Process Development Division 2016

Core Programming Area at the 2016 AIChE Spring Meeting and 12th Global Congress on Process Safety

Houston, Texas, USA
10 - 14 April 2016

ISBN: 978-1-5108-2497-3

Printed from e-media with permission by:

Curran Associates, Inc.
57 Morehouse Lane
Red Hook, NY 12571

CURRAN ASSOCIATES INC. proceedings .com

Some format issues inherent in the e-media version may also appear in this print version.

Printed by Curran Associates, Inc. (2016)

For permission requests, please contact AIChE
at the address below.

AIChE
120 Wall Street, FL 23
New York, NY 10005-4020

Phone: (800) 242-4363
Fax: (203) 775-5177

www.aiche.org

Additional copies of this publication are available from:

Curran Associates, Inc.
57 Morehouse Lane
Red Hook, NY 12571 USA
Phone: 845-758-0400
Fax: 845-758-2633
Email: curran@proceedings.com
Web: www.proceedings.com

TABLE OF CONTENTS

442035 Scaling Liquid Flow through Packed Beds from the Laboratory to the Commercial Plant

Abstract:

In 1952, Sabri Ergun published an article reconciling all the data pertaining to flow through packed columns at that time. The equation presented in that paper has since become known as the Ergun equation. It has been used with varying degrees of success since its publication. The Ergun equation has various problems. First, it uses a Reynolds number based on particle diameter, not fluid conduit diameter. Thus, the Reynolds number used in the Ergun equation does not describe fluid flow. Second, the Ergun equation was developed for spheres. However, myriad odd-shaped packing exists; thus, the Ergun equation must be modified when using non-spherical packing. That modification is generally encapsulated in a dimensionless "sphericity factor" … a cheat factor, in other words. Third, the Ergun equation violates geometric similarity, thus it cannot be used for scaling up or scaling down.

In this paper, we address these problems via Dimensional Analysis. We use Dimensional Analysis to formulate a new method for characterizing liquid flow through packed columns; i.e., for characterizing filtration. We also propose a new method for characterizing the packing material. Since our results obey the laws of similarity, our analysis can be used for up scaling and down scaling filtration units. This method is applicable to liquid flow through packed beds in chemical plants as well as oil flow through fractured rock.

Scale-up of Oscillatory Helically Baffled Reactors

Safaa Ahmed

Supervisors:
Dr. Anh Phan
Prof. Adam Harvey

➢ Outline

- ➢ Introduction
- ➢ Aim of study
- ➢ Devices and procedure
- ➢ Results
- ➢ Conclusions
- ➢ References

➢ Background

<u>**Oscillatory baffled Reactor (OBR)**</u>

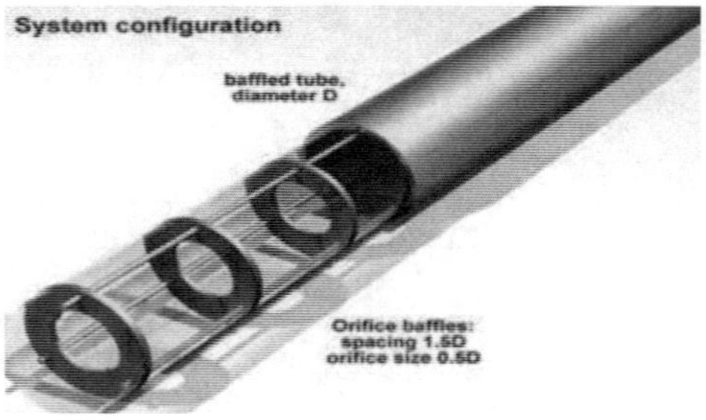

➢ Aim of study

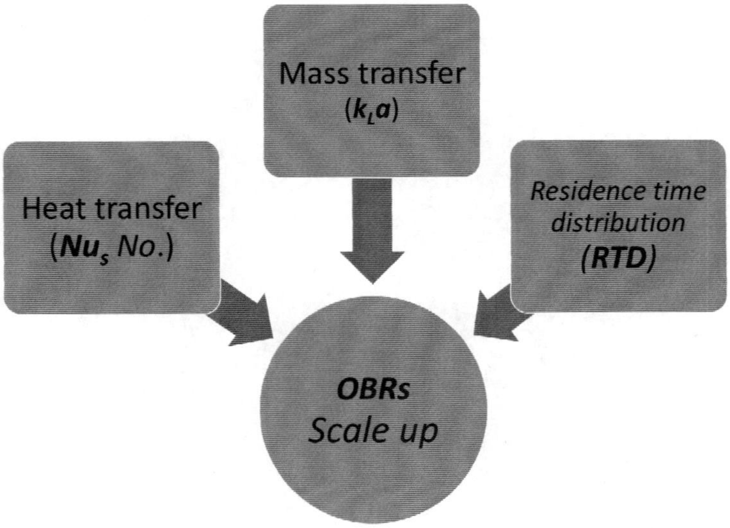

Residence time distribution (RTD)

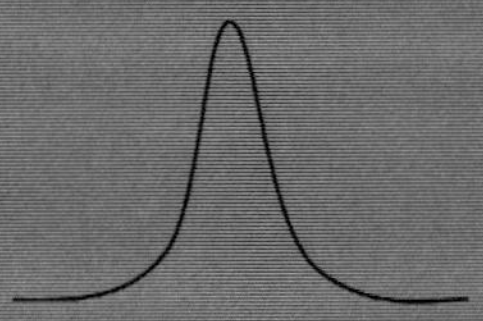

Tanks-in-series model

$$E(\theta) = [N(N\theta)^{N-1}/(N-1)!]\, e^{-N\theta}$$

- **<u>Baffle designs:</u>**

✓Orifice baffles

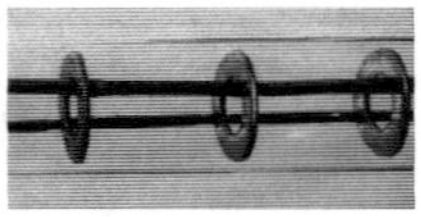

✓ Helical baffles

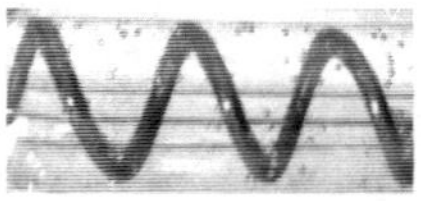

✓ Integral baffles

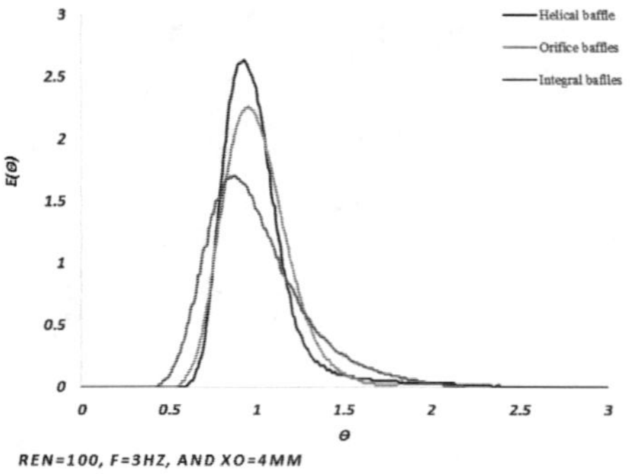

REN=100, F=3HZ, AND XO=4MM

➢ Devices and Methods

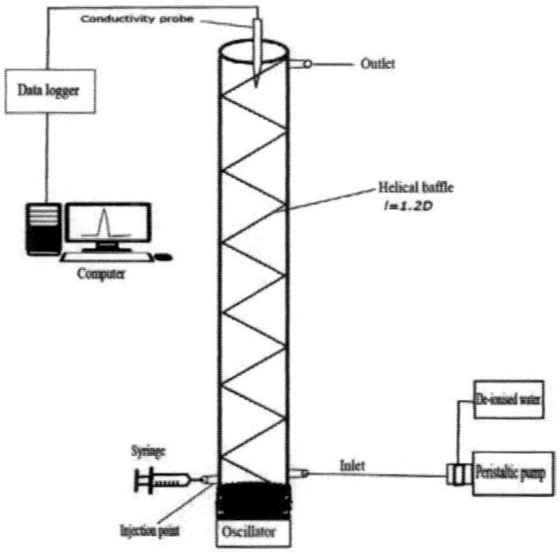

➢ Scale-up limitations

- **Geometric similarity**:
  - ✓ Baffle spacing *(l/D)=1.2*
- **Dynamic similarity:**
  - ✓ Strouhal Number (*Str*): 0.37, 0.27, 0.17
  - ✓ Velocity ration (ψ) = *Reo/Ren*

➢ OHBR scale-up results

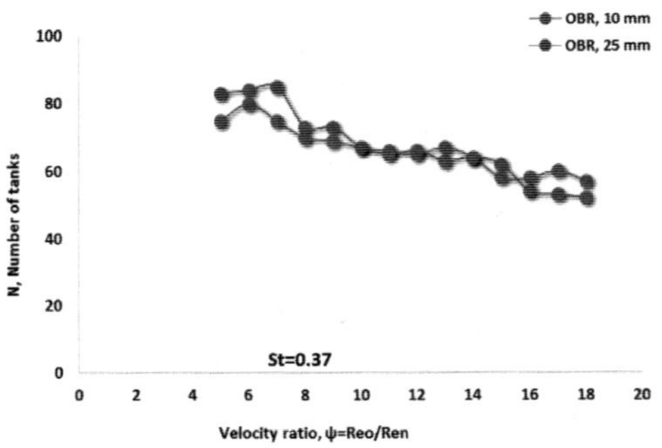

RTD behaviour at *Str*=0.37

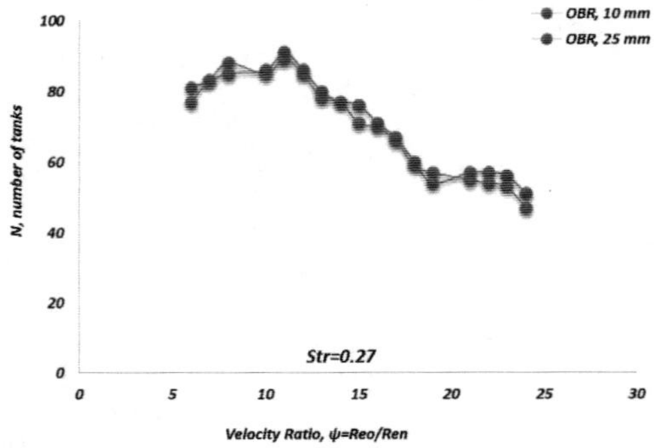

RTD behaviour at *Str*=0.27

➢ Scale up correlation

- $N = 25\,Str^{-0.3}\,\psi\,e^{-0.13\psi} + 30.2$

$$N = \frac{41.5 + aRe_n - bRe_n^{2}}{1.36}\left\{\left[\left(\psi + 1.8 - cRe_n^{-d}\right) + 1\right] e^{-0.35\left(\psi + 1.8 - cRe_n^{-d}\right)}\right\} + 1 \quad [1]$$

Comparison between the experimental data and correlation data

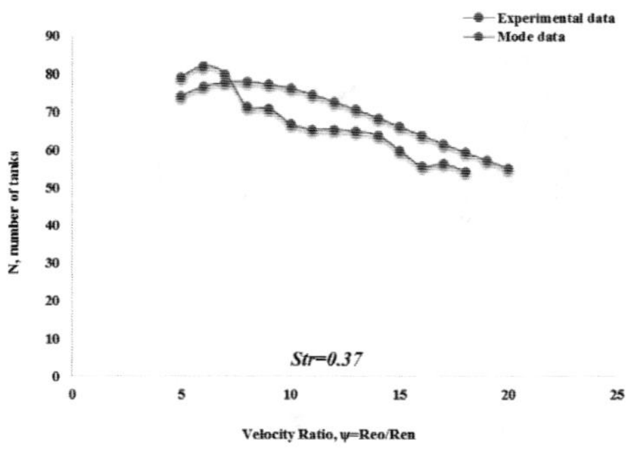

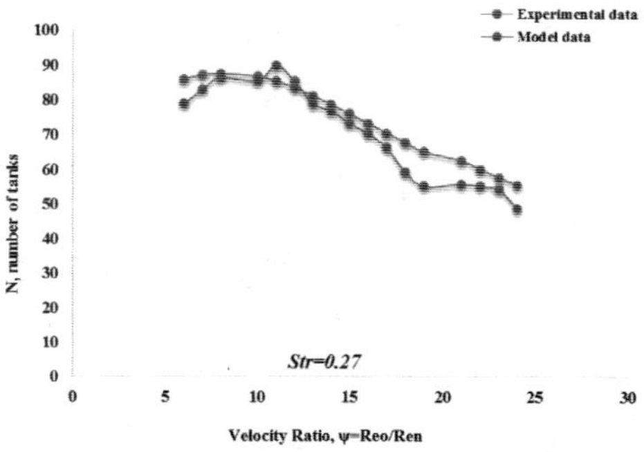

➢ Validation

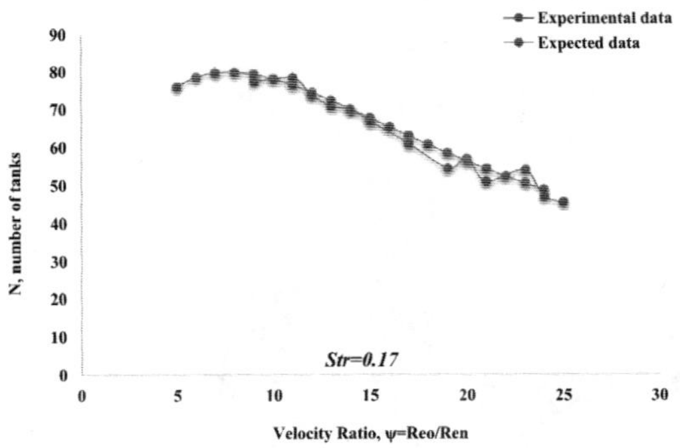

$R^2 = 0.98$

Visualisation of flow pattern inside meso-OBR

(PIV technique & CFD simulation)

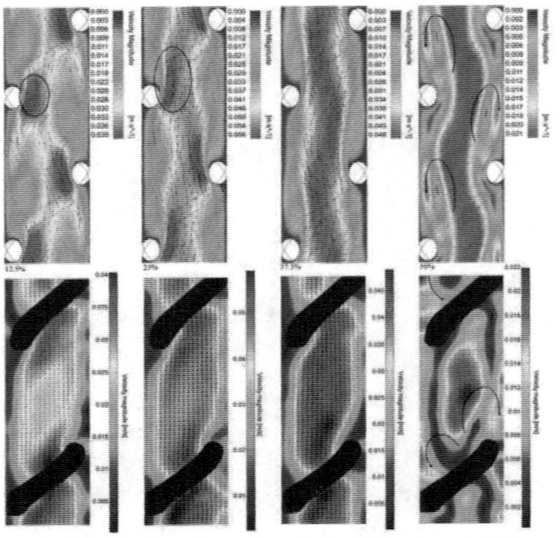

Oscillation conditions: *Ren = 0, Reo = 126, St = 0.2 (xo = 2 mm, f = 2 Hz)*

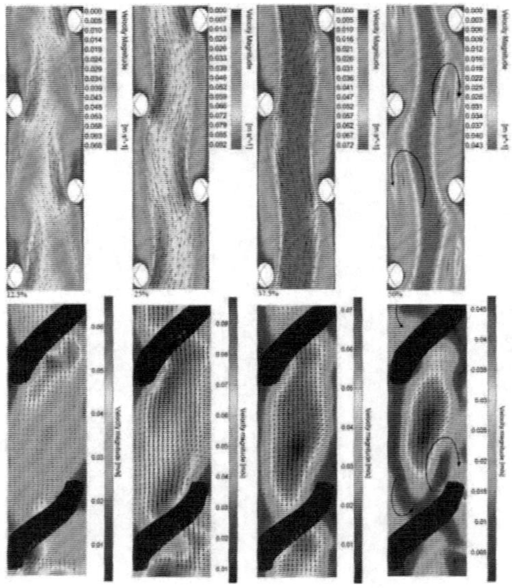

Oscillation conditions: *Ren = 0, Reo = 188, St= 0.13 (xo = 3 mm, f = 2 Hz)*

OBRs scale up based on Mass Transfer Coefficient

(Air-Water system)

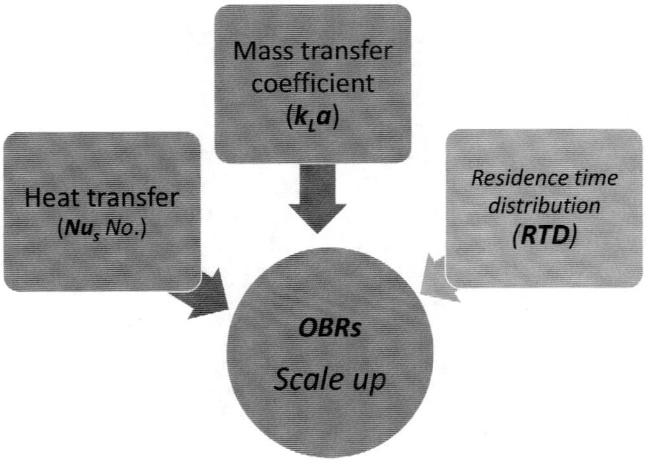

➢ Aim of this step

- Choose the best design of OBR (based on *kLa* results)
- Gas holdup calculation for each scale.
- Mass transfer coefficient (***kLa***) calculation for each scale.
- Determine a model of ***kLa*** for each scale.
- Compare the results of all scales.
- Validate the models.

➢ OBRs designs

✓Single-orifice design

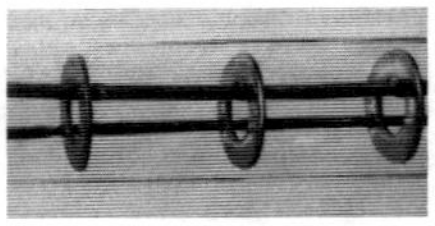

✓ Helical design

✓ Smooth Periodic Constrictions design (***SPC***)

➢ *kLa* comparison (10 mm scales)

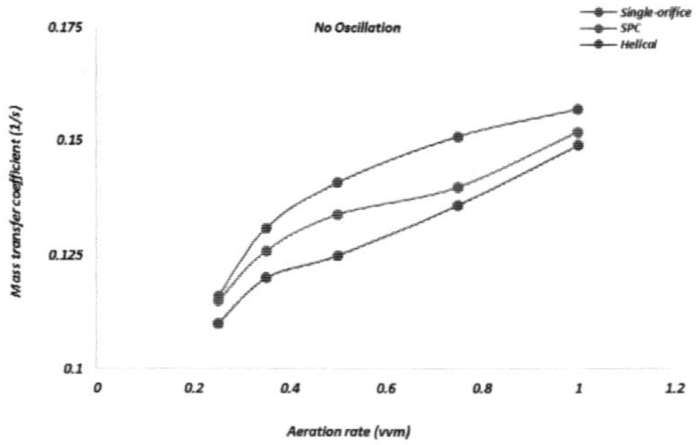

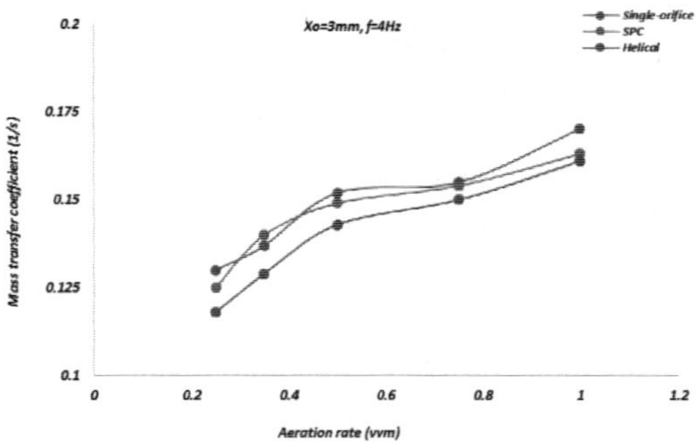

- **Baffles modifications**

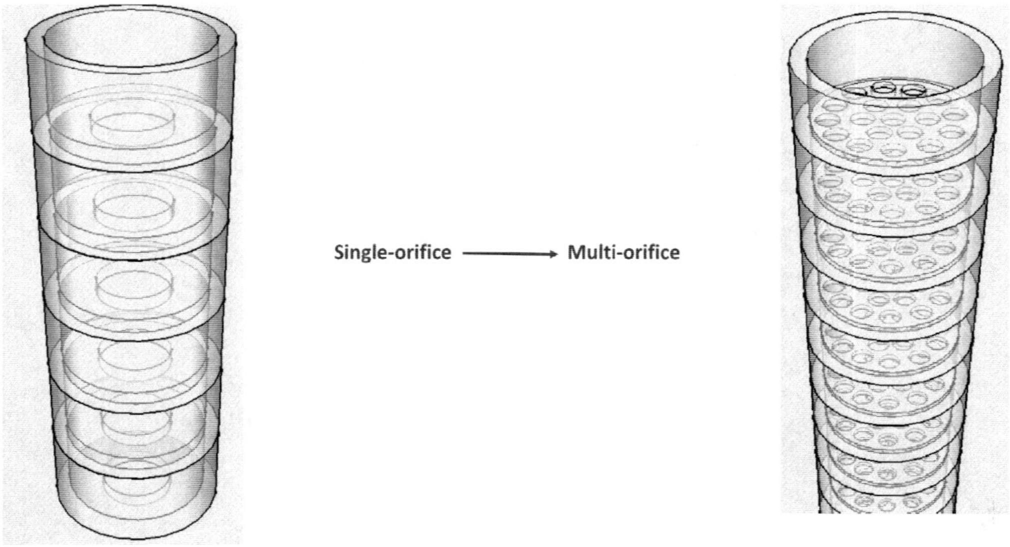

- # Scale up limitations (Similarity)

1- Geometrical parameters:

- Open cross sectional area *(α)= 56%*
- Baffle spacing *(l/D)= 0.5*

2- Dynamic parameters:

- Maximum velocity *(2πfxo): 0 - 1000 mm/s*
- Aeration rate (*vvm*)*: 0.1-0.5 l/min*

➢ Conclusions

- The OBRs were scaled up successfully based on RTD .
- The scale-up correlation was developed.
- The scale-up correlation was validated.
- Flow pattern inside meso-OBR was visualised using PIV technique and Ansys software.
- A best design of OBR has been chosen and modified.

➢ References

[1] Stonestreet, P. and Van Der Veeken, P.M.J. (1999) 'The Effects of Oscillatory Flow and Bulk Flow Components on Residence Time Distribution in Baffled Tube Reactors', Chemical Engineering Research and Design, 77(8), pp. 671-684.

[2] Phan, A.N. and Harvey, A. (2010) 'Development and evaluation of novel designs of continuous mesoscale oscillatory baffled reactors', Chemical Engineering Journal, 159(1-3), pp. 212-219.

[3] Pereira, F.M., Sousa, D.Z., Alves, M.M., Mackley, M.R. and Reis, N.M. (2014) 'CO2Dissolution and Design Aspects of a Multiorifice Oscillatory Baffled Column', Industrial & Engineering Chemistry Research, 53(44), pp. 17303-17316.

[4] Smith, K.B. and Mackley, M.R. (2006) 'An Experimental Investigation into the Scale-up of Oscillatory Flow Mixing in Baffled Tubes', Chemical Engineering Research and Design, 84(11), pp. 1001-1011.

Thanks

Strategy for Designing a Successful Pilot Plant

Anil J. Mehta, The Dow Chemical Company, Freeport, TX, U.S.A.

Abstract

Testing a chemical or a petrochemical production process at a pilot scale is a vital step in scaling up the new process to commercial scale successfully. The author has extensive experience in scaling up new chemical and petrochemicals production processes successfully. The author presents an effective strategy and approach for designing and operating a pilot plant with examples to achieve the desired results. This could be an interactive discussion.

1. Introduction

One thing we all can count on in life is change. The planet and the world we live-in are in a state of change constantly. Our needs are always changing driven by technology changes or challenges such as climate change or other factors. To meet the ever-changing needs, new materials and processes to manufacture them are needed. Sometimes, the same products need to be made using different raw materials which also requires developing a new production process for the product. Thus, new manufacturing processes are always needed.

New materials and processes usually are invented in laboratories at a very small scale. If the product or the process is found promising from initial testing then the process needs be scaled up to produce the product in a much larger quantity at an affordable price to satisfy demand for the product.

Scaling-up a new manufacturing process from a small laboratory scale to a commercial scale requires designing and building a commercial plant, a highly complex job. Commercial manufacturing plants can be very expensive, costing anywhere from a few millions to billions of dollars, depending on the scale and complexity of the process.

2. Why test the new process at a pilot scale?

Scaling-up a new process from a laboratory scale to a commercial scale directly is very risky because many things can go wrong in the scale-up process. The new commercial plant may not work as intended, either not being able to meet product specification or operating at a rate far less than the design rate or both. The process industry is full of examples where a new process scaled-up failed to produce the quality product or failed to operate at a rate close enough to the design capacity. If the new plant fails to produce the intended product at an economical price, improvements to the plant are needed, requiring additional capital and time. In the worst case, the investment may have to completely written off, resulting in a total loss of capital which could be hundreds of millions of dollars and time.

Testing the new process at a pilot scale before commercializing minimizes such risks present in scaling up the process to commercial scale [1-4]. Pilot scale implies a scale much smaller than the desired commercial scale [5]. The pilot testing serves as an insurance for scaling up the new process to commercial scale because although the pilot testing is expensive, it is still much cheaper to invest in a pilot program than experience failure at the commercial scale.

3. Pilot testing strategy

Building a pilot plant is expensive; it can cost millions of dollars. Therefore, if the pilot testing can be done in an existing facility either in-house or at a contract manufacturer's facility, it can save the cost of building the new pilot plant. However, this depends on the type of the process to be scaled-up.

The new production process to be tested may be a continuous process, a batch process, a semi-continuous process or a combination of these. Batch processes tend to be simpler; hence, it is usually easier to find facilities for pilot testing a batch process. Complexity of a process increases as one goes from a batch process to a semi-continuous process, a continuous process and a continuous process with a recycle or even multiple recycle loops. The new process may be for making an existing product using a new synthesis route or using a different raw material e.g. producing ethanol from algae or it may be for a new product.

If the new process is for a new product, in addition to developing the production process, the product also has to be tested extensively for the intended end-use application and the market for the product developed. Launching a new product in the market brings additional and usually even bigger risk. Only a few of many products that get launched, gain sufficient acceptance in the market. Therefore, if the new production process is for a new product, it is best to minimize the pilot plant expenses by testing the process at an existing facility, either within or outside the company such as at a contract manufacturer's facility, until the product gains sufficient acceptance in the market. If a new process is for an existing product, is continuous, has one or more recycle loops, it will likely require designing and building a dedicated pilot plant as such customized pilot plants are hard to find in the market.

3.1 Before committing to pilot testing

A decision to test a process at a pilot scale is a major step in the process development. It requires committing significant capital and resources. Therefore, it is expected that the decision to test the process at a pilot scale follows rigorous testing of the process at a bench scale including safety and toxicology studies. It is expected that by this time all process technology options have been evaluated from techno-economic and safety aspects and there has been a rigorous internal debate for the technology choice before committing to the pilot testing.

3.2 Designing a successful pilot plant

3.2.1 Pilot plant objectives

There must be a clear understanding of the purpose of the pilot plant before designing it. Define in detail the goals for the pilot testing. Prepare a detailed list of the objectives and the data to be generated and how the data will be used – e.g. data on reaction rate, conversion, yield, selectivity, effect of flow regime, recycle, separation – distillation / extraction, product quality, process control, materials of construction etc. Ensure management input in this process to make sure nothing will be missed.

3.2.2 Scale-down to scale-up

One of the purposes of the pilot plant is to test at a much smaller scale the production process to be used at the commercial scale. Therefore, as far as possible, the pilot plant should have the same kind of key equipment and materials of construction that will be used in the commercial plant. It is often not necessary to use the same type of a pump, filter or a heat transfer equipment as in the commercial plant if such unit operations are well understood and not key part of the pilot testing.

The best way to ensure that the pilot plant will have the same kind of key equipment that will be used in the commercial plant is to first complete preliminary design of the desired commercial plant. This requires preparing material and energy balance for the commercial plant. Commercially available process simulation programs such as Aspen Plus or others can be used for the same.

Physical properties data form the backbone of a process simulation and preparing a process model will help identify missing physical properties data if any. Importance of accurate physical properties data can not be over-emphasized for scaling up a process successfully. Search literature extensively for the missing data. If the missing data can't be found in the literature then it needs to be generated within or outside the company.

Determine the type of equipment to be used in the plant and their preliminary size based on the material and energy balance for the process. The next step is to scale down the material and energy balance. At this point, a question arises – what should be the scale of the pilot plant?

3.2.3 What scale for the pilot plant?

The pilot plant should be of the smallest possible and workable scale while ensuring that it provides the high quality scale-up data required for designing the commercial plant, the scale-up data for various unit operations used in the plant.

The smallest possible size minimizes the pilot plant cost which includes cost of equipment, raw materials, waste generation and disposal. It also minimizes handling of raw materials, waste and potential exposure to hazardous chemicals, process inventory and the process safety risk.

Determining the minimum scale for the pilot plant requires determining the minimum scale for each unit operation to be tested such as reaction, distillation, extraction etc. In Dow, we can scale-up distillation columns from as small as a 2" diameter column [6]. In one of the process development project I led, we performed scale-up analysis for the reactor and determined that 1-liter reactor size was sufficient to generate the scale-up data for the commercial scale CSTR. However, this was an extremely corrosive process requiring a glass-lined reactor. The smallest glass lined reactor available was 1 gallon in size. Therefore, the pilot plant was designed based on the 1 gallon reactor size. The smallest size for piloting the epoxidation step required an order of magnitude larger feed rate.

Scale-up factors, the ratio of the commercial scale to the pilot scale, can vary over a very wide range depending on the type of unit operation and complexity of the process. For example, we scaled up successfully an absorber by a factor of over 1500 and we scaled up successfully improvements in a slow batch reaction from about a 100 grams scale to over a million pounds scale.

It is worthwhile to seek input from the subject matter experts for the unit operations being tested or used in the pilot plant such as reaction engineering, mixing, distillation, metallurgy. Pilot testing is expensive so it is imperative that the best expertise available within or outside the company be used to ensure that the correct decisions are made in designing the pilot plant.

3.2.4 Design for abnormal operation scenarios

Once the pilot plant scale is determined, the next step is to prepare material and energy balance for the pilot plant. Simulate various pilot plant operation cases such as start-up, shut-down, low throughput rate, high throughput rate etc. and ensure that the pilot plant is designed for these situations.

3.2.5 Testing materials of construction

If the process is corrosive, requiring corrosion resistant materials, make provision to test in the pilot plant the materials of construction you plan to use in the commercial plant and collect material performance data on long term exposure to the process.

3.2.6 Pre-construction review of the design

Once, key parts of the pilot plant design are completed, review the design with the larger team involved in the project, appropriate experts and supervision before ordering equipment and beginning construction of the pilot plant. It is much easier to make changes in the design at the design stage then after constructing the pilot plant. This will minimize potential for changes after pre-startup review and save the time in making the changes before the pilot plant can be started up, ultimately speeding up the process.

3.2.7 Process Safety

Most chemical processes involve dealing with hazardous materials. Exposure to a hazardous material can result in a serious safety incident. As scientists and engineers, it is our utmost responsibility to make the pilot plant as safe as possible. Safety must be an integral aspect of the design and operating discipline – how the pilot plant is started-up, operated and shut down. We can not depend upon personal protective equipment as they are the last layer of defense. Integrating safety in design builds the first layer of defense. Ensure input of process safety experts in the pilot plant design. Conduct process hazard analysis of the design. Identify worst case scenarios and design safe guards to protect against them.

Reactive chemicals testing should be done before even conducting laboratory scale experiments. Risk of a run-away reaction increases significantly as the scale of the process increases from a laboratory scale to a pilot scale. The surface area available for cooling from heat loss to the surrounding decreases rapidly per unit volume of the process materials. The highest temperature in a distillation column is usually found in the column's reboiler. If the liquid mixture in the reboiler is reactive, temperature for a potential run-away reaction should be determined and the reboiler should be operated well below the potential run-away reaction temperature. The run-away temperature is the temperature at which the rate of temperature rise exceeds the rate of cooling for the reboiler.

Reactive chemicals testing such as accelerated rate calorimetry can provide self-heating rate for the mixture. See the example in Figure 1 [7]. Cooling rate for the given equipment can be measured under real conditions as shown in Figure 2.

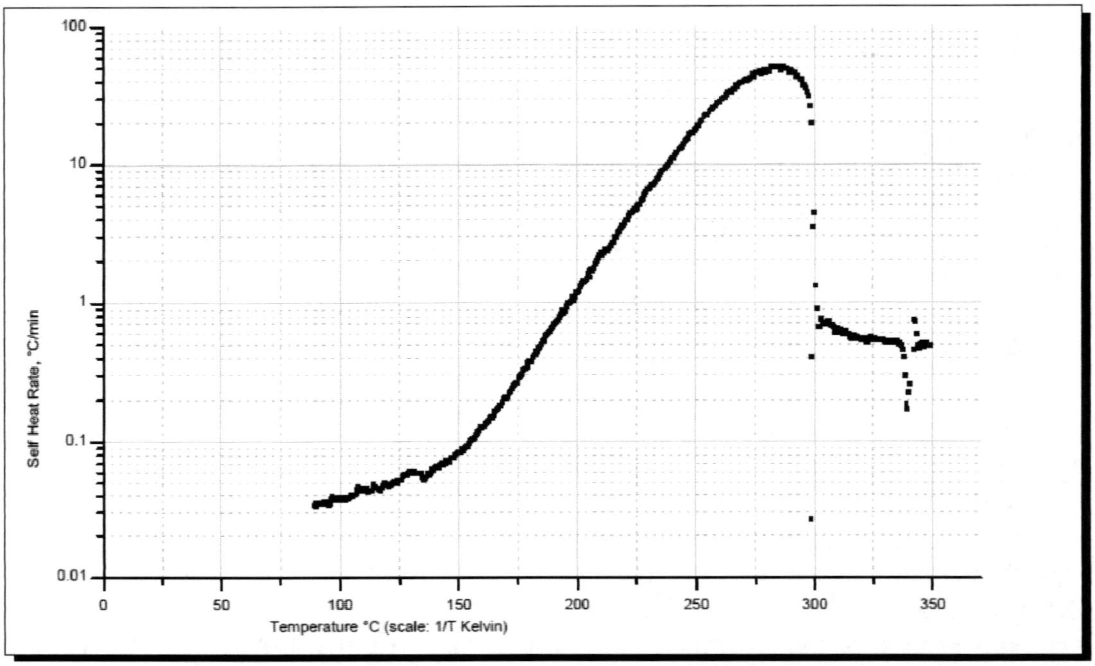

Figure 1: Self heating rate of a process mixture

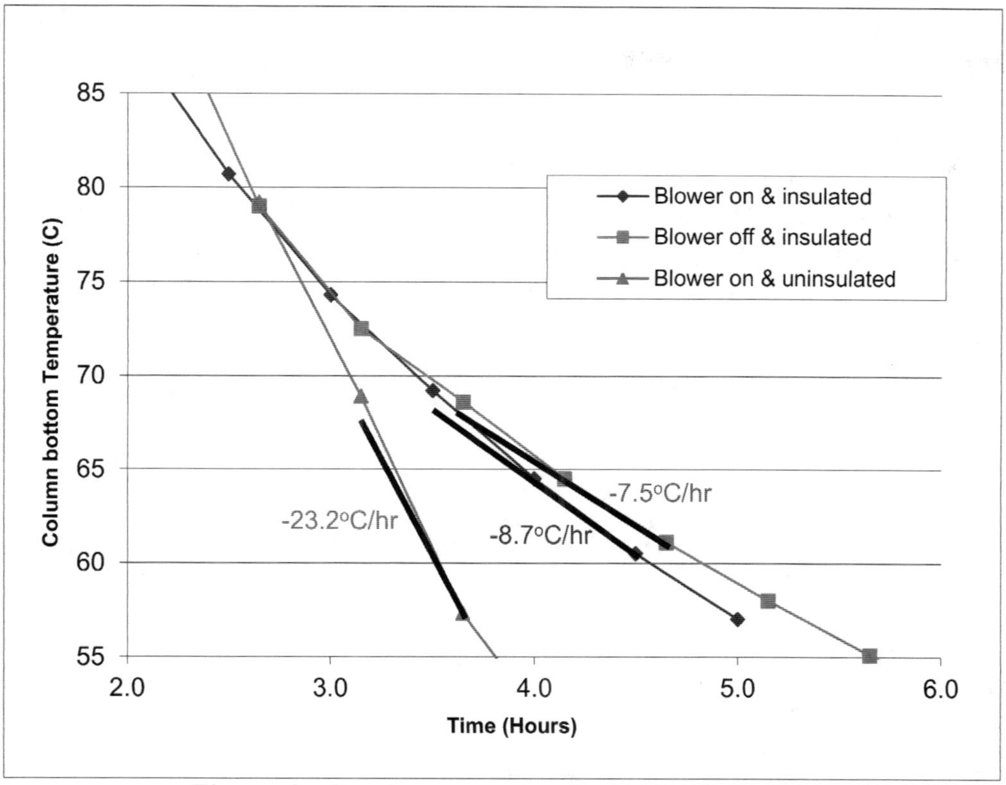

Figure 2: Cooling rate of a pilot plant reboiler

Safety must be a top priority, especially the safety of the people. When dealing with hazardous chemicals, do not take them lightly. Do not take shortcuts to meet the timeline. Conduct pre-startup review and correct deficiencies before starting up the pilot plant.

3.3 Process Samples Analysis and Data quality

One of the purposes of a pilot plant is to generate scale-up data required for designing the commercial plant, in addition to demonstrating the process at the pilot scale. Getting data on process streams composition at key locations in the process is critical.

Identify sampling points in the process during the design phase. Prepare a comprehensive process sampling and analysis plan describing all process sample points, sampling frequency and methods to be used for analyzing the samples. Set-up required analytical capability – set up the hardware such as gas chromatographs, titrators etc. Develop and validate the methods for analyzing the samples before starting up the pilot plant. Define the experiments to be conducted in the pilot plant and have an experiments plan ready.

Extract maximum possible information from each experiment. Each process sample is a result of a real physical phenomenon. Even a sample from a process upset or from a startup or a shutdown period presents an opportunity to learn about the process. It takes a lot of investment of time and resources to design, build and operate the pilot plant. Therefore, it is very important that the experiments are not wasted from lack of inadequate analytical capability.

High quality analysis of the process samples goes a long way in achieving the objectives of the pilot plant. The two graphs in Figure 3 show an example of a high quality analysis from one of the projects I led in the past. The two charts show sample composition accountability for two separate sample points in a continuous glycerine hydrochlorination pilot plant operating 24 hours a day, 7 days a week. The samples from both sample points were analyzed using the same instruments. Each data point in the graphs is a sum of the composition from 3 separate analyses – organic species analyzed on a gas chromatograph (GC), acid by titration and water by Karl-Fisher titration. The sum of the three needed to be 100 ± 5% for the analysis to be valid.

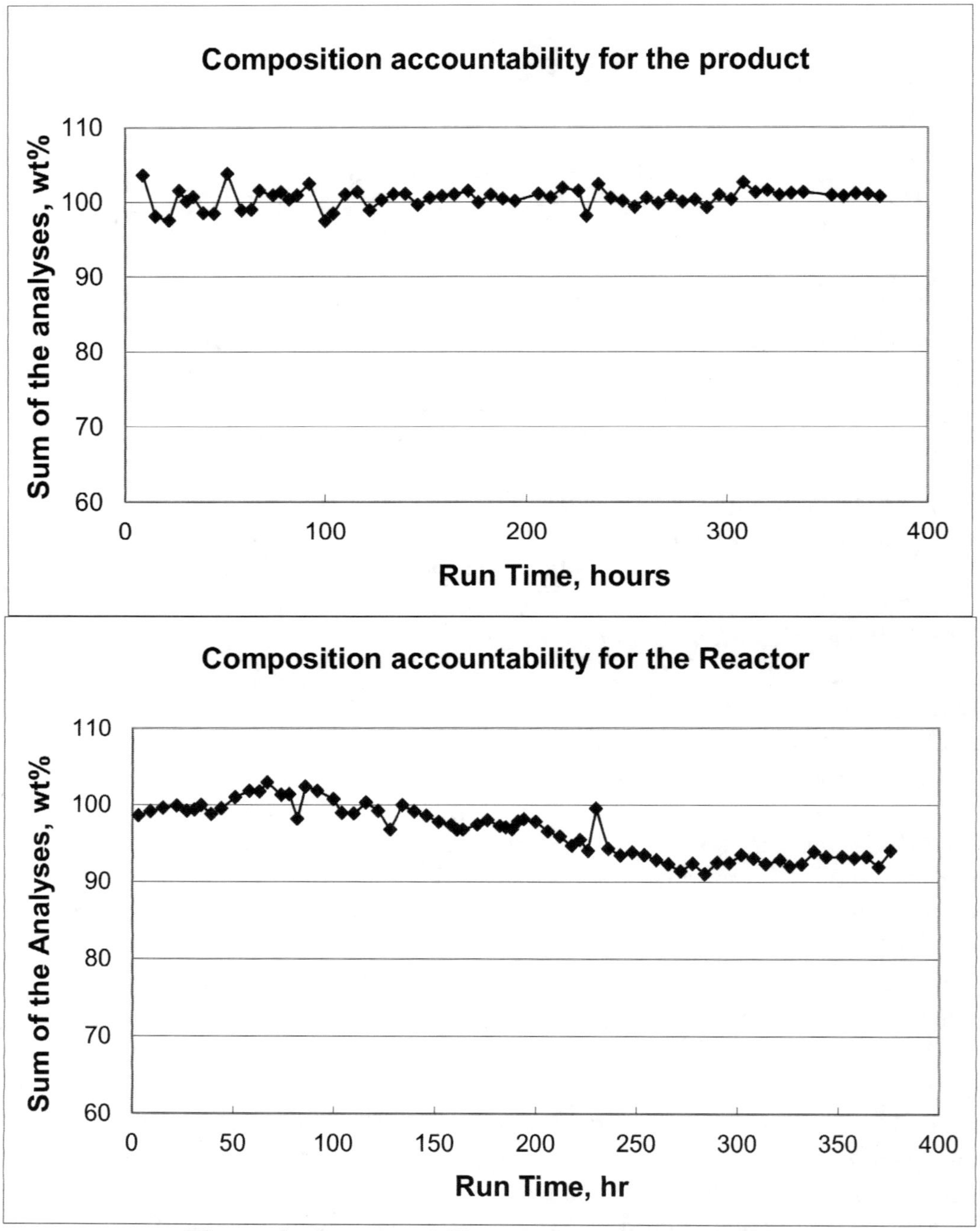

Figure 3: Accountability of the samples analyses

The first chart in Figure 3 which is for the product stream, shows that the sample composition obtained by using the three separate and independent analytical methods was usually within +/-2% for over 350 hours of continuous operation. The second chart which is for a reactor stream shows that the sample composition was within +/-2% for the first 100+ hours of operation and then it gradually decreased to 90-95% range. Excellent accountability for the first sample point confirmed that the three analytical instruments and the analysis methods worked well. The lower accountability later on for the second sample point could not result from the instruments but must have been a result of a real phenomenon. After further investigation, we concluded that the decrease in the accountability was a result of gradual build-up of heavier components in the process which did not elute from the GC column. Thus we were able to learn about the rate and extent of heavies build up in the process as a result of having the high quality analysis of the samples.

3.4 Staffing, training, start-up and operation

Determine in advance the number of persons required to operate the pilot plant, analyze process samples, dispose the waste etc. and staff the project accordingly. Train personnel on the process and samples analysis before starting up the plant. In the case of a complex process, a training program may be needed for the pilot plant operation. Start-up the pilot plant with water if possible and train personnel before introducing chemicals to the process. It may not always be possible to operate the plant with water first.

3.5 Data management- storage, retrieval and learning from the data

Plan ahead for data storage, set up data historian for the process and the analytical data for easy data retrieval at a later date. Analyze the data, learn from the data, refine the process model and the design of the commercial plant. Redo the techno-economic evaluation of the refined commercial process and decide accordingly. This is usually a major decision stage for the project on whether to commercialize the new process.

3.6 Document

It is very important to document in detail all significant aspects of project. Document design and construction of the pilot plant including the reasons for the choices made. Document the data generated, analysis of the data and learning. Document any model refinement achieved as a result of the new data and the results of the refined techno-economic evaluation of the process.

4. Conclusion

Pilot testing of a new production process is a vital step in scaling-up the process from a small laboratory scale to a commercial scale. There are many aspects of the process requiring attention. An effective strategy with examples has been presented in this paper for avoiding the pit-falls and testing the new process at the pilot scale successfully.

5. References

1. "Pilot plants" Kirk-Othmer Encyclopedia of chemical Technology (2005)
2. Ploufe, Patrick, Arturo Macchi and Dominique Roberge; "Chemical Process Development and Scale-up challenges" in Flow Chemistry, Vol. 1 – Fundamentals (2014)
3. Distillation: Equipment and Processes (2014), Chapter 10: Laboratory Distillation and Scale-up
4. Cusack, Roger; "Pilot Plants confirm Process Validity" Chemical Engineering, June 1998
5. Oxford Dictionary of Chemical Engineering (2014), P-282
6. From past consultations with Dow Chemical's distillation experts
7. From a test conducted in Dow Chemical's reactive chemical laboratory

441847 Design of Multiple Extractors for Optimizing Liquid-Liquid Extraction Processes

Abstract:

Separation processes involving liquid-liquid extraction cannot always be solved by a single, counter current extraction column. In some applications, two columns or a column with mixer/settler in series should be considered and may offer the best solution from a process optimization and/or economic standpoint. Fractional extraction is an obvious process whereby two columns in series can be considered in lieu of a single column. However, other processes such as columns with backwash, extractions with reaction, unfavorable density gradient and presence of solids can also involve more than a single column. In any of these types of processes, pilot plant testing is critical to the optimal design of the system. This paper covers a number of different applications where a single column was considered vs two extractors in series. The results from pilot plant testing are presented along with the eventual design for scale up to the commercial processes.

CARBONYL REMOVAL IN STEAM CRACKERS & FCC UNITS

The Merits of Carbonyl Pre-Treat (CPT) Technology

Dr. Sabah Kurukchi
and
Joseph Gondolfe
Janus Technology Solutions, LLC
Houston, TX (USA)

16AIChE
Spring Meeting, Houston, TX

JANUS

Presentation Outline

- **Statement of Problem Definition**
- **Technology Overview (SC Plus FCCU)**
- **Experimental Results from Absorption**
- **The Perfect Technology Solution for SC**
- **Experimental Results from L/L Extraction**
- **The Perfect Technology Solution for FCCU**
- **Conclusions**

16AIChE
Spring Meeting, Houston, TX

JANUS

Statement of Problem Definition
Oxygenates in Steam Crackers and FCC Units

- Hydrocarbons at high temperatures undergo decomposition to C and H_2
- In the presence of steam, some of the carbon converts catalytically at the operating temperature to CO and H_2 on the Steam Cracker radiant steel coil surfaces containing Ni and Cr.
- Similar reaction mechanism for the formation of CO & H_2 is true within the Fluid Catalytic Cracking Unit (FCCU) over its aluminosilicate catalysts.
- Oxygenates are catalytically formed in SC and FCC Units including organic acids, alcohols and carbonyls in concentrations ranging from 10-1000 ppm depending on the type of feedstock, catalyst, ratio of dispersion steam-to-hydrocarbon (feed) at operating temperature.

16AIChE
Spring Meeting, Houston, TX

JANUS

Technology Overview
SC Plus FCCU Key Operating Parameters

	FCC	DCC	SC
Reaction temp/°C	500–550	530–590	760–870
Reactor pressure/barg	1–3	1–2	1–0
Residence time/s	1–5	1–10	0.1–0.2
Catalyst/oil ratio (wt/wt)	4–8	10–15	–
Dilution steam (%)	1–3	5–30	30–80

16AIChE
Spring Meeting, Houston, TX

JANUS

Effects of Aldol Condensation in Caustic Services

Aldehydes and Ketones (acetone) present in Hydrocarbon streams polymerize by Aldol Condensation in the presence of caustic (acting as catalyst) in the formation of Red Oil polymer.

Red Oil causes fouling of Steam Cracker Caustic Tower and the spent caustic stream.

Red Oil causes fouling of FCC Caustic Contactor and Oxidizer

Red Oil Formation Leads to Fouling

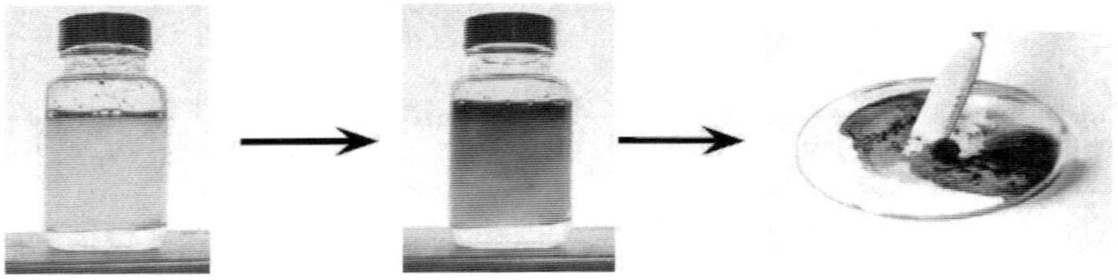

Yellow Oil → **Red Oil** → **Sticky-Viscous Solid Resin**

16AIChE
Spring Meeting, Houston, TX

JANUS

Caustic Equipment Fouling by Aldol Polymer

Consequences of Red Oil Formation

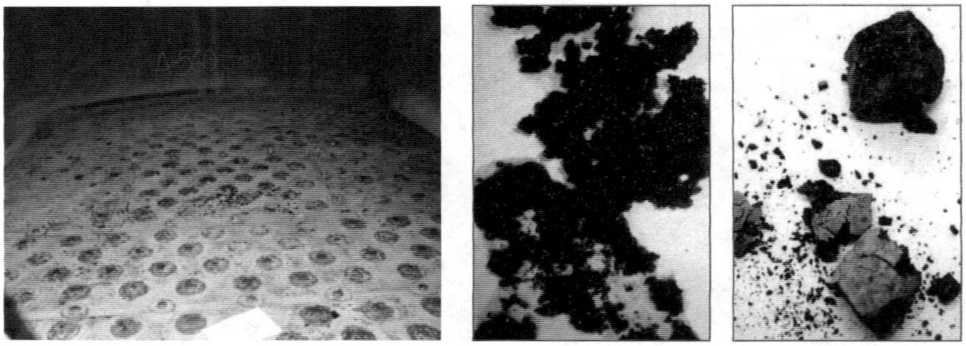

16AIChE
Spring Meeting, Houston, TX

JANUS

Preventing Red Oil Polymer Formation in Caustic Towers

- Aldol Polymer (Red Oil) present technologies deal with treating the symptoms by:
  - Filtering the large polymer particles from the recirculating caustic solution
  - Additive(s) addition to disperse the formed polymer

The perfect solution is to prevent Red Oil formation by removing the Carbonyls upstream of the Caustic Contactors.

16AIChE
Spring Meeting, Houston, TX

JANUS

Mechanism of Aldol Condensation Reactions

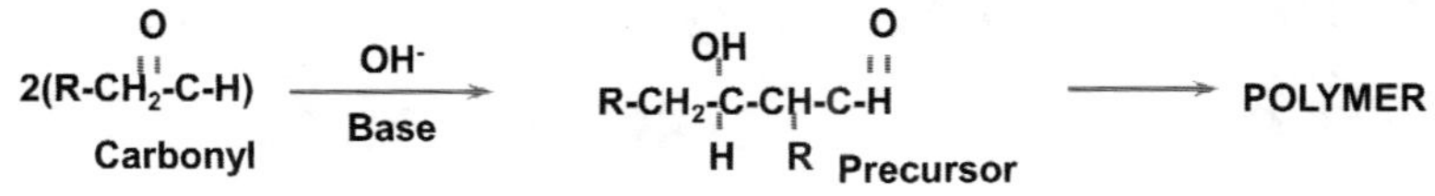

Step [1] Formation of a nucleophilic enolate

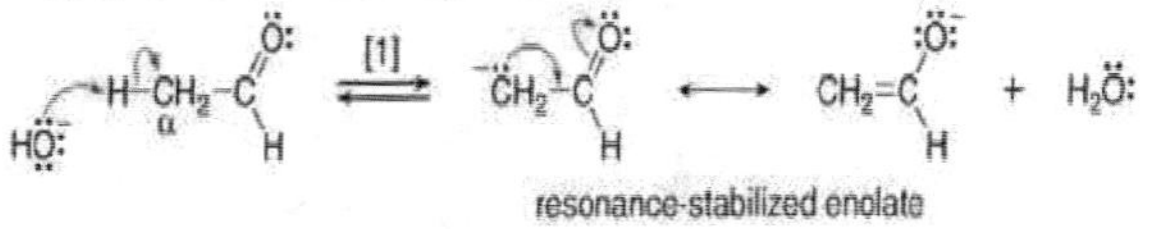

- In Step [1], the base removes a proton from the α carbon to form a **resonance-stabilized enolate.**

Steps [2]–[3] Nucleophilic addition and protonation

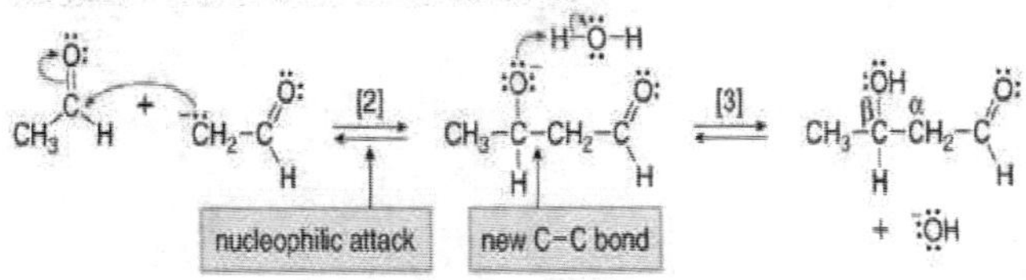

- In Step [2], the nucleophilic enolate attacks the electrophilic carbonyl carbon of another molecule of aldehyde, thus forming a new carbon–carbon bond. **This joins the α carbon of one aldehyde to the carbonyl carbon of a second aldehyde.**
- Protonation of the alkoxide in Step [3] forms the **β-hydroxy aldehyde.**

16AIChE
Spring Meeting, Houston, TX

JANUS

Typical CG Caustic Tower Feed Composition

Typical stream contaminants are:

- CO_2 200 - 400 mppm
- H_2S 20 -100 mppm
- Acetaldehyde 100 - 200 mppm
- Acetone 5 - 10 mppm

- Acetaldehyde is very reactive in the presence of caustic solution and react to form Aldol Polymer.
- Acetone is less reactive and present in very low concentrations; thus, cannot promote fouling tendency.

Feed to Caustic Tower		Liquid Cracker	Gas Cracker
Composition, mol %	Comp. MW	mol%	mol%
H2O	18.01	0.400	0.300
HYDROGEN	2.02	17.000	36.000
CO	28.01	0.100	0.054
METHANE	16.04	28.000	7.500
ACETYLN	26.04	0.500	0.166
ETHYLENE	28.05	29.100	32.800
ETHANE	30.07	7.000	21.000
MACETYLN	40.06	0.350	0.006
PROPDIEN	40.06	0.300	0.006
PROPYLEN	42.08	11.000	0.783
PROPANE	44.10	0.600	0.094
13BUTD	54.09	3.000	0.434
C4's	56.11	1.100	0.181
C5's	66.10	0.550	0.177
C6-C8 NONA	84.16	0.400	0.000
BENZENE	78.11	0.500	0.352
TOLUENE	92.14	0.040	0.044
CO2	44.01	0.02 to 0.04	0.02 to 0.04
H2S	34.08	0.002 to 0.01	0.002 to 0.01
Acetaldehyde	44.05	0.01 to 0.02	0.01 to 0.02
Acetone	58.08	0.0005-0.001	0.0005-0.001
Mole Wt		25-26	18.5-19
Flow Rate Ton/h		300-400	300-400
Pressure Bars		14-17	14-17

16AIChE
Spring Meeting, Houston, TX

JANUS

Experimental Results from Absorption Metal Hydrogen Sulfite Solution Removes Carbonyl from Cracked Gas Stream

Extensive laboratory testing showed Metal Hydrogen Sulfite solution contacted with the cracked gas in absorption tower removes the Aldehydes and Ketones down to negligible ppm levels.

16AIChE
Spring Meeting, Houston, TX

Carbonyl Removal Packed Absorber Rig Sketch

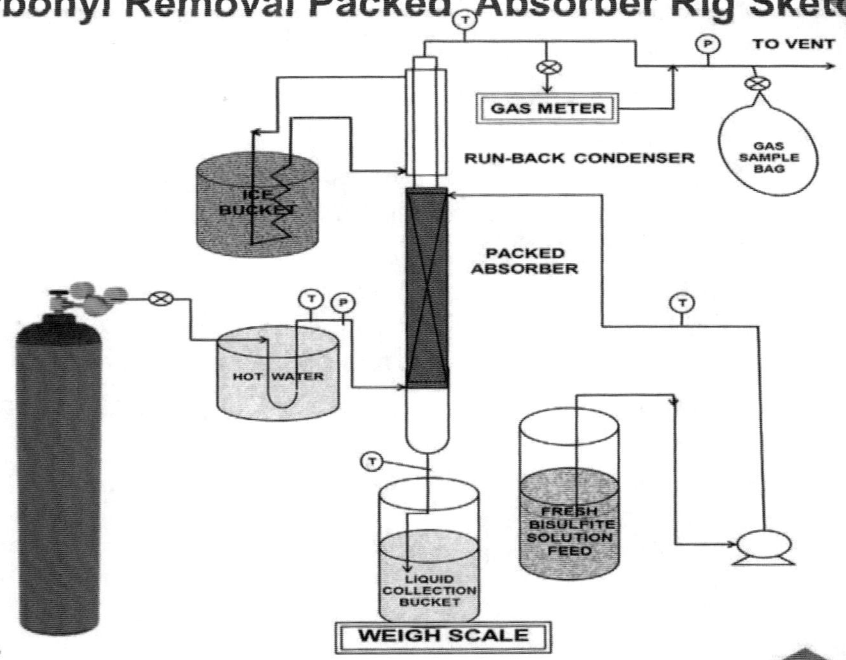

16 AIChE
Spring Meeting, Houston, TX

JANUS

Experimental Carbonyl Removal Absorber Results Summary

Run Set #	Metal Bisulfite wt%	Inlet Carbonyl Conc, wppm	Exit Carbonyl Conc, wppm
1	10	200	<1 (Detection Limit)
2	5	200	<1 (Detection Limit
3	1	200	<3

16 AIChE
Spring Meeting, Houston, TX

JANUS

Carbonyl Pre-Treat (CPT) Technology Fouling Prevention Advantages

- Pretreating the cracked gas with Metal Bisulfite Solution removes >99% of the Carbonyl.
- The amine absorber and /or caustic tower treating the cracked gas will operate without fouling; thus less expensive tower internals can be used.
- The spent caustic will essentially contain sodium salts and dissolved hydrocarbons that can be cold stripped with nitrogen or methane to obtain hydrocarbon (HC) free spent caustic.
- HC free spent caustic is saleable to Pulp & Paper plants and Leather Tanneries.
- HC free spent caustic eliminates fouling in spent caustic treatments:
  1. WAO and Catalytic WAO
  2. Incineration
  3. Deep Neutralization - Stripping

16AIChE
Spring Meeting, Houston, TX

JANUS

Conventional CG Caustic Treating Diagram

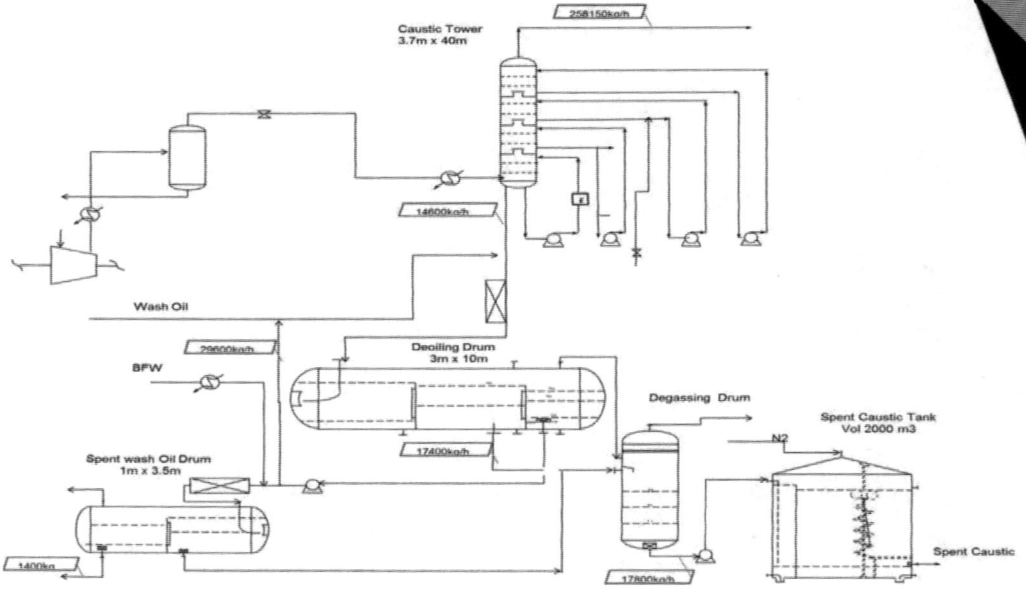

16AIChE
Spring Meeting, Houston, TX

JANUS

The Perfect Solution for SC Caustic Treating with CPT Technology

Eliminate:

- **Deoiling Drum**
- **Spent Wash Oil Drum**
- **Use of Wash Oil**

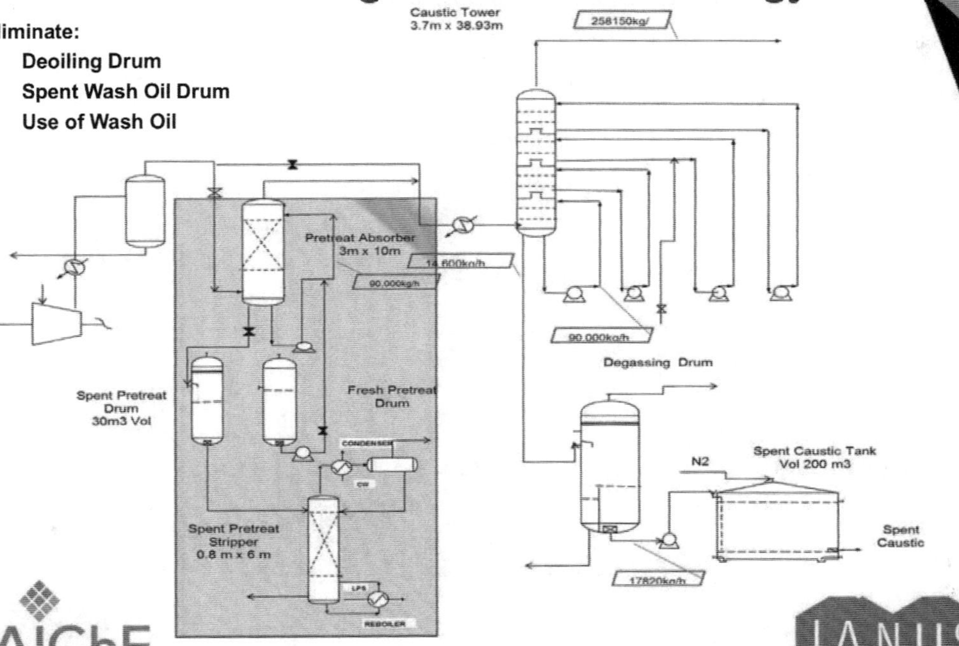

16AIChE
Spring Meeting, Houston, TX

JANUS

Experimental Results from L/L Extraction Metal Hydrogen Sulfite Solution Removes Carbonyl from FCC LPG Streams

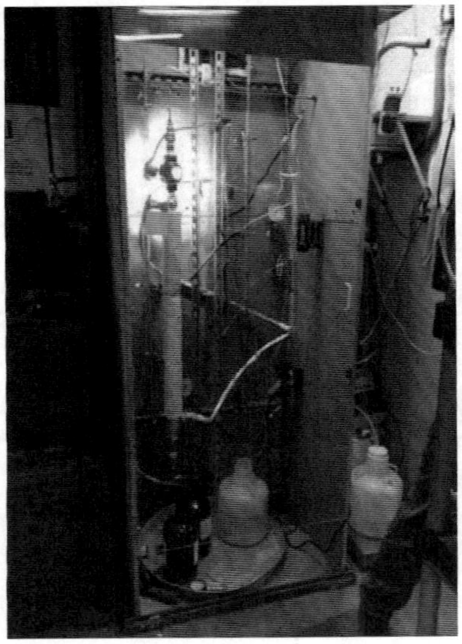

Extensive laboratory testing showed Metal Hydrogen Sulfite solution contacted with the Liquid HC stream in L/L Extractors removes the Aldehydes and Ketones down to negligible levels.

16AIChE
Spring Meeting, Houston, TX

JANUS

Carbonyl Removal Packed L/L Extractor Rig Sketch

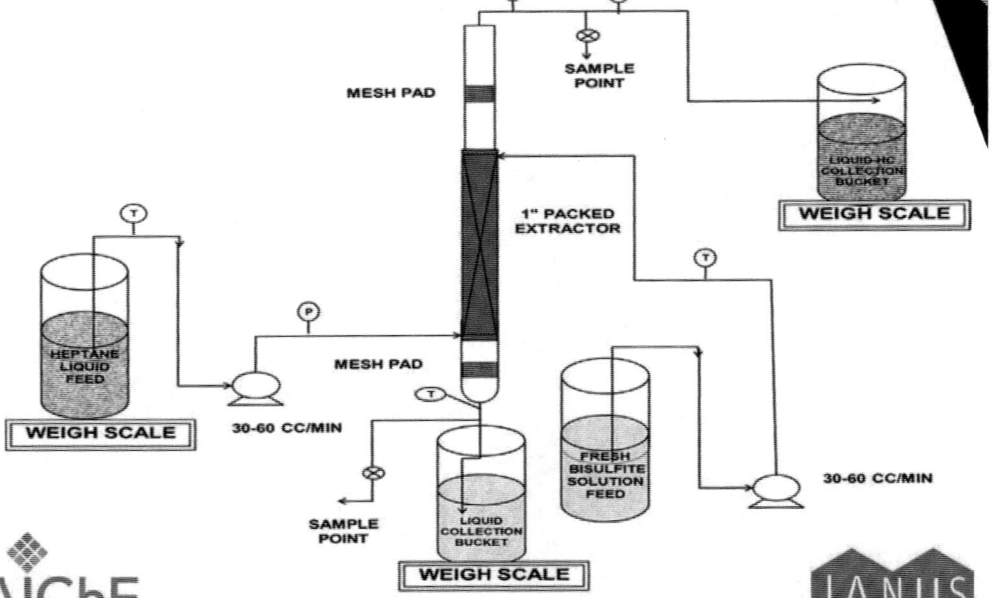

16AIChE Spring Meeting, Houston, TX

JANUS

Experimental Carbonyl Removal L/L Extractor Results Summary

	Wt% MHS Conc.	Inlet Acetaldehyde Conc. wppm	Inlet Acetone Conc. wppm	Outlet Acetaldehyde Conc. wppm	Outlet Acetone Conc. wppm
1	10	200	200	<0.1	0.5
2	10	506	496	<0.1	0.6
3	1	200	200	0.3	9
4	1	500	500	<0.1	10

16AIChE Spring Meeting, Houston, TX

JANUS

The Perfect Technology Solution for FCCU LPG Treating with Carbonyl Pretreat Unit

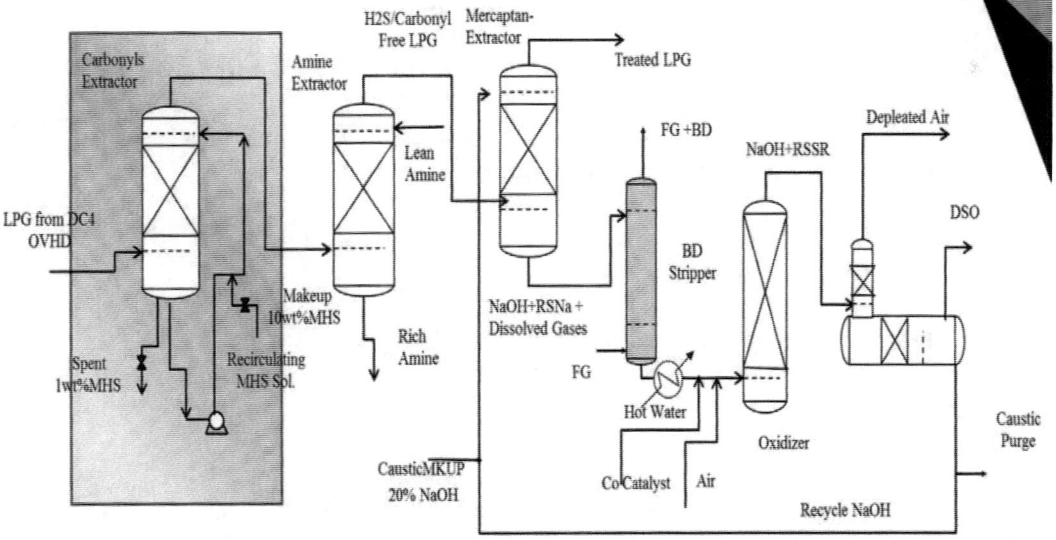

16AIChE
Spring Meeting, Houston, TX

JANUS

FCCU LPG Stream Carbonyl Pre-Treat (CPT) Fouling Prevention Advantages

- **Pretreating the LPG stream with Metal Bisulfite Solution removes >99% of the Carbonyl.**
- **The amine Extractor and /or caustic RSH Extractor treating the LPG stream will operate without fouling; thus more efficient sulfur Removal is achieved**
- **The treated LPG will be free of carbonyls and with lower sulfur content.**
- **Carbonyl free LPG will meet requirements of :**
  - **Alkylation Feed**
  - **Hydrogenation Feed**

16AIChE
Spring Meeting, Houston, TX

JANUS

Conclusions

- Oxygenates are formed in both steam crackers and FCC units.
- Carbonyls in steam crackers and FCC Units once contacted with a caustic solution form RED OIL that foul both process equipment and render caustic and spent caustic streams to be "highly fouling" by their continued reactivity in solution.
- Removal of carbonyls from hydrocarbon streams can be achieved by washing with metal bisulfite solution.
- Innovative "Patent Pending" Cost Effective methods have been devised to totally remove the carbonyls from hydrocarbon streams upstream of caustic contacting equipment in an Carbonyl Absorber or Carbonyl L/L Extractor.
- CPT Technology provides a "perfect solution" to eliminate problematic Red Oil formation in Steam Crackers and FCC Units.

16AIChE
Spring Meeting, Houston, TX

JANUS

CARBONYL REMOVAL IN STEAM CRACKERS & FCC UNITS

The Merits of Carbonyl Pre-Treat (CPT) Technology

Contact Information

JANUS

Dr. Sabah Kurukchi, VP & CTO
JANUS Technology Solutions, LLC

Tel (W): 713-7836128
Tel (M): 713-203-8746
Email: skurukchi@janusts.com
5820 Doliver Drive
Houston, Texas 77057-2455 (USA)

16AIChE
Spring Meeting, Houston, TX

JANUS

444758 Combustion Pyrolysis of Light Hydrocarbons

Abstract:

Combustion pyrolysis of light hydrocarbons

Balamurali Nair, Pankaj Gautam, and Sreekanth Pannala

SABIC technology Center, Sugarland TX-77478

The growing abundance of natural gas resources and their competitive advantage with respect to other light paraffinic hydrocarbons make the production of chemical intermediates and other high value chemicals from natural gas an attractive option. At SABIC (Saudi Basic Industries Corporation), we are committed to efficient and optimal utilization of natural resources. One method to convert the natural gas resource to high value products is through combustion pyrolysis. Combustion pyrolysis involves partial oxidation of hydrocarbons to produce smaller unsaturated molecules which can be used as building blocks for commodity chemicals. In addition, combustion pyrolysis can also be used to upgrade low value off gases/ feed streams to produce valuable products instead of using these streams as fuel or for flaring. Although combustion pyrolysis technology is five decades old, it is not widely practiced commercially due to several reasons including operational and scalability challenges. The homogeneous nature of the reaction without catalysts, simple reactor design and favorable economics could make this process attractive in the coming decades. In this work, a brief review of the existing combustion pyrolysis technology followed by experimental results on a pilot unit will be discussed.

The performance of a two stage combustion pyrolysis pilot unit for the production of chemical intermediates acetylene and ethylene- from a variety of hydrocarbon feeds is elucidated here. Combustion pyrolysis uses oxy-fuel combustion for heat generation. The extreme temperatures prevalent in the combustor (2500-2800 C) make the process development challenging. The hot gases from the combustor are mixed with the reactor feed and fed into a reactor where pyrolysis reactions takes place. The important parameters which determine the performance of a combustion pyrolysis unit are the combustor firing rate, residence time, and type of feedstock. Some of the key aspects of this technology include feed flexibility and modularity. The process development challenges include the heat management, combustor material stability, and scalability to industrial sizes. The variation of acetylene and ethylene yield with firing rate, residence time and type of feedstock will be discussed. The results of piloting help to scaleup the process to commercial level.

444990 Utilizing Pilot Plants for Process Development to Bridge the Gap Between Laboratory and Industry

Abstract:

Pilot plants have historically been constructed to manage the risk of scaling up commercial plants using laboratory data or simulated models. Pilot plant operating data can be correlated to process specific scale-up factors to be used in modeling as a more robust tool for predicting commercial yields. The capital investment to construct and operate a pilot plant is substantially less than a commercial demonstration plant, which allows systematic operational changes to pinpoint the desirable operating ranges. This presentation will illustrate real-world examples of how to bridge the gap between laboratory testing and commercial plants, showing the need for a variety of technical expertise, relatively low capital investment, and site infrastructure.

444424 Incorporating Process Safety Risk in Design and Optimization of Hazardous Chemicals Supply Chain

Abstract:

Incorporating process safety risk in design and optimization of hazardous chemicals supply chain

2,3Nitin Roy, 1Dr. Fadwa-Eljack, 3Dr. Mahmoud El-Halwagi, 2,3Dr. M. Sam Mannan

1Department of Chemical Engineering

Qatar University

Doha, Qatar

2Mary Kay O'Connor Process Safety Center

Texas A&M University System

College Station, Texas 77843-3122, USA

3Artie McFerrin Department of Chemical Engineering

Texas A&M University System

College Station, Texas 77843-3122, USA

ABSTRACT

Life-cycle analysis of a product includes the study of the overall cycle of the product from its raw materials to the disposal of the materials. Supply chain management and optimization can be considered as one part of life-cycle analysis. Industries have now very tight relations and connections with the suppliers and distributors. Any small disturbances may have huge implications in the different parts and parameters of the supply chain and the cascading effect can affect industries that are situated in different parts of the world. Supply chain optimization is very important in decision making, which gives the power to stakeholders to assess the gigantic supply chains. It has always been very hard to estimate the effect of an event on the supply chain. Supply chains may be defined as an integrated network of facilities and transportation options for the supply, manufacture, storage, and distribution of materials and products. One of the most important factors of supply chain disruption is the process safety. This part of the uncertainty is often overlooked. Quantifying process safety has been one of the main goals for the decision makers for fast and intelligent strategic decisions. Different techniques are there to analyze and quantify the risks in a facility, either storage or manufacturing and the transportation. The numbers of these techniques and quantitative indices have grown a lot during the last decade. A good analysis of these indices and techniques is necessary to intelligently choose the right risk analysis tool for the approach. Application of these risk analysis techniques will make the decision making more fast and accurate.

A mathematical model of a detailed and practical supply chain has been formed. This model has production units, storage units and transportation routes connecting them. All the production units are simulated in Aspen Hysys for getting information for economic analysis and risk analysis. The whole problem is formulated as Mixed-Integer Linear Programing (MILP) formulation. The problem was formulated in GAMS and optimized using CPLEX.

Keywords: Process Safety, hazard potential, hazardous chemical risk, supply chain network, modeling and optimization

Sustainable Productivity *and* Zero Harm *in Practice**

Is it Possible?

Rao Akula P.E., STS
Senior Member AIChE

Arcadia CA 91007, USA
narasimharaoakula2014@gmail.com

* In the oil/gas/energy industry operations and engineering/construction projects
Legend: New, Emerging

Prepared for Presentation at
American Institute of Chemical Engineers
2016 Spring Meeting
12th Global Congress on Process Safety
Houston, Texas
April 11-23, 2016

Sustainable Productivity *and* Zero Harm *in Practice**

Is it Possible?

Rao Akula P.E., STS
Senior Member AIChE

Arcadia CA 91007, USA
Narasimharaoakula2014@gmail.com

Keywords: Sustainable productivity, zero harm in practice, emerging practices, out-of-the-box, innovative thinking, group culture, internal/personal controls

Abstract

With each passing generation, the world seems to become a better place to live in. What seemed to be an unimaginable goal/impossible task yesterday, appears to have become more feasible today, with a very real possibility to turn into a reality by tomorrow. How? Why?

In spite of all the historical workplace safety issues and challenges, today's companies aspire for, and even accomplish *injury free* operations, and execution of projects and operations. Unfortunately zero harm outcomes along with sustainable productivities are *not* commonplace, even today. There still is a wide-spread belief/informed concern in the oil and gas industry that no one can actually prevent all injuries, and aspire for an outcome like "zero injuries" *in practice* on a sustainable basis, considering the nature and risks of the industry and the complexities of the projects/operations undertaken across the planet. Some experts of today even firmly (and privately) think or feel concerned that the world of (oil & gas and energy) operations/projects have the realities of productivity targets, profit and loss objectives, which may limit (or even contradict) the efforts to prevent injuries.

So the natural question to ask ourselves is "*Is it possible to accomplish **Sustainable Productivity** and **Zero Harm outcomes** in Practice*" in our operations/projects?

The answer is yes! What follows is philosophical perspective, based on experience driven narration of facts & logic, and a passionate/optimistic vision of the future, and some "out-of-the-box" thinking, where these outcomes could become realities and commonplace. Let us get started, and systematically think through, to our future vision!

* In the oil/gas/energy industry operations and engineering/construction projects
Legend: New, Emerging

1 How (and Why) Did We Get Here?

I think there is a lot of potential for progress in the future, based on what we have seen so far. In order to systematically visualize the future potential, we need to honestly analyze the past and present. In the context of safety and productivity management in our industry, we can do that in the following series of thoughts:

Each new generation appears to be better than the previous ones. Why?

- Humanity's accomplishments are based on cumulative learnings to date
- Each generation gets to stand on the shoulders of all the previous generations combined. Therefore they may appear to be taller/bigger/better!
- We are relentless in our pursuit of improvements.
- It is inherent human nature to be challenged, by the challenges human mind can think of, and the *compulsive need* to rise above them
    - Show us the second tallest mountain in the world, and we want to climb it
    - Show us the tallest, and we *need* to!
    - Give a place a no-go name like Death Valley and, well, that's nothing but a dare.

What was unusual/uncommon yesterday, is commonplace today. For example,

- Video phones were unimaginable 50 years ago; They are commonplace today
- Sustainable productivity *and* zero harm outcomes were rare in the past hydrocarbons engineering projects/programs & industrial operations. Today, these are not just feasible, but a reality.
- But this outcome is *not commonplace* in the industry, even today (yet).
- It takes nothing less than *Passionate Human Innovation and Thinking* to drive the possibilities of converting what is a difficult outcome today into what would be commonplace tomorrow.

This paper is about Sustainable Productivity *and* Zero Harm outcome (through innovative HSE/Safety Management) in the oil & gas/hydrocarbon processing business, addressing the following:

- Can we get Sustainable Productivity *and* Zero Harm outcomes in practice, in our business?
- Can we apply some *Out of the Box, & Passionate/Innovative Thinking* to get there *in Practice/Reality*? To make it commonplace?

Yes! ☺

Let us take a quick (but systematic) journey through time, in the world of safety management, before we get to the innovative parts

- Where were we yesterday?
- Where are we today?

- Where *are we headed* in the future?
- What has changed/*is changing*?
- What *can be changed*?
- How *can we control* the outcomes better?

2 Safety Management in the Past

2.1 The Golden Gate Bridge Project in California

Figure 1. Golden Gate Bridge

I picked this example, because it illustrates the importance as well as the impact of the "out of the box" thinking on safety outcomes. It also brings out the "common place" aspect of our thinking and how it changes with time.

In the 1930s, when the Golden Gate Bridge at San Francisco was just a project, there was an industry perception that "each million dollars spent in a project of that nature, it would be reasonable/pragmatic to expect one fatality at work". At the end of the project, there were 11 fatalities (10 in one fall, 1 in a scaffold collapse), and the project spent $37 million dollars. In addition, 19 lives were saved by some unique safety innovations instigated and implemented by a very passionate project manager (Joseph B. Strauss).

The Project manager decided to achieve "unheard of safety performance", distinguish himself from his contemporaries, find a "new point of view" and care for his workers. He made everyone in the project wear a hard hat (!), and installed safety nets to catch falling workers since the highest safety exposure area was – possible injury due to fall from heights. If he had done nothing, 30 would have died, and the project would still have been considered successful by the standards/expectations of the day. The 19 grateful survivors formed a "Halfway to Hell Club".

3 Safety Management Today

3.1 Safety/Hazard Management Matrix Today

Since the highest exposure for safety incidents is during our field activities, let us look at a typical safety/hazard management matrix in use today in a typical oil & gas operation/project:

Table1: Safety/Hazard Management Matrix Today

Hazard/ Energy Source	**Description/ Example**	**Exposure Control/ Protection Measures (Besides ISD,SCDM*)**
Body Mechanics	Pinch Points, Access Issues, Slips, Trips & Falls	Eliminate/PPE
Chemical	Acids, Corrosion Inhibitors, Poisonous gases like H2S, Asbestos, lead-based paint	Eliminate/PPE
Electrical	Lockout/Tag out Permit issues, Electrocution	Substitute (use PPE)
Gravity	Fall from heights, dropped objects, permit violations, poor scaffolding	Admin Controls
Mechanical	Exposure due to lack of protection guards from rotating equipment	Admin Controls
Noise	Hearing loss from high decibel surroundings	PPE
Pressure	Isolation problems, release of hydrocarbons and other hazardous conditions (e.g. steam, compressed air)	PPE
Radiation	Radiation exposure (e.g. coke drums level instruments)	Isolate
Thermal	Lack of insulation on hot pipes, fires, explosions	Engineering Controls
Biological	Wild life in idle/long isolated equipment or elsewhere (like Off-sites)	Engineering Controls

ISD = Inherently Safer Designs; SCDM = Safety Critical Design Measures

3.2 Safety Work Practices Today

A typical list of work practices we come across these days at the workplace may look like this:

Obligation to Stop Unsafe Work - Everyone has an obligation to stop work that is unsafe be it work on a design that is unsafe, or physical work being performed in any location whether it be in the design office, the field, at home, …ANYWHERE

1. **Permit to Work** - Be mindful of routine and non-routine activities and design out potential hazardous situations; Develop work plans that minimize simultaneous operations; "Inherently Safe Design"; Future tie-in flanges
2. **Working at Heights** - Be mindful of routine and non-routine activities requiring working at height and provide appropriate platforms designed for appropriate loads
3. **Energy Isolation** - Provide a means to safely de-energize /de-pressure. Provide means to lockout systems to prevent inadvertently re-energize / re-pressure
3. **Vehicle Safety** - Be mindful of vehicle access / egress movements / requirements and provide appropriate clearances and traffic control devices e.g. markers, guard rails, and etcetera's; Practice defensive driving; Don't use your cell phone while driving
4. **Ground Disturbance** - Minimize or eliminate trenching, drilling, and excavation as appropriate
5. **Confined Space Entry** - Minimize / eliminate the need for confined space entry; Internal coatings and claddings; Internal cleaning devices; Requirements for external inspection
6. **Lifting Operations** - Be mindful of routine and non-routine lifting activities and provide appropriate lugs, lifting beams, trolleys, and cranes designed for appropriate loads
7. **Management Of Change (MOC)** - Strictly adhere to the "design freeze" concept (remind the client that the concept applies to them as well); Generate MOC documentation and obtain approvals before proceeding with design changes; Document and obtain approval for spec adaptations

3.3 Current Limitations on Sustainability of Outcomes in Safety Performance and Productivity

While the plans and practices such as those described above for safety and productivity management sound good on paper, gaps continue to exist between the goals and reality when it comes to outcomes. I think we all agree that sustainable productivity and zero harm outcomes are hard to accomplish in a group. Why? Some impediments explaining the potential challenges in filling such gaps (between plan and reality) are presented below:

- Untethered, human nature naturally leans towards freedom/ rights; and moves away from responsibilities
- Externally imposed influences and controls are less effective than self-imposed
- Business outcomes regarding productivity and zero harm objectives may be viewed (by some) as contradicting each other
- No one is blessed with all the human relationship skills at birth, nor are they taught (well) in our education systems. They are acquired skills, as we grow. Same applies to safety management skills and perception controls
- Therefore, good companies spend enormous amount of resources in training, & resolution of conflicts. Focus on preventing them is usually a budget controlled one.
- Sustainable productivity requires sustainable peace and harmony within the work groups/company. Many (if not all) countries on the planet have yet to accomplish sustained overall peace and harmony. One reason for this situation could be the top-down management method we use for control of issues. Same applies to our industry's safety performance/statistics, and their management of personnel, & our personal lives.
- Common sense is not that common!

Here are some of my explanations & (suggested) solutions in this context:

- Is expecting "sustainable outcomes" as "utopian thinking" as accomplishing "sustainable world peace"? (I don't think so). Many projects, operations, and tasks we execute today already accomplish these outcomes. The key word here is sustainable (not the outcome).
- We have to make progress in our abilities to accomplish sustainable objectives in our personal lives, and working in groups – (think decades or longer)
- Our focus should be on developing abilities for self-regulation (besides the practices/tools we already have) in our groups
- The team with most self-discipline is the most likely to succeed
- How do we do this?

4 Innovative Thinking for the Future

4.1 Extension of Emerging Practices

Based on what we have seen so far, and examining the emerging practices in our industry, let us think about the areas and controls that are physical, and those *beyond physical,* with focus on emerging practices and potentially new practices:

- Health – (Physical, Mental)
- Safety – (Plant/Process, People, Engineering/Design)
- Environment – (Physical, Solid, Liquid, Gas, Culture, Work, Family)
- Control (& Influence)
- Laws (Local, Regional, National, etc.) - Enforcement
- Regulations – Audits (Internal, External)
- Management Policies (Company, Business Unit, Client)
- Expectations/Requirements (Company/Boss, Client, Personal) KPIs (Performance – Personal, Group, Perceptions – Ours & Theirs)
- Work Practices (ours, theirs, industry's)
- Affluence and Influence (consequences, rewards, incentives)
- Values & Excellence (meet vs. exceed expectations)

4.2 Out-of-the-Box thoughts for Future

If we reverse the focus from outward management /controls to self- imposed personal controls on an individual and on harmoniously working groups, radically different focus areas emerge:

- Health Objectives – Physical, mental, social (beyond work)
- Well-Being - Personal, Family, and Community
- Happiness – Personal, Group, Groups
- Will Power - Personal, Group, Groups
- Consciousness - Personal, Group/Family, Groups/Community
- Vitality – Personal, Group

- Peace – Personal, Group
- Sustainable Group outcome – Zero Harm, Productivity

5 Philosophy of Peace and Happiness

More often than not, productivity of a group is negatively impacted by personal and personality conflicts within groups based on gaps between declared and real objectives of individuals (aka dishonest communications). Some conflicts could also arise from unfair assignment of credits and blames, rewards and consequences within the groups. It is also not uncommon to come across individual (as well as group) actions for their own benefit, at the expense of others. Therefore, it is critically important that groups working together should ensure means to attain and sustain peace within the group. Safety and productivity may not appear to be directly related to each other. However, superior and sustainable performance in both areas is strongly connected to peace and happiness within the group. Given below is a philosophical perspective on peace and happiness:

5.1 What and Why?

- The most important ingredient of success in business is business acumen. Philosophy is an integral part of business acumen (& part of our values).
- Peace is the Foundation (think sustainable peace). It
  - Is not just absence of conflicts (think prevention)
  - Is not just calmness (be actively calm & calmly active with conscious will, health, and vitality)
  - Includes freedom (think freedom/choice of attitude (& gratitude!))
- Happiness is the Mansion we build on the Foundation (positive and tangible)
  - It comes from small events - Examples (UK Survey top 4)
    - Craftsman whistling over a job well done
    - Boy building sand castles
    - Mother bathing a baby
    - Doctor who saved a life
  - A project completed with actual zero harm outcome *and satisfied employees, customers, & supply chain*
  - The only way to multiply happiness is by dividing it
  - If you like to be happy
    - For an hour – take a nap
    - For a day – go fishing
    - For a year – build a fortune
    - For a life time – work for someone else's happiness
- Happiness is an *inside* job

5.2 Suggestions for Sustainable Outcomes

Sustainable positive outcomes in productivity and zero harm outcomes are made possible by motivated competent managers and workers who practice consistently fair and honest communications, work practices, accountability, rewards and consequences to all within the context of their business. If you want to accomplish sustainable positive outcomes in your productivity & zero harm objectives in your groups (Project(s), Office, Location, Region, etc.), consider these points:

- Embrace peace & avoid/elevate (negative/destructive) conflicts and disagreements – (also think behaviors)
- (Constructive conflicts/disagreements (from the diversity and inclusion perspective) aimed at continuous improvement are to be handled differently – with efficient systems and mechanisms to convert into opportunities & realize them)
- The only way we can change others is by reforming ourselves
- Focus on sustainable peace within the group – not short-term profits/results at the expense of long-term values/ relationships
- Be able to agree to disagree, respectfully
- Begin a culture of peace within your group. Here are some characteristics of the peaceful and productive culture at the workplace:
    - An unmistakable ability to enjoy every task & moment in the group
    - A loss of interest in judging others
    - A loss of interest in judging yourself
    - A loss of ability to worry
    - Frequent overwhelming episodes of appreciation
    - Content feelings of connectedness with others and group outcomes
    - Frequent attacks of smiling
    - Increased susceptibility of caring experienced by others & passing it on
    - Peace and happiness are visible in the group members – They seem to carry a song in their heart & a spring in their step!
    - A pleasant taste, that remains in everyone's memory *and grows* with completion of each successful project

6 Conclusion

It is clear that in order to control outcomes on performance objectives in safety and productivity, we will continue to need innovative and out-of-the-box thinking. It is nice to see winners in this area, but there is no reason why these outcomes are to be limited to few. An individual's success is more likely when one looks inward for improvements, instead of any outward quest for improving others, and systems. Any excuses, and practices like assigning blame to circumstances and other persons can never properly justify any lapses in anyone's performance. In fact, an individual's chances for success (in their performance and productivity) are greatly enhanced when they accomplish peace, harmony and happiness within themselves and with others. Similarly, a group can succeed if their focus is on their group culture, harmony, peace, and happiness. When it comes to sustainability of the group's performance, our focus should go beyond the physical means to the behavioral ones, top down external controls to internal/ personal controls, and beyond personal discipline to group culture and social values. At the end of the day, out-of- the-box thinking, smart and fair systems are only part of the solutions to the puzzle. The oil/gas/energy industry's sustainable performance for productivity and zero harm outcomes in practice can be made possible only by people's actions. Their sustainable success should be enabled by motivated competent managers and workers who practice consistently fair and honest communications, work practices, accountability, rewards and consequences to all within the context of their business.

About the Author:

Rao Akula is a Senior Member of AIChE, and a California Registered Professional Chemical Engineer, with 30+ years work experience in three countries. Rao has a Master's degree in Computer Information Systems, and a Bachelor's degree in Chemical Engineering. Rao has worked in a wide variety of settings in the industry - Management, HSE, Process Safety, Environmental Permitting, Remediation/Clean-up & Compliance, Advanced Controls, Process Design (petroleum refineries, ammonia and ethylene plants); Government, Operating Petroleum Refineries, Engineering & Consulting. Rao's past employers include WorleyParsons (Arcadia, California), BP (Long Beach, California), State of California Department of Toxic Substances Control (Glendale & Sacramento California), Brown & Root (Alhambra, California), Kuwait National Petroleum Company (Mina Abdulla Refinery, Kuwait), and Engineers India (New Delhi, India).

443716 Energy Efficient Vacuum Pressure Swing Adsorption System

Abstract:

Energy Efficient Vacuum Pressure Swing Adsorption System

A high energy efficiency vacuum pressure swing adsorption system design with direct drive centrifugal compressor has been developed and will be presented.

Vacuum Pressure Swing Adsorption (VPSA) technology is commonly used to supply oxygen to a wide range of industries. Glass, waste water treatment, pulp and paper, mining and cement, metal fabrication, pharmaceutical, and fermentation are a few examples. Many of these applications use oxygen rather than air which includes a substantial portion of nitrogen for environmental sustainability and productivity improvements. The competitive cost of oxygen supply is an important consideration factor when choosing an oxygen supply system.

Electric energy cost is the largest operating cost for VPSA system since the raw material, air, is free. Typical VPSA oxygen plants are designed to be operated with minimum operator assistance. With increasing electricity costs, it becomes more important to develop technology to reduce VPSA energy consumption, and therefore reducing the total oxygen cost to the customers.

Conventional VPSA technology uses displacement rotary lobe types of blowers operating at fixed speeds to compress air to the desired process pressure. These machines generally have low efficiency with no capability of turn down. Up to 40% of the energy supplied to compressor is wasted through various inefficiencies. In addition to high energy requirements and poor utilization, these blowers are generally very noisy. In order to meet 85 DBA requirements, state of the art silencers and a building is typically required to mask the excessive noise to protect the operators hearing.

It is desirable to develop a new VPSA compressor technology which has higher energy efficiency and is quieter.

Comparing to displacement rotary blowers, centrifugal compressors have a demonstrated much higher energy utilization (efficiency). The primary difference of utilizing a centrifugal compressor in a VPSA process is that the vapor pressure swing process requires routine rapid pressure changes. In order to maintain centrifugal compressor peak efficiency, the compressor drive system is required to match the head pressure requirement of both the pressurizing and evacuating step of the adsorbent beds by continuously varying the speed.

A direct drive, high speed permanent magnetic motor with magnetic bearings was developed to provide the centrifugal compressor with the variable speed required by VPSA process. In contrary to the conventional induction motor and gearbox system, the direct drive centrifugal compressor is capable of accelerating and decelerating rapidly to meet the short cycle time requirement for VPSA system, and also eliminates the losses from gearbox, conventional bearings, and lube oil system.

Comparing to the conventional VPSA there are many benefits attributed to the direct drive centrifugal VPSA system:

1. Electric power usage is significantly lower, up to 20% power reduction is expected.
2. System layout is more compact, since the motor and compressor are smaller and there is no gearbox and lube oil system.
3. The system is significantly quieter therefore the building is not required.
4. Seal water and associated piping are no longer needed.

445014 CFD Simulation of a Lab-Scale Bubble Column

Abstract:

Bubble columns are widely used in the chemical, petrochemical and biochemical industry for gas liquid contact due to their simple construction and absence of moving internal parts. The prediction of the fluid flow behavior, gas hold-up and bubble size distribution is of major importance for the development and design of such reactors. These characteristics directly affect the performance and eventually economics of the process. Experimental methods to investigate bubble columns are common. But numerical methods are also widely used to study bubble columns, since the numerical simulations has made significant progress in the last several years for such multiphase flow.

In this contribution we have numerically investigated a 3D lab scale bubble column to validate the predictive capabilities of the commercial STAR-CCM+ CFD solver by CD-adapco. For the simulation of the two phase flow, an Eulerian-Eulerian approach was used. The influence of several forces which define the interactions between the continuous water phase and the dispersed gas phase is investigated, which includes, amongst others, drag force, lift force, turbulent dispersion force, virtual mass force and turbulence induced by bubbles. Bubble size distribution as well as bubble break-up and coalescence is incorporated by using the S-gamma model proposed by Lo[1].

Validation is done with respect to the gas plume oscillation frequency, liquid and bubble velocity and gas holdup against published data by Pfleger[2] and Diaz[3]. A reasonable agreement between the numerical and experimental results was found. Further the importance of the different interaction forces could be shown. A mesh study was conducted to compare calculation time and accuracy of the results with respect to the used cell types, hexahedrons and polyhedrons.

In the second part of this contribution, the oxygen transfer from the gas bubbles to the continuous water phase is investigated. The overall mass transfer coefficient is compared with literature data and also shows a reasonable agreement.

[1] Lo, S. & Zhang, D. (2009); J. Comp. Multiphase Flows (1), pp. 23-38

[2] Pfleger, D.; Gomes, S.; Gilbert, N. & Wagner, H.G. (1999); Chem. Eng. Sci. (54), pp. 5091-5099

[3] Diaz, M.E.; Montes, F.J. & Galan, M.A. (2008); Chem. Eng. Proc. (47), pp. 1867 - 1876

443736 Sublimation: Solids Do Volatilize

Abstract:

Chemical engineers typically do not encounter processes involving sublimation. However, solids do sublimate. Chemical engineers become involved with sublimation in laboratories and pilot plant areas where various solids may accumulate. If these solids are hydrocarbon-based, then the chemical engineer supervising the area must calculate the exposure limits for personnel working there. The same calculation must be done for any volatile inorganic substances in the laboratory or pilot plant as well.

We generally calculate these exposure limits by assuming a CSTR approximates the laboratory or pilot plant area. However, such models require a rate constant --- a rate constant that describes the generation of the sublimating chemical.

This presentation describes a method for measuring a sublimation rate constant and discusses the exposure level calculation using that rate constant. The experiment described in this presentation is the sublimation of naphthalene.

443786 A Multi-Functional Circulating Fluidized Bed Pilot Plant for Catalytic Process Development

Abstract:

Catalytic processes are widely used in chemical or energy industries for producing either chemicals or liquid fuel. The Dalian Institute of Chemical Physics (DICP) has long been involving in the catalytic processes development. One of the recent achievements is the fluidized bed methanol-to-olefins technology (DMTO), which has been successfully commercialized and is becoming one of the main streams for light olefins production in China. This presentation will introduce the multi-functional circulating fluidized bed pilot plant in DICP. This pilot plant includes a regenerator and three different reactors: a riser reactor (10 m high), a multi-staged turbulent fluidized bed reactor (15 cm diameter), and a bubbling fluidized bed reactor (40 cm diameter). Any one of these three reactors can be used either alone or in combination with the other two upon the requirement for different processes. The unit can process 1-3 t/d methanol feed when used for DMTO process development. In addition to the reactor-regenerator system, the unit also has feed pre-treating, product separation, and dry gas combustion system. The presentation will illustrate the optimal design and operation of this unit by use of the development of DMTO and some other catalytic processes as examples.

443484 Cavitation Refining of Junk Oil

Abstract:

Introduction

The PHCR Process includes pressure, heating, cavitation and reaction. Inputs include junk oil for enhancement and recovery. (knickle). The process involves breaking of bonds. The input depends upon the junk oil but typically might involve transport ships bottoms. The bottoms can then be blended with a lighter hydrocarbon such as diesel oil and with water. Experimental results have indicated this mixture can create an equivalent to number two fuel oil. The degree of cracking would add an important analysis to this process.

When a chemical reaction occurs, molecular bonds are broken and other bonds are formed to make different molecules. Bonds of two water molecules are broken to form hydrogen and oxygen.

$$2H\,2\,O \rightarrow 2H\,2 + O\,2$$

Bonds do not break and form spontaneously but require an energy change is required. Bond energy is the energy input required to break a bond. The shorter the bond length, the greater the bond energy.

Although each molecule has its own characteristic bond energies, some generalizations are possible. For example, although the exact value of a C–H bond energy depends on the particular molecule, all C–H bonds have a bond energy of roughly the same value because they are all C–H bonds. It takes roughly 100 kcal of energy to break 1 mol of C–H bonds, so we speak of the bond energy of a C–H bond as being about 100 kcal/mol. A C–C bond has an approximate bond energy of 80 kcal/mol, while a C=C has a bond energy of about 145 kcal/mol.

In a chemical reaction the atoms in the reactants rearrange their chemical bonds to make products. The new arrangement of bonds does not have the same total energy as the bonds in the reactants. Therefore, when chemical reactions occur, there will always be an accompanying energy change.

For an endothermic chemical reaction, energy is absorbed as reactants are converted to products. A procedure would be to first look at the molecules determine what bonds exist and then examine the new molecules to determine the bond breakage located in the reactant side. Then we can find the net change of the reaction. Energy is required to break bonds. For this question simply look at the number you calculated as your enthalpy of reaction. Is it positive or negative? It is positive so this means that it is in fact endothermic. It requires energy in order to create bonds.

Bond Energies

Molecules have different C-C bond energies but usually they are similar and the enthalpy change for breaking bonds is **positive.** We can calculate a pseudo Ea in the Arrhenius Equation to determine temperature dependence. By knowing bond energy at two different temperatures and using the Arrhenius equation. C-C and H-C and H-O in hydrocarbons are covalent bonds. Covalent bonds share electrons.to form a molecule – covalent bonding. Bond Dissociation Energy (Bond Energy) is the energy need to break one mole of bonds. Bond energy is a measure of the amount of energy needed to break apart one mole of covalently bonded gases. The SI units used to describe bond energy is kiloJoules per mole of bonds (kJ/mol).

Organic Reactions and Cracking

Cracking is the process whereby complex organic molecules such as kerogens or other heavy hydrocarbons are broken down into simpler molecules such as light hydrocarbons, by the breaking of carbon-carbon bonds in the precursors. The rate of cracking and the products are strongly dependent on temperature and presence of catalysts. Cracking is the breakdown of large alkanes into smaller, more useful alkanes and alkenes. Simply put, hydrocarbon cracking is the process of breaking a long-chain of hydrocarbons into short ones.

Bitumen is an oil-based substance. It is a semi-solid hydrocarbon product produced by removing the lighter fractions (such as liquid petroleum gas, petrol and diesel) from heavy crude oil during the refining process. (Refined bitumen) At ambient temperatures, bitumen is a stable, semi-solid substance. Bitumen is created according to the requirements of its end use specification. This is achieved either by the refining process or by blending.

Bitumen is too viscous to pump at room temperature. It must be kept hot or blended and heated.
Bitumen can either be cracked or diluted to reduce viscosity. The bitumen can be diluted with light ends such as diesel. (up to 20% diluent). This will be referred to as diluted bitumen.

Cavitation

There are two primary types of cavitation including **1.** Vaporization Cavitation; and
2. Internal Recirculation Cavitation

According to Bachus, vaporization cavitation accounts for nearly 70 percent of all cavitation cases, with internal recirculation cavitation accounting for 15 percent of cases

In chemistry, a radical (more precisely, a free radical is an atom, molecule, or ion that has unpaired valence electrons. With some exceptions, these unpaired electrons make free radicals highly chemically reactive towards other substances, or even towards themselves: their molecules will often spontaneously dimerize or polymerize if they contact each other. Most radicals are reasonably stable only at very low concentrations in inert media or in a vacuum.

A notable example of a free radical is the hydroxyl radical (HO•), a molecule that has one unpaired electron on the oxygen atom. In contrast, the hydroxyl ion (HO−) is not a radical, since the unpaired electron is resolved by the addition of an electron.

Creation of free radicals occur in a number of ways, including synthesis with very dilute or rarefied reagents, reactions at very low temperatures, or breakup of larger molecules. Any process that puts enough energy into the parent molecule can affect a large hydrocarbon molecule. Input energy can include ionizing radiation, heat, electrical discharges, electrolysis, pump energy and chemical reactions. Indeed, radicals are intermediate stages in many chemical reactions.

Following recent nomenclature revisions, a part of a larger molecule is now called a functional group and "radical" now implies "free". However, the old nomenclature may still occur in the literature.

A familiar free-radical reaction is combustion. Combustion consists of various radical chain reactions that the singlet radical can initiate. The flammability of a given material strongly depends on the concentration of free radicals that must be obtained before initiation and propagation reactions dominate leading to combustion of the material. Once the combustible material is consumed, termination reactions again dominate and the flame dies out. When a hydrocarbon is burned, a large number of different oxygen radicals are involved.

The dimensionless **Cavitation number** (**Ca**) is used to characterize the relationship between the difference of a local absolute pressure from the vapor pressure and the kinetic energy per volume, and is used to characterize the potential of the flow to cavitate.

Where

1. is the density of the fluid.

2. is the local pressure.

3. is the vapor pressure of the fluid.

4. is a characteristic velocity of the flow.

Experimental Analysis

Analysis of beginning and final samples of the process can be made by application of GC-VUV technology, vacuum ultraviolet (VUV) detector for gas chromatography (GC). The percentage cracking of the process can then be obtained.

References

1. Knickle, H. et al "Innovative Hydrocavitation Technology Process to Create Composite Fuels" AIChE Spring Meeting, Session 141, Houston TX (2012)

2. Petrucci, Ralph H., Harwood, William S., Herring, F. G., and Madura Jeffrey D. General Chemistry: Principles and Modern Applications. 9th ed. Upper Saddle River: Pearson Education, Inc., 2007.

3. Carruth, Gorton, Ehrlich, Eugene. "Bond Energies." Volume Library. Ed. Carruth, Gorton. Vol 1. Tennessee: Southwestern, 2002."Bond Lengths and Energies."

4. UWaterloo, Nov 2010. http://www.science.uwaterloo.ca/~cch...20/bondel.html

5. *Anal. Chem.*, **2014**, *86* (16), pp 8329–8335, **DOI:** 10.1021/ac5018343, July 31, 2014

444848 Application of Self-Heat Recuperation to Industrial Methanol Synthesis Process

Abstract:

Many methanol synthesis processes have reactant recycle system with product separation to achieve high conversion ratio of the whole process, as well as other chemical processes. For this product separation from reactant recycle stream, a gas-liquid separator has often been used and distillation column has often been used for product purification. However, this gas-liquid separator and distillation column are well known as energy consuming processes. In fact, a huge amount of coolant is required for the cooler located between reactor and the gas-liquid separator in the methanol synthesis process. Furthermore, there are two distillation columns are normally located in the methanol synthesis process to purify the crude methanol and these columns consumed a large amount of energy. In addition, some of the exothermic reaction heat are utilized for the preheating of the reactant stream, leading to reduction of recovered heat amount by steam generation.

Authors have developed self-heat recuperation technology. In this technology, each process unit is divided into functions to analyze the process energy balance and not only the latent heat but also the sensible heat of the process stream can be circulated without any heat addition, leading to huge reduction of energy requirements in several chemical processes. This self-heat recuperation technology is suitable for thermal and separation processes and is applied to several chemical processes for energy saving as case studies.

In this research, we investigated the feasibility of applying self-heat recuperation technology to a methanol synthesis process using industrial conditions and data and developed an innovative process for methanol synthesis process from the energy saving point of view. By installing the self-heat recuperation technology to the methanol synthesis process, the energy consumption of the process can be greatly reduced. In addition, all of the exothermic reaction heat of the methanol synthesis can be recovered by steam generation.

443748 $Mn(OAc)_2$ as an Effective Catalyst for Synthesis of Dimethylhexane-1,6-Dicarbamate (HDC) from 1,6-Hexanediamine(HDA) and Dimethyl Carbonate (DMC)

Abstract:

The catalytic synthesis of dimethylhexane-1,6-dicarbamate (HDC) from 1,6-hexanediamine (HDA) and dimethyl carbonate (DMC) over a series of homogeneous catalysts (e.g., metallic nitrates and acetates) were investigated. $Mn(OAc)_2$ was screened as the most effective catalyst for this reaction. The effects of reaction temperature, DMC-to-HDA molar ratio, catalyst concentration and reaction time, were investigated. Under the optimum conditions (reaction temperature of 363K, DMC-to-HDA molar ratio of 4:1, catalyst concentration of 7wt.% (based on HDA), and reaction time of 5 h), HDC yield can reach as high as 98.0% with 100% HDA conversion. Online in situ Fourier transform infrared (FTIR) were used to investigate the interactions between $Mn(OAc)_2$ and substrate. The results indicated that the new species, which was formed by the reaction of HDA with $Mn(OAc)_2$, attacked carbonyl carbon of DMC to generate intermediate HMC and finally to HDC, Density Functional Theory (DFT) calculation support the results from FTIR.

The Principle of Intrinsic Reality
Solving the World's Woes through Chemical Engineering and Economics

Jed Moyano Bellen, SCPM, PMP
Hyundai Engineering & Construction Co. Ltd.
9 Temasek Blvd., #27-02 Suntec Tower 2, Singapore 038989
jed.bellen@hdec.co.kr

Prepared for Presentation at
American Institute of Chemical Engineers
2016 Spring Meeting
11th Global Congress on Process Safety
Houston, Texas
April 10-14, 2016

UNPUBLISHED

The Principle of Intrinsic Reality Solving the World's Woes through Chemical Engineering and Economics

Jed Moyano Bellen, SCPM, PMP
Hyundai Engineering & Construction Co. Ltd.
9 Temasek Blvd., #27-02 Suntec Tower 2, Singapore 038989
jed.bellen@hdec.co.kr

Keywords: Principle of Intrinsic Reality, Chemical Engineering, Economics, Sustainability

Abstract

Human beings in profound rationality always seek for explanations of the phenomena around them. To understand these phenomena, human beings have developed systematized body of knowledge as a practical way of dealing with life's challenges. Among the systematized body of knowledge developed are chemical engineering and economics.

An avant-garde way of looking at each of these disciplines leads to the Principle of Intrinsic Reality. This principle relates the fundamental laws of the two disciplines to glean insights on the constraints imposed by conservation laws on economic systems at both the macro and micro levels.

It can be argued that if mass and energy in the universe are constant, then logically diminishing returns and marginal utility are bound to be observed. This logic is quite evident in a microorganism culture which follows the Monod growth model.

It can also be thought that conservation laws imply scarcity of resources. As such, it enjoins everyone to find optimal and efficient use of resources in the context of trade-offs, production decisions and sustainability. It even calls for the development of new technologies to produce more outputs from less inputs (i.e. shifting of production possibility frontier to the right). In this way, the Principle of Intrinsic Reality leads to the imperative for sustainable optimization.

Furthermore, it is deemed that the conservation laws can explain the phenomenon called "polarity of wealth." As mass and energy when harnessed create wealth, then the constancy of mass and energy implies boundary for wealth creation. In effect, wealth can neither be created nor destroyed but can only shift from one entity to another thereby creating polarity in wealth distribution. This assertion can be the key towards more equitable restructuring of economic systems to close income gap and thereby result to stable and resilient populace.

In conclusion, this paper envisions for a paradigm shift from traditional money-centric view of chemical engineering and economics to a synergistic convergence of disciplines which can powerfully tackle the world's woes and thereby effect sustainability.

1. The World in which We Live Today

The past few years saw unprecedented global challenges which have put into question existing economic models and scientific paradigms. Among these challenges are the dystopian global economy, climate change, environmental degradation and terrorism. The magnitude of these challenges has grown to a profound level that newer ways of thinking and perspectives become necessary. With that end in mind, this paper attempts to look at the fundamental laws of chemical engineering and economics in new ways to recognize our intrinsic reality from which these pressing challenges can be tackled at large.

2. The Principle of Intrinsic Reality

It is an observable fact that advancement in science increasingly blurs boundaries of various disciplines in tackling proactively the pressing global issues or problems. Within this purview, two seemingly diverging disciplines namely chemical engineering and economics can be taken together to glean insights on our intrinsic reality and what it means for our economic and ecological systems on a larger scale.

In the author's contention, intrinsic reality is grounded on the fundamental laws of chemical engineering and economics. These laws are the Conservation Laws (Chemical Engineering) and Law of Diminishing Returns (Economics). It can be argued that the Law of Diminishing Returns is observable because of the constraints imposed by the Conservation Laws. As such, this leads to the Principle of Intrinsic Reality which can be stated thus;

> *In any system, both at the macro- and micro-levels, the limiting factor for growth is the constancy of potential inputs.*

Potential inputs mean not the actual inputs to processes but the constant material or energy available which can be harnessed to create wealth. The reasoning or proof of this principle is presented in subsequent discussions.

2.1 Proof Based on the Monod Growth Model

The best illustration to understand the Principle of Intrinsic Reality is the Monod Growth Model. Monod in his 1949 seminal paper presented an empirical relation between exponential growth rate and concentration of essential nutrient. The data was fitted into hyperbolic equation:

$$R = R_K\left(\frac{C}{C_1 + C}\right)$$

In this equation, R is the growth rate during the exponential phase; R_K is the rate limit for increasing concentrations of C; C is the concentration of the nutrient; and C_1 is the concentration of nutrient at which the rate is half the maximum.

Figure 1: Graphical Representation of Monod Growth Model

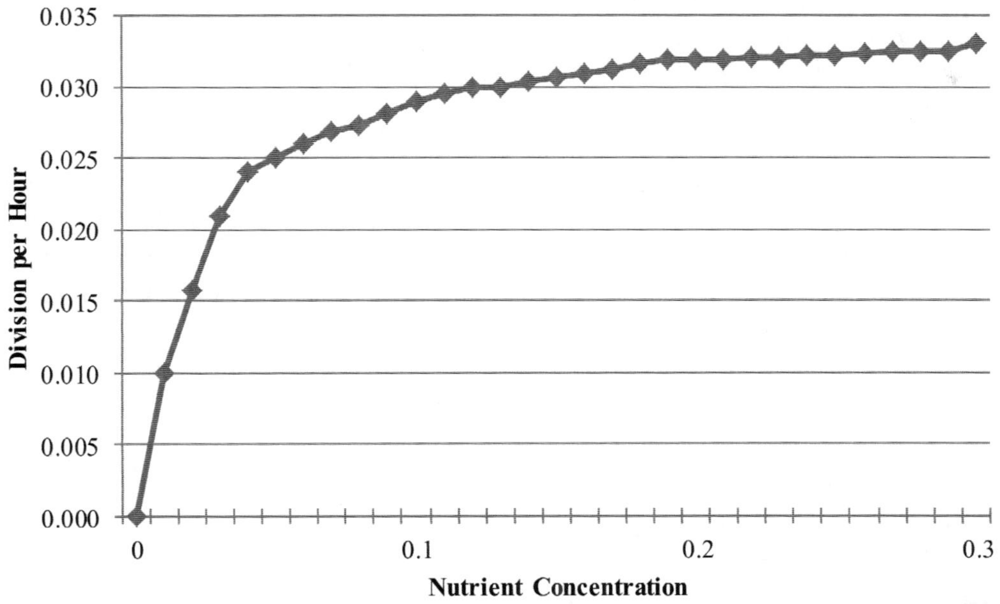

Monod (1949) further pointed out that the limiting factors commonly found are exhaustion of nutrients, accumulation of toxic metabolic products and changes in ion equilibrium.

The foregoing discussion has been observed at the micro-level with the bacterial culture. At the macro-level (i.e. the world at large), analogous phenomena are also observed to occur because of the same constraint imposed by the constant quantity of resources (i.e. material and energy). One example of these phenomena is that human beings cannot grow indefinitely. If they do, then how can the world sustain that growth with limited available resources (e.g. foods) to support it?

Another analogous phenomenon is in population growth. Thomas Malthus (1789) predicted that when population growth exceeds agricultural production, the population level is forced to return to subsistence-level conditions. This phenomenon as Malthus (1789) believed is due to the restriction of available resources supporting population growth, which others have claimed to be exploding as the earth becomes over populated.

In a surprising manner, a similar phenomenon can be observed in production processes in which the increase in a single factor of production assuming all other factors constant, results not to more marginal output but to a reduced one. This was attributed by classical economists, Thomas Malthus and David Ricardo, to the decreasing quality of inputs while neoclassical economists attributed this to the disruption in the entire production process as additional units of input are added to a fixed amount of capital. But in view of the Principle of Intrinsic Reality, the phenomenon is viewed intrinsic to the production system in which case infinite amounts of mass or energy cannot be processed with perfect efficiency. Hence, the law of conservation of mass and energy comes into the fore.

It can be deduced from the discussion that there is an underpinning intrinsic reality to the more obvious phenomena in the world from the micro-scale (bacterial growth) to the macro-scale (human growth and production processes).

2.2 Proof Based on Diminishing Marginal Utility

Another way to prove the Principle of Intrinsic Reality is through the diminishing marginal utility observable among humans and perhaps other organisms as well. Marginal utility, as defined in economics, is the gain from an increase or the loss from a decrease in the consumption of goods or services. The diminishing marginal utility is an observable phenomenon in which the utility or satisfaction an individual gains diminishes as he or she consumes more of the same unit of goods or services. Graphically, this can be represented as in Figure 2.

Figure 2: Graphical Representation of Diminishing Marginal Utility

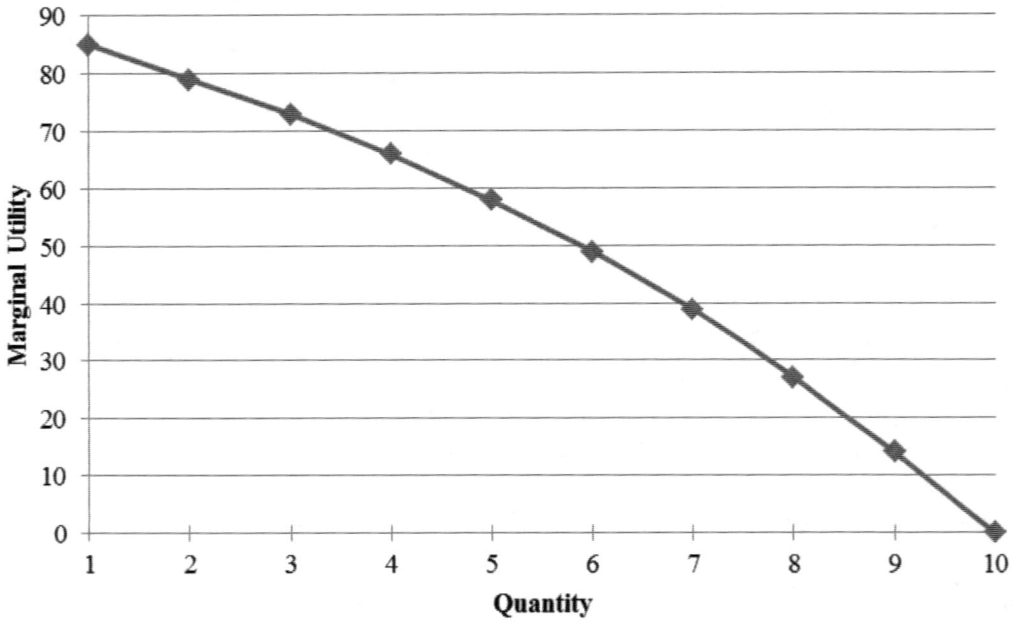

It can be deduced from this phenomenon and in support of the discussion in section 2.1 that human beings cannot grow indefinitely due to the fact that in our intrinsic reality, there is always a limit so as the constancy of available resources (i.e. mass and energy) can support our existence.

This means that in view of our intrinsic reality, we cannot have infinite greed otherwise an upsurge of unintended consequences are bound to happen (i.e. environmental degradation, terrorism and eventually demise of humanity).

Therefore, recognizing our intrinsic reality is the first step towards solving the world's woes and thereby resulting to sustainability and security.

2.3 Expanding the Production Possibility Curve through R&D

The caveat here is that the constant mass or energy (i.e. the limits) leads to trade-offs. Trade-offs is encountered when personal or business decisions are made. In making these decisions, one has to ensure that he or she becomes better off with his or her chosen option. We also realize in our life that we cannot get all that we want and we have to make trade-offs so long as it places us in a better or preferred position. This is more of a philosophical discourse, but it goes to show that the concepts of chemical engineering can be as useful in life and in business as it is in chemical processing.

To tackle these trade-offs, the Production Possibility Frontier in economics can be used. In view of the Principle of Intrinsic Reality where growth is limited by the fixed amount of material and energy, the Production Possibility Frontier needs to shift to the right as shown in Figure 3.

Figure 3. Graphical Representation of Expanding the Production Possibility Curve

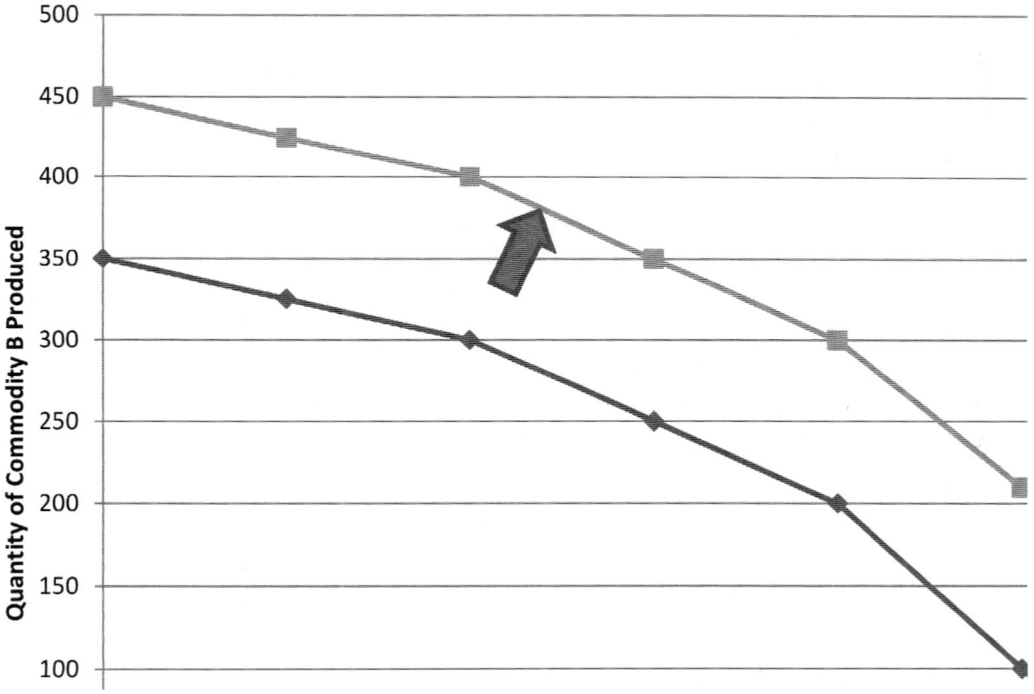

This shift has an implication on the practice of chemical engineering as Research and Development (R&D) is needed to achieve the said shift. R&D enables the economy to produce more of the two commodities (i.e. Commodity A and Commodity B) resulting in economic growth. The key challenge here to Chemical Engineering, particularly on Chemical Engineering R&D, is to look at the bigger picture where economic growth emanating from R&D output is pivoted on sustainability; considering its impact to the environment and the society at large.

3. Paradigm Shift in the Practice of Chemical Engineering

"Money" or "Profit Maximization" has been at the center of the practice of chemical engineering (Smith 2005). Various mathematical optimization methods are used to design systems, processes or components of processes with the end in mind of optimizing profits. This mental model however, will no longer work in our situation today as we face economic, ecological, social and spiritual crisis (Scharmer 2010).

To address this, Scharmer (2010) pointed out seven acupuncture points to shifting capitalism into a regenerative ecosystem economy. Within this purview, the practice of chemical engineering has a greater role to assume. For example, Scharmer (2010) suggested that the design of all production and consumption cycle be completely earth-to-earth, i.e. without necessitating for landfill and that it has to co-evolve with the natural ecosystem. In this regards, the way we do optimization should be revamped into sustainable optimization. By sustainable optimization, it means optimization of the entire system and adding in environmental as well as societal objectives to the equation.

In this definition, two things need to be considered. The first one is systems thinking which emphasizes the understanding of the entire system so the interaction among components will be known and unintended consequences mitigated (Senge 2006). This is a holistic approach to problem solving and finds enormous use in process design.

It should be noted however that the construction of processing plant, whether it is a greenfield or a brownfield project, is rooted on projected market demand. In some instances, economic volatility necessitates companies to experiment with multiple pathways to survive. In this circumstance and within the purview of systems thinking, design thinking comes into the fore. Design thinking is a collective principle emphasizing empathy for users, doing prototyping and tolerance for failure (Kolko 2015). This principle finds enormous use in chemical engineering designs such as in product development, process development, systems improvement and the like. This is also useful for simplifying or humanizing chemical engineering designs with the end in mind of contributing to sustainability. It also has to be noted that of paramount importance in using this tool is imagining the future with the view in mind of status quo and the likely outcome for sustainable development.

The second consideration is the inclusion of environmental and societal objective functions to the objective functions (economic criteria) of profit, cost, energy and yield (Edgar, et al. 2001) in optimization. The environmental objectives can be the impact or risk to environment and the societal objectives can be the impact or risk to society or common good.

These are just few examples of the needed paradigm shift in chemical engineering practice in consideration of the principle of intrinsic reality. But there can be more. A picture of the envisioned new paradigm for the practice of chemical engineering is shown in Figure 4.

Figure 4. New Paradigm in the Practice of Chemical Engineering

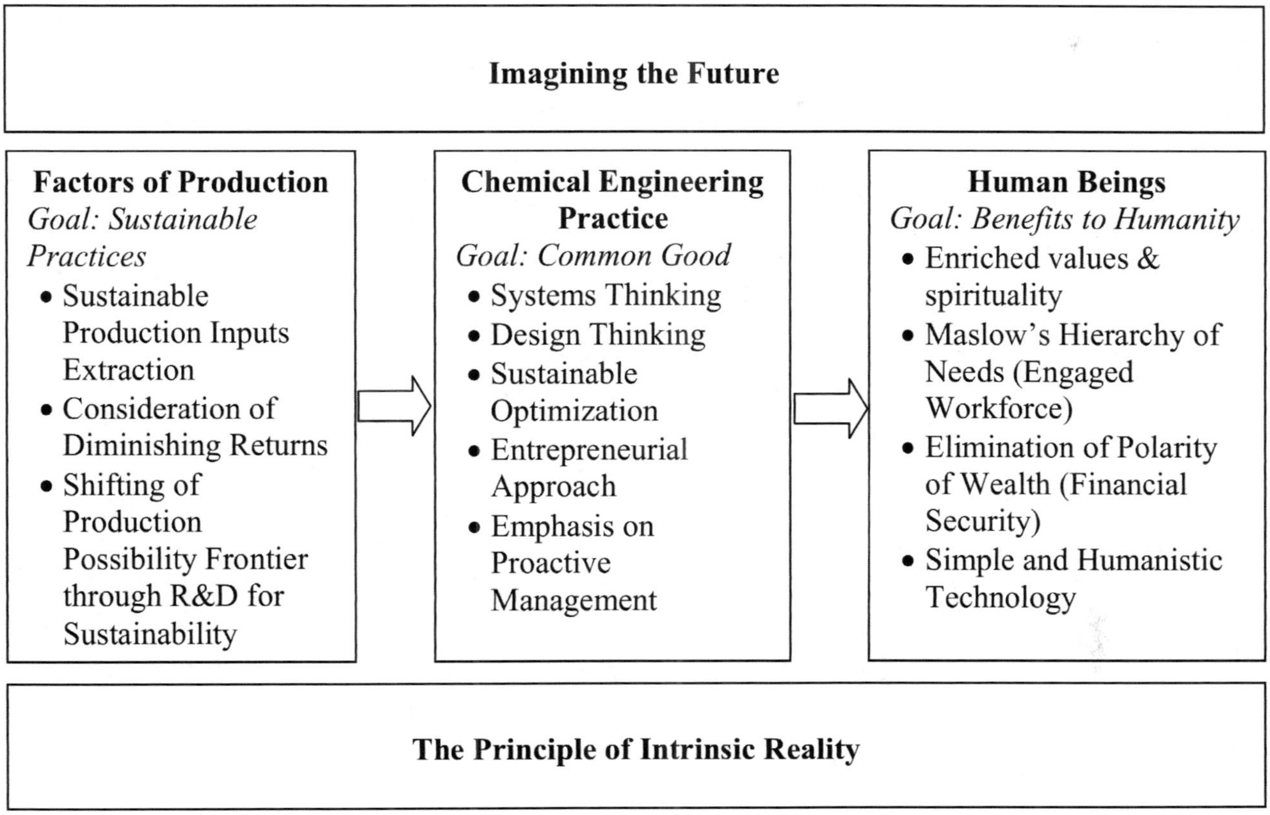

It should be realized that at the core of our economic systems is chemical engineering which can drive the economy into positive direction with the least unintended consequences. Related to that is capitalism in which upgrading of its operating systems can affect the practice of chemical engineering to a substantial degree. A discussion of this is provided in the following section.

4. Wealth Creation and Its Imperative for Sustainability

The development of this paper was inspired by the book "Freakonomics" in which the authors Steven D. Levitt and Steven J. Dubner (2005) asserted that economics can explain almost everything. It came to mind that if economics as a basic science can explain almost everything, then why not chemical engineering which was founded on chemistry, the central science?

Reflecting on it deeper, while explanations clear our doubts, we need more than that so we can tackle the world's woes. As such, this author thought that perhaps combining the two disciplines of chemical engineering and economics can be a potent force towards this end. How this can possibly be done is discussed in this section.

The starting point here is a review of the economic history of the world. In his book "A Farewell to Alms", Gregory Clark (2007) reviewed the economic history of the world in which he divided it into two periods; the period before 1800 and the period after. This is graphically represented in Figure 5.

Figure 5. Shares of Major Economies in World GDP through Time

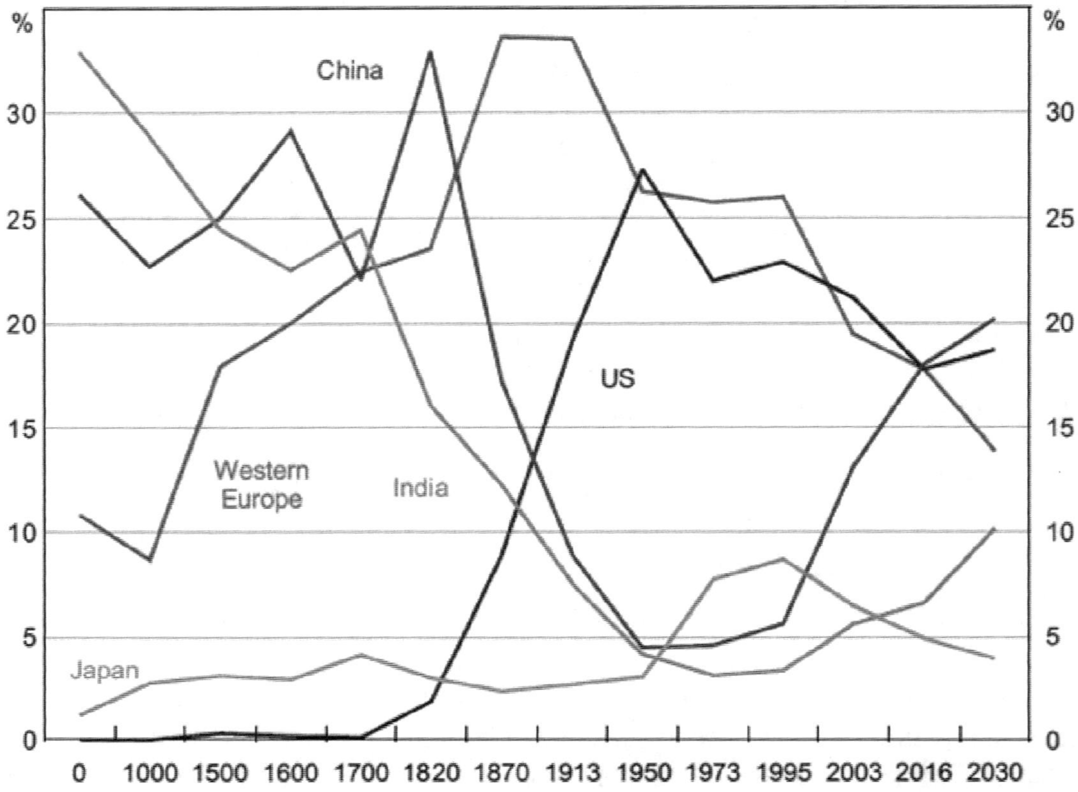

Note: Based on purchasing power parity (PPP) valuation of country GDP

Sources: International Monetary Fund (IMF), *World Economic Outlook* database, April 2011; Maddison (2006)

Thompson (2012) interpreted this graph by stating that "everything to the left of 1800 is an approximation of population distribution around the world and everything to the right of 1800 is a demonstration of productivity divergences around the world." In other words, everything to the left before Industrial Revolution depicts economic power by the sheer size of population and everything to the right after the Industrial Revolution shows the effect of mass production on country's economic output.

Clark (2007) on the other hand called the era before the industrial revolution, i.e. ca. 1800, as Malthusian era in which income per person varied across societies and epochs but growth is not observed. As Clark (2007) explained, this was due to a mechanism called Malthusian trap in which short-term increase in income through advancement in technology is dragged down by population growth. He further explicitly stated:

> "Economics as a discipline arose in the dying decades of the Malthusian era. Classical economics was a brilliantly successful description of this world. But the torrent of goods unleashed by the Industrial Revolution not only created extremes of wealth and poverty across nations, it also undermined the ability of economic theory to explain these differences."

In view of this statement, Figure 5 is reinterpreted in this paper. The contention of this author is that *polarity of wealth* can be gleaned from the graph. *Polarity of wealth* means the uneven concentration of wealth in the global economic system as some parts of the world are richer than others. The polarity can also be observed to shift as some richer parts of the world decline in economic power and some poorer parts of the world rise in economic wealth.

A historical example can be gleaned from Figure 5. India and China up until 1500 CE had 50 and 60 percent share of the world's economy, respectively. In 1 CE, China had a population of almost 60 million, while the area of the present day United States had a population of 680,000. In the late 18th century, the Industrial Revolution rendered countries with large population, just countries with the largest populations.

Looking at North America, it took them 1800 years to overtake China's economic output. However, by 1950, the US economic output was three times as great as China, even though its population is three times less. Comparing China with UK in 1913, the UK had a population of 45 million and China had 437 million but the economic output was almost the same. Analyzing what happened in Europe and later in the United States, Europe was on the lead even before 1800 (Nye 2012).

In 2010 according to data, China overtook Japan as the second biggest economy, though the numbers may not be strictly comparable. Japan's data are seasonally adjusted while those for China were not. Angus Maddison suggested that China and India were the biggest economies in the world for most of the past 2000 years. That is why economists find it a mystery why they fell so far behind and now they are flourishing ('China's economy' 2010). This may not be mysterious considering the *polarity of wealth* as was mentioned earlier and that this polarity can be argued to emanate from the Principle of Intrinsic Reality.

The reasoning here is that the world has constant mass and energy. Assuming that all these mass and energy can be harnessed to create wealth, it could mean that there is a limit for wealth creation. Theoretically, the limit can be expanded indefinitely through R&D but experience demonstrates that this was not achieved yet or getting anywhere near being achieved. As such, the constant mass and energy vis-à-vis the limit for wealth creation can be thought of as the reason for the polarity of wealth, i.e. wealth can only shift from one entity to another.

The implication of this assertion can be demonstrated by the fact that the 1% richest have accumulated more wealth than the rest of the world combined (Oxfam International 2016). The world has never seen 100% of its population to be wealthy. As such, time and again, equitable distribution of wealth is called for and mechanisms to achieve such_should be made a priority in government policies.

In a more concrete way, Scharmer (2010) suggested measures to address this growing inequality which is consistent with the mindset espoused by the Principle of Intrinsic Reality. These measures are as follows:

- Creation of a New Class of Common Property Rights

The creation of common property rights is intended to address the cost to society of private economic undertakings. For example, a private company is extracting natural resources from a certain place. While the private company can profit from that endeavor, the community in that place might suffer the undesirable consequences (e.g. flooding due to forest denudation). To ensure sustainability and institutionalize the voice of future generations in this case, a trust can be established which will own the rights and distribute funds (e.g. from cap-and-trade system for carbon emissions) to citizens (one share per person), communities and the government for investment in public infrastructure.

The trust is an effective means because it does not function like a company (driven by profits) or government (driven by special interest groups) but that it is accountable for long-term sustainability of the specific commons it manage for future generations.

- Creation of a Human Right to Basic Income

One way to ensure equitable distribution of wealth is to create a human right to basic income for every human being. This right when combined with free or affordable access to healthcare and education would create a playing field where everyone has a chance to pursue aspirations and dreams. In other words, it enables individuals to put their real creativity for the benefit of larger community. Another intention of this measure is to free up paid labor (a form of dependence) so that human beings can enhance their creativity and entrepreneurial skills.

- Inclusive Capitalism

To eradicate the *polarity of wealth*, capitalism should be made inclusive so that money and capital flows are directed to all sectors of the economic system. Currently, our money system leads to accumulation of oversupply of money and capital in high financial but low environmental and social gain areas. At the same time, there is an undersupply of money and capital in areas that address societal and community needs. It implies that there is too much money in areas where it is not needed but not enough on areas where it is needed. To address this imbalance, it is recommended to redesign the system so that the current extraction-based financial system is replaced by institutions which can support the real economy and thereby serve the ecology and community.

It should be noted that shifting of wealth is also observed in various geographical regions. When this occurs, geo-political instability can result. This goes to show that if we want lasting stability and security within the limits of our intrinsic reality, then we should not tolerate unabated greed. Otherwise, *polarity of wealth* will continue to persist and we cannot realize our full potential as individuals and collaborators for common good.

5. Way Forward…

Recognizing our intrinsic reality leads us to greater understanding of our very nature bounded by the constancy of mass and energy. It shed light on why growth stabilizes after addition of more inputs. Within this purview, trade-offs are also encountered which necessitates for R&D. R&D initiatives however should be underpinned on sustainability.

Looking at the practice of chemical engineering from the lens of intrinsic reality, the approach should be futuristic while considering relevant aspects related to factors of production and human beings.

The bigger picture shows that Engineers particularly Chemical Engineers are major players in driving the real economy. As Scharmer (2010) aptly puts it, "The essence of technology and knowledge creation, and the ultimate driving force of the economy, is the same: it is the human capacity to create." As such, if sustainability is to be achieved, Chemical Engineers have a greater responsibility to innovate for regenerative and social technologies. In this way, we can stem the polarity of wealth by providing affordable technology to enable disadvantaged individuals enhance their creativity, innovation and entrepreneurship. This can help them partake in the benefits of inclusive capitalism.

As demonstrated in this paper, the synergy emanating from the convergence of Chemical Engineering and Economics can powerfully tackle the world's woes. As such, there is a good reason for chemical engineers and economists alike to collaborate particularly on sustainability initiatives.

6. References

[1] 'China's economy overtakes Japan's in real terms' 2010, *The Economist*, 16 August, viewed 5 March 2016, http://www.economist.com/node/16834943

[2] Clark, G 2007, *A Farewell to Alms: A Brief Economic History of the World,* Princeton University Press, Princeton, NJ.

[3] Edgar, T, Himmelblau D & Lasdon, L 2001, *Optimization of Chemical Processes*, McGraw-Hill, New York.

[4] Huang, Y & Wang, B 2011, 'From the Asian Miracle to an Asian Century? Economic Transformation in the 2000s and Prospects for the 2010s', in H Gerard & J Kearns (eds), *Proceedings of the Reserve Bank of Australia annual conference*, Reserve Bank of Australia, Sydney, viewed 30 January 2016,

http://www.rba.gov.au/publications/confs/2011/huang-wang.html

[5] Kolko, J 2015, 'Design Thinking Comes of Age', *Harvard Business Review*, pp. 66-71, viewed 24 February 2016, https://hbr.org/2015/09/design-thinking-comes-of-age

[6] K.N.C. 2012, 'Mis-charting economic history: more 2,000 years in a single graphic', *The Economist blog*, web log post, 20 June, viewed 29 February 2016,

http://www.economist.com/blogs/graphicdetail/2012/06/mis-charting-economic-history

[7] Levitt, S & Dubner, S 2005, *Freakonomics: A Rogue Economist Explores the Hidden Side of Everything*, William Morrow, New York, NY.

[8] Malthus, T 1798, *An Essay on the Principle of Population*, J. Johnson in St. Paul's Church-yard, London.

[9] Maslow, A 1943, 'A Theory of Human Motivation', *Psychological Review*, vol. 50, pp. 370-396, viewed 24 February 2016, http://psychclassics.yorku.ca/Maslow/motivation.htm

[10] Monod, J 1949, 'The Growth of Bacterial Cultures', *Annual Review of Microbiology*, vol. 3, pp. 371-394, viewed 2 February 2016,

http://www.annualreviews.org/doi/abs/10.1146/annurev.mi.03.100149.002103

[11] Nye, J 2012, 'Fascinating new graph shows the 'economic history of the world since Jesus'', *MailOnline*, 23 June, viewed 15 November 2015,

http://www.dailymail.co.uk/news/article-2163610/Fascinating-new-graph-shows-economic-history-world-Jesus.html

[12] Oxfam International 2016, *An economy for the 1%, how privilege and power in the economy drive extreme inequality and how this can be stopped*, Oxford, viewed 5 March 2016,

https://www.oxfam.org/sites/www.oxfam.org/files/file_attachments/bp210-economy-one-percent-tax-havens-180116-en_0.pdf

[13] Samuelson, P & Nordhaus W 2009, *Economics*, McGraw-Hill Higher Education, New York, NY.

[14] Scharmer, O 2010, 'Seven Acupuncture Points for Shifting Capitalism to Create a Regenerative Ecosystem Economy', *Oxford Leadership Journal*, vol. 1, pp. 1-21, viewed 24 February 2016, http://www.oxfordleadership.com/journal/vol1_issue3/scharmer.pdf

[15] Senge, P 2006, *The Fifth Discipline: The Art & Practice of the Learning Organization*, Doubleday/Currency, New York.

[16] Smith, R 2005, *Chemical Process Design and Integration*, John Wiley & Sons Ltd., West Sussex, England.

[17] Thompson, D 2012, 'The economic history of the last 2,000 years in 1 little graph', *The Atlantic blog*, web log post, 19 June, viewed 29 February 2016, http://www.theatlantic.com/business/archive/2012/06/the-economic-history-of-the-last-2-000-years-in-1-little-graph/258676/

Acknowledgment

The author would like to thank Hyundai Engineering and Construction Co. Ltd. for the support in this endeavor. The author is also thankful to Ms. Sofia Ayu for meticulously editing this manuscript and Mr. Marcial N. Barroga for the poster design.

445030 Heavy Metal Detection in Treated Water Via a Fabric Nanocomposite Sensor

Abstract:

We present the conception of an electrochemical sensor for on-site/portable water analysis. This flexible sensor was developed to quantify the amount of heavy metals (in real-time), alleviating the wait time, large sample size, possible contamination, and expensive analytical equipment associated with current procedures for water analysis. This sensor is composed of a multiwalled carbon nanotube (MWNT) functionalized polymer mat, for a flexible yet conductive sensor. The MWNTs are then functionalized with a cyclo-oligomeric calixarene, which is shown to selectively form a supramolecular complex with metal ions. Upon complex formation, the charge carriers are drawn away from the carbon layer and the detection is prevalent at appropriate levels. The optimization of MWNT and calixarene functionalization, as well as the selectivity and sensitivity of the sensor will be discussed.

444086 Optimization of Layered Graphene Oxide Nanostructures through Controlling the Membrane Deposition Rate Towards Size-Sieving Separation

Abstract:

Single-layer graphene oxide (GO) flakes were coated on anodized aluminum oxide (AAO) substrate through a simple vacuum filtration process, the effect of the deposition rate on the nanostructures of obtained membrane was investigated though pure water flux and organic vapor diffusion, which indicated GO membrane assembled with slow deposition rate might have more ordered lamellar structure and, correspondingly, have smaller effective pore size. In contrast, under fast deposition rate, GO flakes assembled in a more random way. Therefore, through controlling the membrane deposition rate, the nanostructure of layered GO membrane could be optimized for size dependent separation. We found after thermal reduction, 100 nm GO membrane (prepared @ 0.02 nm/s) has effective pore size smaller than 6.1 Å, which exhibit great potential in n-hexane/2,2-dimethybutane (DMB) separation.

444376 Ozone Source Apportionments of Local Anthropogenic Emissions

Abstract:

Ozone is one of six principal air pollutants designed by US EPA. NOX and VOC emissions from anthropogenic sources are the most important factors to control ozone pollutant.

In this report, photochemical dispersion models are used to estimate the effect of ozone from local anthropogenic emission sources. EKMA-type diagrams are generated for NOX- and VOC- effects. Furthermore, APCA (Anthropogenic Precursor Culpability Assessment) technology is used to estimate ozone source apportionments.

EKMA-type diagrams show that 10-30% emission reductions could reduce ozone concentrations to meet standard. Further APCA results show the average ozone contributions from regions and emission categories. APCA results explain different effects for different monitoring stations.

444142 Robust Design of a Radiant Wall Burner for a High Temperature Environment Utilizing Computational Fluid Dynamics Simulations

Abstract:

Radiant Wall Burners are often utilized in Ethylene Cracking Furnaces. The conditions and fuel gases for a specific unit can vary considerably within the unit and across similar units. With burner air preheated to high temperatures, premixing the fuel can result in dangerous and destructive flashback, while complete nozzle-mixing of the fuel can result in burner instability at startup or turn-down conditions. These challenges can be solved through the generation of a robust radiant wall burner design, which is facilitated by executing Computational Fluid Dynamics simulations of many different geometric configurations.

444397 Flare Minimization and Emission Characterization during Start-Ups of Chilling Train in Ethylene Plants

Abstract:

Flaring is a crucial safety operation to protect chemical plant personnel and equipment during plant start-up. However, it generates huge amounts of off-spec products, and causes negative environmental and social impacts, as well as results in tremendous raw material and energy losses which could be saved to generate much more needed products. Thus, cost-effective start-up flare minimization strategies which is a double-win practice, simultaneously benefiting both the industrial and environmental sustainability, are becoming more and more important and attractive to the industries.

The chilling train is a very important section for the ethylene plants and accounts for more than 50% of the total start-up time and flare emissions of the whole ethylene plants. Within this section, the charge gas from the upper stream process is refrigerated to a very low temperature (lower than -160 °C) to separate hydrogen and methane. Meanwhile, the left charge gas is liquidized for cryogenic separation. The chilling train system consumes over 40% of the refrigerating capacity of an ethylene plant. During the chilling train start-up, different operational conditions with different compositions and flowrate of precooling process will affect both product separation efficiencies, start-up stability time and flare emission.

Traditional start-up scenarios are developed based on experience and improved by trial and error, which is inefficient and dangerous. In this study, a general methodology on flare minimization for chilling train start-up operations via rigorous dynamic simulation is presented. The methodology starts with setup and validation of chilling train steady-state and dynamic simulation models. The initial state of start-up in the real plant is achieved through systematically transforming from the validated dynamic model, and several operation conditions with different compositions and flowrate of precooling stream are examined to identify an optimal initial state of start-up based on the obtained initial state. Thereafter different flare minimization strategies for the chilling train start-up are conducted on the optimal initial state model. Also, emission source characterization under different flare minimization strategies for chilling train start-up is investigated. Deep insights of the start-up strategies, emission source distribution and dynamic emission profiles are provided. This study provides detailed technical support for both the industry and environmental agencies on evaluating and developing cost-effective flare minimization strategies in the future.

439811 Purification of Ibuprofen Enantiomers By a New Technique Combining Distillation and Crystallization

Abstract:

Stripping crystallization (SC) is a new separation technique which combines distillation and crystallization. Basically, SC is operated at a triple-point condition, in which the liquid is simultaneously vaporized and crystallized due to the three-phase equilibrium. In essence, the process is continued until the liquid phase is completely eliminated and only the pure solid crystals remain in the feed. Thus, crystal washing is not required since only pure solid crystals remain in the feed and no impurities are adhered on the crystal surfaces at the end of SC. In the previous research, SC has been successfully applied to separate p-xylene from the mixed xylenes by Shiau et al. (2005, 2008, 2013).

Ibuprofen, also called 2-(isobutylphenyl)-propionic acid, is a non-steroidal anti-inflammatory drug widely used to treat headaches and minor pains. Although it is known that S-ibuprofen is responsible for the anti-inflammatory effects while R-ibuprofen might cause some long-term side effects, ibuprofen is still sold as a racemic mixture in the market due to the complexity involved in the separation of enantiomers. A thermodynamic model is developed in this work to simulate the three-phase equilibrium conditions during the SC operation. The experiments show that seeding with ultrasound mixing or magnetic stirring can be efficiently employed to purify S-ibuprofen from ibuprofen enantiomers. The experimental results, including the final enantiomeric purity and recovery ratio of S-ibuprofen, are consistent with the simulation results predicted by the model.

References:

1. Shiau, L. D.*, Wen, C. C. and Lin, B. S. "Separation and purification of p-xylene from the mixture of m-xylene and p-xylene by distillative freezing", *Ind. Eng. Chem. Res*, **44(7)**, p2258-2265, 2005.

2. Shiau, L. D.*, Wen, C. C. and Lin, B. S. "Separation of p-xylene from the multicomponent xylene system by stripping crystallization", *AIChE Journal*, **54(1)**, p337-342, 2008.

3. Shiau, L. D.*; Liu, K. F. "Investigations into the effects of the cooling rate on stripping crystallization", *Ind. Eng. Chem. Res*, **52**, p1716-1722, 2013.

442557 Flow Field Simulation and Distribution Performance Analysis of a Gas Pipe Distributor in Fccu

Abstract:
A fluidized catalytic cracking (FCC) is an important petroleum refinery process, which, aided by highly selective zeolite catalysts, cracks large-molecule low-value heavy oil feedstock into small-molecule highvalue
products. When spend catalyst is fed into the regenerator for reactivating, it is fluidized by the gas distributor at the bottom of regenerator. The gas distributor is designed to uniformly distribute gas to maintain the fluidized state of the fluidized bed and to provide the reaction gas to guarantee the burning coke deposited on the spend catalyst. Therefore, the gas distribution performance of the gas distributor influences the quality of initial gas distribution directly. So the pipe distributor used in a industrial FCC unit, which is composed by main pipes and a number of branch pipes with many nozzle tubes, was investigated by means of flow field calculation. The gas distribution performance of the pipe distributor was analyzed based on the simulated flow field. The simulation result shows that there are a obvious nununiform
gas distribution from the different nozzle gas flow rate. The static pressure increases gradually along the flow direction in the nozzle tube while the axial flow rate decreases, which leads to the nozzle gas flow rate increasing gradually. Furthermore, an obvious bias flow occurs at the inlet of the nozzle tube, which influences the gas flowing into the nozzle. The pressure drops performance of nozzle and pressure varying in the branch pipe are analyzed comprehensively. Due to the nonuniform
gas distribution by the gas pipe distributor in the practical usage, the problems such as overlarge temperature difference among the fluidized bed layers in the regenerator have often happened, which will not only reduce burning coke efficiency and but also decrease activity of the regenerated catalyst. Meanwhile, the erosion of the distributor also occurs, and in consequence the function of the distributor fails, which even causes the nonscheduled shutdown of the FCC unit and leads to tremendous economic loss. The support is provided for optimizations of both structural parameters and operation parameters of the gas pipe distributor and improvement of the gas distribution performance. Furthermore, improvement suggestions were provided based on the above reasoning. The results are expected to be useful for the distributor design.

441535 Hydrogen Disposal Flare System Hazard Identification and Mitigation Analysis

Abstract:

A hazard assessment on a hydrogen disposal flare system was conducted in preparation for a hydrogen turbopump test at the Air Force Research Laboratory (AFRL), Aerospace Systems Directorate at Edwards AFB, CA. The test operations required the disposal of large quantities of liquid and gaseous hydrogen through a commercial flare system to prevent the formation of an explosive environment. As part of the assessment, the primary steps in the test operation process were evaluated for the formation of a potential hazard by looking at each associated failure mode and the resulting affect on personnel and facilities. Evaluation of each operational step was conducted in a Process Failure Modes and Effects Analysis (PFMEA) that identified the probability and severity of the identified effects and associated hazards. If the probability and severity were unacceptable, mitigating steps were applied until the overall hazard risks were reduced to acceptable levels. The recommendations of this assessment were incorporated into both the flare system design and the operational procedures. This paper outlines the results of the PFMEA which include identification of the effects/hazards, available risk-mitigating techniques, and implementation of these techniques used for the mitigation of the identified hazards. Based upon the findings significant design modification to the vender's commercial equipment was needed to comply with recent changes in the code. Most of the identified hydrogen flare system hazards/effects were categorized into four groups: internal explosions, obstructions, external explosions, and maintenance degradation. Many corrective actions were implemented to mitigate both the probability and severity of the identified hazards. In addition, this paper discusses the installation of oxygen monitoring systems on the flare system, determination of appropriate system purge requirements, installation of purge system status alarms, incorporation of a triple redundant flare ignition system with status indicators, and a number of other hazard risk mitigation techniques. This comprehensive description of the hazards and mitigation of a hydrogen disposal system is a plant is vital part to safe operation of both hydrogen commercial and aerospace facilities.
<align="center">

444020 Thermodynamic Modeling of Aqueous Lithium Sulfate and Its Solubility in the Presence of Na+, K+ and Mg2+ Sulfates

Abstract:

This work is an extension of the previous works done by our group towards development of a rigorous and comprehensive thermodynamic model for high salinity produced water that is essential for mass and heat balance calculations and simulation and design of novel treatment processes. Specifically this work presents a thermodynamic model that accurately represents all thermodynamic properties of aqueous lithium sulfate solution including the solubility of Li2SO4 in water and also in the presence of Na+–K+–Mg2+ sulfates. By explicit consideration of solution chemistry of each subsystem, symmetric electrolyte NRTL model is used to correlate available experimental data of all pertinent thermodynamic properties including activity coefficient, osmotic coefficient, excess enthalpy, and liquid heat capacity over the entire concentration range. This facilitates quantification of eNRTL binary molecule–molecule, molecule–electrolyte, and electrolyte–electrolyte interaction parameters. Also, solubility data are used to regress the thermodynamic constants required for modeling the liquid–solid equilibrium of lithium sulfate in water and in the presence of Na+–K+–Mg2+ sulfates. Displaying a significant prediction ability, the model accurately represents the thermodynamic behavior of the system from infinite dilution up to saturation and temperatures from 298.15 K to 573.15 K.

442068 Enhanced Furfural Production By an Innovative Bottom Dividing Wall Column with a Decanter Configuration

Abstract:

Enhanced Furfural Production by an Innovative Bottom Dividing Wall Column with a Decanter Configuration

Nguyen Van Duc Long, Le Cao Nhien, Le Quang Minh and Moonyong Lee†

School of Chemical Engineering, Yeungnam University, Gyeongsan 712-749, South Korea

Furfural, which are used to derive other industrial chemicals, has been identified as one of the major bio-based platform chemicals that can compete with petroleum-based chemicals. However, the current commercial furfural process has low yield and requires intensive energy. This paper reports the simulation, detailed design and optimization of an industrial process recovering furfural from lignocellulosic biomass, a renewable and inexpensive feedstock. The aqueous feed mixture from the acidic hydrolysis of biomass was fed to the separation process to recover furfural as the main product, along with methanol as valuable byproducts. The purification process was intensified by applying an innovative bottom dividing wall column with a decanter configuration (BDWC-D). The response surface methodology (RSM), which allows the interactions between variables to be identified and quantified, was used to optimize the BDWC-D structure. The predictions by the RSM showed satisfactory agreement with the rigorous simulation results. The heat pump technology and heat integration were then employed to improve the purification process. The results showed that the proposed sequence could save the overall energy requirements and total annual cost significantly as compared to the conventional sequence. This sequence can be employed for both grass-root and retrofit designs.

Keywords: Lignocellulosic Biomass; Furfural; Energy Efficiency Improvement; Dividing Wall Column; Response Surface Methodology

Acknowledgments

The authors wish to acknowledge the financial support from the R&D Convergence Program of NST (National Research Council of Science & Technology) of Republic of Korea and KITECH (Korea Institute of Industrial Technology) (ES150001).

*** Formatting Issue - Best Available Paper/Graphic ***

444278 Insight into the Mechanism of Toxic Chromium (VI) from Wastewater Using Magnetite Coated PINE Powder

Abstract:

INSIGHT INTO THE MECHANISM OF TOXIC CHROMIUM (VI) FROM WASTEWATER USING MAGNETITE COATED PINE POWDER

Pholosi A., Ofomaja A., Naidoo E.B.

Biosorption and Wastewater Treatment Research Laboratory, Department of Chemistry,

Faculty of Applied and Computer Sciences, Vaal University of Technology, P. Bag X021, Vanderbiljpark, 1900 South Africa.

INTRODUCTION

A knowledge of the interaction between pollutant molecules or ions and active sites on an adsorbent surface is an important phenomenon in many industrial processes such as catalysis, adsorption and other separation techniques. The use of lignocellulosic materials for Cr(VI) adsorption has gained interest over the years **(Miretzky and Cirelli, 2010)**. The adsorption properties of lignocellulosic materials has been attributed to acidic functional groups on their surfaces which can take part in ion-exchange, complexation, chelation and hydrogen bonding with pollutant species in solution **(Vithanage et al., 2015)**. Two mechanisms for lignocellulosic adsorption of Cr(VI) have been proposed; (i) an electrostatic mechanism in which at low pH, the protonated acidic functional groups on the biomaterial carrying a positive charge attracts the negatively charged Cr(VI) species **(Demirbas, 2005)** and (ii) the adsorption-coupled reduction mechanism in which Cr(VI) is reduced to Cr(III) either in solution (Direct reduction) or after being adsorbed (Indirect reduction) by electron-donor groups of the biomaterial that have lower reduction potential values than that of Cr(VI) **(Gao et al., 2008)**. The Cr(III) ions formed in solution may then form complexes with Cr(III)-binding groups on the biomaterial surface or released into bulk solution by repulsion between Cr(III) and other positively charged groups on the biomaterial **(Miretzky and Cirelli, 2010)**.

METHODS

Sodium hydroxide treated pine coated with magnetite (NTP-NC) was prepared by co-precipitation method. The effect of solution pH was performed in a batch system in which 0.5 g of the adsorbent was contacted in six 250 cm^3 beakers containing 75 cm^3 of 75 mg/dm^3 Cr(VI) solutions set at pH 1, 2, 4, 6, 8 and 10 at 26 °C and agitated at 200 rpm for 2 h and the clear solution analysed for Cr(V/I) left using UV-Vis spectrophotometry at 460 nm

RESULTS

The results of the effect of initial solution pH on the adsorption of Cr(VI) onto NTP-NC is shown in Fig. 1. To confirm the operating mechanism, the change in $[H^+]$ concentration at each initial pH, the oxidation/reduction potential before and after adsorption and the amounts of Total Cr, Cr(VI)and Cr(III) present in solution at each pH was monitored for NTP-NC and the results displayed in Fig1 . The change in $[H^+]$ concentration ($\Delta[H^+] = [H^+]_{final} - [H^+]_{initial}$) at the end of the adsorption process for each initial solution pH was recorded and plotted against the initial solution pH. The results showed that the change in $[H^+]$ concentrations at the end of the adsorption process where higher for lower initial solution pH's (where lower amounts of Cr was found in solution) than for higher initial solution pH's (where higher amounts of Cr was found in solution) and the change in $[H^+]$ concentration became small and

*** Formatting Issue - Best Available Paper/Graphic ***

almost constant after initial solution pH 4. This is consistent with the assumption that the removal of Cr from solution is associated with [H^+] ion consumption and that as the concentration of [H^+] ions reduces in solution, the Cr removal reduces. Cr(VI) ion in acidic solution has a positive redox potential value (above 1.3 V at standard conditions) and in the presence of an electron donor becomes unstable and can be reduced to Cr(III) (**Park et al., 2008**). The results show that the initial ORP values (before adsorption) for the Cr solutions at pH's of 1, 2, 4, 6, 8, and 10 were 339, 280, 165, 55, -53 and -172 mV for NTP-NC respectively. The positive values of ORP at lower solution pH's indicates higher possibility for reducing Cr(VI), while the lower and negative ORP values indicates lower possibility for reducing Cr(VI). On addition of the adsorbents into the solutions, a change in ORP were observed . The values of ORP at the end of the adsorption process became 330, 268, 30, 9 -29 and -61 for NTP-NC respectively. A reduction in ORP at the end of the adsorption stage signifies that reduction has occurred in the system. **Yu et al. (2014)** showed the amount of Cr(VI) removed from aqueous solution in contact with zero-valent iron (nZVI) could be related to the change in ORP and pH of the solution. The authors observed a decrease in ORP from 400 to 100 mV and a corresponding 10 unit pH increase as Cr(VI) was removed from solution.

Finally, analysis of the prepared Cr solutions at the various solution pH's showed that the concentration of Total Cr obtained from ICP and that of Cr(VI) using UV-vis measurements were very similar, suggesting that Total Cr was almost equal to Cr(VI) in solution. After contacting the various solutions with the adsorbents, the concentration of Total Cr in the pH 1 solution was observed to have reduced to 21.56 mg/dm^3 for NTP-NC, while the Cr(VI) concentrations was 10.66 mg/dm^3 for NTP-NC. With solution pH 2, the concentration Cr(III) left in solution was lower than that of solution pH 1 for both samples. The reason is that at pH 1 there is an abundance of H^+ ions in solution which accumulates on the functional groups on the adsorbent making the surfaces of the adsorbent positively charged and repelling away the positively charged Cr(III) ions from the adsorbent surfaces (**Suksabye et al., 2009**). At solution pH 2, it is believed that there is less accumulation of H^+ ions at the adsorbent surfaces which led to reduced electrostatic repulsion and increased adsorption of Cr(III) from solution. This increase in Cr(III) adsorption is also confirmed by the reduction in Total Cr in solution between solution pH 1 and 2. The concentrations of Cr(VI) left in solution at pH 2 was higher than at pH 1 for NTP while for NTP-NC the Cr(VI) concentrations were lower at pH 2. For solution pH's higher than 2 and up to 10, the residual Cr(III) in solution becomes very low and almost constant. This is consistent with the small change in [H^+] in solution which is almost constant at this pH range.

FTIR Evidence for the adsorption of Cr(III)

When Cr(III) is formed in solution via adsorption-coupled reduction mechanism it may undergo hydrolysis and/or complexation depending on the solution pH and the total Cr(III) concentration. Hydrolysis may lead to the formation of various Cr(III) species with an increase in solution pH which gives rise to a reduction in its solubility and increasing the chances of precipitation of the Cr(III) species. At solution pH 1 – 4, Cr^{3+} and $CrOH^{2+}$ are the major species which can be adsorbed by negatively charged ligands on the biomaterial surface. On the other hand, at solution pH range of 4 – 10, Cr(III) species (hydroxo complex), such as $CrOH^{2+}$, and $Cr(OH)_3$, are the dominant forms in the aqueous system and at high pH is readily transformed into soluble complex, . When NTP-NC was contacted with Cr(VI) solution for 2 hr, the final solution pH was increased to pH 2.53. The FTIR spectra for NTP-NC (Fig. 2) shows the presence of characteristic Fe-O peaks of bulk Fe_3O_4 at 553.26 cm^{-1}, 564.36 cm^{-1}, 570.21 cm^{-1} and 576 21 cm^{-1}. Functional groups associated with the pine includes the broad band at 3250.99 cm^{-1} representing –OH of cellulose, 2905.17 cm^{-1} representing C-H stretching, while the peaks at 1627.69 and 1602.87 cm^{-1} shows the COO^- and C=O bonds respectively. After 2 hr contact with Cr(VI) solution, the peak at 564.34 cm^{-1} was observed to have split into several new peaks (Fig. 3b (ii)). These emerging peaks are characteristic of Fe-O bonds of maghemite (γ-Fe_3O_4), for example the characteristics peaks of maghemite was observed at 595.21, 685.65 cm^{-1} and 582.70, 634.91 cm^{-1} suggesting a change in the iron oxide phase (Fe_3O_4 to γ-Fe_2O_3) due to the conversion of Cr(VI) to Cr(III) by magnetite nanoparticles. According to **Peterson et al. (1997)**, Cr(III)(hydr)oxide produced is precipitated on the magnetite surface. In this study, α-Cr_2O_3 peak were observed at 604.63 cm^{-1} indicating that Cr(III) was formed during the adsorption process. Finally it was also observed that the carboxylate group at 1635.55 cm^{-1} was shifted to 1637.73 cm^{-1} while a new peak was observed at 1737.72 cm^{-1} suggesting that complexation between Cr(III) and Carboxylate ions also occurred.

REFERENCE

Demirbas, A. Energy Source 27 (2005) 1449–1455.

Miretzky, P., Cerelli, A.F. A review. J. Hard. Mater. 180 (2010) 1-9.

Vithanage, M., Rajapaksha, A.U., Ahmad, M., Uchimiya, M., Dou, X., Alessi, D.S., Ok, Y.S. J. Environ. Manage. 151 (2015) 443-449.

D. Park, C.K. Ahn, Y.M. Kim, Y.-S. Yun, J.M. Park, J. Hazard. Mater. 160 (2008) 422–427.

P. Suksabye, A. Nakajima, P. Thiravetyan, Y. Baba, W. Nakbanpote.. J. Hazar. Mater. 161 (2009) 1103–1108.

R.F., Yu, F.H., Chi, W.P., Cheng, J.C. Chem. Eng. J. 255 (2014) 568–576

M.L. Peterson, A.F. White, G.E. Brown, G.A. Parks Environ. Sci. Technol. 31 (1997) 1573–1576.

Fig. 1 Effect of solution pH on the adsorption of Cr(VI)

***** Formatting Issue - Best Available Paper/Graphic *****

Fig, 2: FTIR of NTP-NC

443101 Thermodynamic Modeling of HNO_3-H_2so_4-H_2o Ternary System with Symmetric Electrolyte NRTL Model

Abstract:

Nitric acid concentration/sulfuric acid concentration (NAC/SAC) process has been widely used for concentrating dilute aqueous nitric acid and recovering spent sulfuric acid. Nitric acid is concentrated by employing sulfuric acid to bind water and break nitric acid-water azeotrope at approximately 68 wt. % nitric acid. To support heat and mass balance calculations and process simulation for NAC/SAC processes, we develop a comprehensive thermodynamic model for nitric acid-sulfuric acid-water ternary system based on previously published thermodynamic models of nitric acid-water [Wang et al., Fluid Phase Equilibria, (2015) (DOI: 10.1016/j.fluid.2015.04.015)] binary and sulfuric acid-water [Que et al., Journal of Chemical & Engineering Data, 56 (2011) 963-9772] binary systems developed with symmetric electrolyte NRTL (eNRTL) activity coefficient model. The present work takes into account one-step hydration of nitric acid molecules, two-step hydration of sulfuric acid molecules, and further hydration of hydronium ions. With up to three temperature coefficients for each eNRTL binary interaction parameter and up to two temperature coefficients for each chemical equilibrium constant, the ternary model is validated with boiling point data from 383.15 to 471.45 K and activities at 273 K. Furthermore, contour plots of boiling points, vapor phase composition, and specific heat capacity of the ternary system are generated for engineering use.

Keywords:

Nitric acid, Sulfuric acid, Electrolyte NRTL model, Vapor-liquid equilibrium, Boiling point, Heat capacity, Nitric acid concentration/sulfuric acid concentration (NAC/SAC)

444551 Continuously Adjustable, Molecular-Sieving "Gate" on 5A Zeolite for Distinguishing Small Organic Molecules By Size

Abstract:

Zeolites/molecular sieves with uniform, molecular-sized pores are important for many adsorption-based separation processes. Pore size gaps, however, exist in the current zeolite family. This leads to a great challenge of separating molecules with size differences at ~0.01 nm level. Here, we report a novel concept, pore misalignment, to form a continuously adjustable, molecular-sieving "gate" at the 5A zeolite pore entrance without sacrificing the internal capacity. Misalignment of the micropores of the alumina coating with the 5A zeolite pores was related with and facilely adjusted by the coating thickness. For the first time, organic molecules with sub-0.01 nm size differences were effectively distinguished via appropriate misalignment. This novel concept may have great potential to fill the pore size gaps of the zeolite family and realize size-selective adsorption separation.

444791 Dissolution and Precipitation of Minerals Under Different Alkali Concentrations in a Reactive CO_2 Batch System

Abstract:

The primary focus of this research is to study the dissolution and precipitation of minerals and salts from Berea sandstone and brine samples taken from the Great Salt Lake in a reactive CO_2 batch system. Various experiments were constructed to measure dissolution and precipitation reaction rates under different alkali concentrations. Changing the alkali concentrations of the Great Salt Lake brine samples produced changes in dissolution and precipitation of minerals from the Berea sandstone and the Great Salt Lake brine samples. Data from ICP-MS was used to quantify mineralogical changes in the samples and Micro-CT was used to measure the porosity change in the Berea sandstone samples. The results from these analytical instruments showed patterns of mineralogical dissolution and precipitation in the post-experimental fluid that correlate to the change in alkali concentrations.

***** Formatting Issue - Best Available Paper/Graphic *****

443365 Technical Reconstruction of Double Effect Distillation for Methanol Process

Abstract:

Abstract: Methanol is one of important basis raw materials of chemical industry and also an excellent environment-friendly fuel, which make it as a significant commodity in the global market. Distillation is an essential unit operation for purified production of methanol. Product quality, production cost, environmental impact and energy consumption are largely related to distillation process.

Our research effort, under this circumstance, has been devoted since 2002 to energetically optimize 4-column methanol distillation system that is widely adopted in methanol units from coal, with reduced cycling water requirement. This patented process technology saw its industrial application in Chinese largest single unit of 1800Kt/a of Baofeng Group located at North China, with the lowest steam consumption of 0.9/1 t(steam)/t(methanol) compared with normally 25% higher for both simulative case study and on-site operation. Methanol product is in accordance with O-M-232G and the recycling water is reduced significantly from 67 t(CW)/t(methanol) to 52 t(CW)/t(methanol).

The flow diagram of this advanced double effect distillation process is showed in Fig.1. Its advantages arise for limitations found in current units: (1) heat exchange network failed to fully recover energy causing large CW flow to remove extra heat; (2) for pressured column, bubble point reflux need larger reflux ratio than the sub-cooled; (3) the heat of distillate from pressured column was normally ignored; (4) the low concentration of methanol of feed inlet to recovery column leads to higher reflux radio, responsible for unsatisfied unit energy efficiency.

Fig.1 The double effect distillation process.

(Column1.Light ends column; Column 2.Pressured column;

Column 3.Atmospheric column; Column4.Recovery column)

As a solution, a new flow diagram is built up and showed in Fig.2. (1) HEN totally reorganized (not completely show); (2) sub-cooled reflux to pressured column caused 10% reduction to the reflux ratio, economizing steam consumption; (3) top product of pressured column give heat to crude methanol before feeding light ends column; (4) change product and by-product position of recovery column to give efficient unit energy consumption.

Fig.2 The reconstruction of distillation process.

(Column1.Light ends column; Column 2.Pressured column;

Column 3.Atmospheric column; Column4.Recovery column)

It is estimated that the production of methanol will be over 60 million tons in 2014. If this reconstruction could be applied in the whole methanol industry, 1.3 million tons of standard coal would have been saved. This is equivalent to reduce 5 million tons emission of carbon dioxide, so the environmental impact can be largely decreased. Field operation shows reconstruction perfect substitution worth to spread to the whole methanol industry, energetically efficient and low environmentally impacting.

444353 A Combined Rheology and Stability Study on Emulsified Acid for High Temperature Applications

Abstract:

HCl emulsion has been a popular formulation to stimulate carbonate reservoirs due to the various advantages that emulsified HCl has over ordinary HCl; including but not limited to deeper penetration and less corrosion damages. Emulsified HCl has even more potential advantages if utilized properly. The reaction kinetics, stability and rheology of emulsified HCl plays a significant role on achieving successful acidizing treatment.

The objective of this paper is to experimentally study the factors affecting the stability of emulsified HCl. The emulsion was prepared using 15wt% HCl and diesel with cationic emulsifier. The acid solution-to-diesel ratio was 70:30. The effect of stirring intensity, concentration of the emulsifier, addition of corrosion inhibitor, temperature, addition rate of acid phase to oil phase and mixing time of the emulsion have been studied. Stirring intensity was varied from 600 to 1500 rpm. Emulsifier dosage ranged between 5 to 20 gpt. Temperature used for heating, to mimic reservoir condition, was as high as 120°C.

The results showed that there is an optimum condition for emulsifier dosage and for mixing time of the emulsion. The optimum emulsifier dosage was found to be 10 gpt with 5 minutes of mixing time. As for the addition rate of acid phase to oil phase, 1.4 ml/min of drop-wise addition was found to give a very stable emulsion. The effect of stirring intensity on the emulsion stability was proportional, the higher the rpm speed the better the stability. Corrosion inhibitor, on the other hand, played a role as a destabilizer. With regards to heating temperature, it was found out that the higher the temperature, the faster the emulsion breaks.

The novelty of this study is the comprehensive approach in combining stablity and rheology of emulsified acid for high temperature applications. The particle size analysis and solution morphology help in better understanding HCl emulsion and how it works.

443503 A Computational Analysis of Fire Scene for Safety in a Compartment

Abstract:

This article details numerical prediction of fire spread carried out to study the distribution of carbon monoxide concentration, smoke dispersion, soot concentration and temperature in laboratory compartment, using a commercial fire modeling software tool. The heat release rate curve predicted from the simulation was compared with literature values carried out in a road tunnel. The time variations of carbon monoxide transportation, smoke concentration and temperature spread in compartment room are reported to evaluate the effect on humans. The soot concentration was described and used throughout this study as an indicator of safety in the compartment room. The effects of vent dimension and ventilation capacity were analyzed by smoke detector response times. This work provides improvements derived from reinterpretation of actual fire scene considering the ability of mechanical exhaust system and ventilation in compartment room to mitigate the hazard of fire.

444359 Development of Novel Emulsified Acid Using Waste Oil

Abstract:

In carbonate reservoirs, hydrochloric acid (HCl) is commonly used for stimulation purposes. The main disadvantage for using HCl is its high and rapid reactivity allowing for less control on acid reaction with formation. In addition, HCl may react with the tubing causing severe corrosion and damages. Several slow-reacting acid systems, though HCl-based, have been developed to retard the HCl reaction and control its reactivity to allow deeper penetration in carbonate reservoirs before it breaks out and reacts with formation rock. However, the selection of the hydrocarbon phase of emulsified acid is critical to the creation and stability of the emulsified acid system. Diesel oil, crude oil and solvents such as xylene have been reported to be successful in the field.

In this work, alternative and less expensive source of oil is developed to replace the current practices. Emulsified acid is a water-in-oil emulsion where oil is the continuous phase and acid is the dispersed one. Industrial and filtered waste oil was used to prepare the emulsified acid of this work. Extensive work has been carried out to study the chemical composition of the waste oil, namely SARA and to select the proper emulsifier. Emulsions were prepared using 15% HCl with emulsifier concentration ranging from 0.5 to 2.0 vol% and 70:30 of acid-to-oil ratio. Droplet size distributions were measured and used to benchmark with other emulsions. Emphasis was given to the stability of the emulsion especially at high temperatures. In order to apprehend the effect of corrosion inhibitor on the emulsion stability, its concentration was varied (up to 0.3 vol%). Later, rheological analyses of the waste oil emulsion were compared to the commonly used diesel-based emulsified acid.

Results showed the potential use of waste oil to prepare high temperature emulsified acid (up to 100°C). The addition of corrosion inhibitor to the emulsified acid caused a negative effect on stability resulting in a less stable emulsion. On the other hand, addition of cationic emulsifier helped to formulate a more stable emulsion. Apart from that, the optimum concentration of the emulsifier was at 2.0 vol%. The rheological data confirms the stability of the new emulsified formulation at higher temperatures. This paper summarizes the findings of using waste oil emulsion and recommends it for field applications. The newly proposed emulsified acid introduces waste oil as a replacement of diesel.

The novelty of using waste oil will help to significantly reduce the expenses of formulating emulsified acid. Since waste oil is basically waste it has a good impact on the environment, that being said, instead of throwing it, the industry will be able to recycle it.

443764 Optimization of Heat Exchanger Network in Steam Assisted Gravity Drainage Oil Sands Surface Facilities

Abstract:

Steam assisted gravity drainage (SAGD) is the most viable and environmentally safe recovery technology for extracting heavy oil and bitumen. It is a thermal in-situ production process where parallel wells are drilled horizontally into an underground bitumen reservoir. Steam is produced at the central processing facility (CPF) and injected into the reservoir through the wells. The steam heats the bitumen to a point where gravity allows it to flow down to the lower well where the mixture of bitumen and water is then pumped to the surface. The water and bitumen are separated at the CPF. Typically, more than 90% of the energy consumed in producing a barrel of bitumen using SAGD is used to generate the steam that is injected into the bitumen reservoir. Fossil fuels is employed in the process, which causes GHG (greenhouse gas) emissions. The usage of hot and cold utility is reduced by increased heat exchange and optimization of the heat exchanger network, which results in reduction of GHG emissions. In this work, pinch analysis is performed to minimize energy consumption.

443855 Modeling the Thermal Behaviors of an Ultracapacitor Module

Abstract:

Ultracapacitors, also known as supercapacitors, have the potential to meet the increasing power requirements of the sustainable energy systems for the grid-scale energy storage systems and electrified vehicles. As compared to batteries, ultracapacitors offer a higher power density, higher efficiency, and longer shelf and cycle life. Because the performance of an ultracapacitor depends on temperature, it is important to calculate accurately the thermal behavior of a single ultracapacitor cell and the module composed of multiple ultracapacitor cells for the efficient and reliable systems integration of an ultracapacitor in grid and transport applications.

In this work, modeling is performed to study the effect of the layout configugation of the ultracapacitor cells on the thermal behaviors of an ultracapacitor module. The ultracapacitor module is subject to the charge/discharge cycling with constant-current between the nominal voltage of the ultracapacitor module and the half of this voltage. The charge/discharge current values examined are 50, 100, 150, and 200 A. The validation of the modeling approach is provided through the comparison of the modeling results with the experimental measurements.

443856 Modeling the Electrolyte Flow Behaviors of a Zinc/Bromine Flow Battery Module

Abstract:

The zinc/bromine (Zn/Br_2) flow battery is a promising rechargeable power source for energy storage systems because of its inherent chemical simplicity, high degree of reversibility at the electrodes, good energy density, and abundant low-cost materials. The Zn/Br_2 flow battery system, intended for the grid-scale energy storage system, is to be made in modular form. A Zn/Br_2 flow battery module is composed of the multiple stacks consisting of bipolar carbon electrodes. Since the stacks require electrolyte circulation on both the negative and positive sides, two plumbing loops are provided. The flow loops contain an electrolyte storage tank and a circulation pump. A heat exchanger may be included for temperature control in the flow loop. The design of the electrolyte circulation plumbing determines the distributions of electrolyte flow and pressure within the stacks and the efficiency of the stack is dependent on the flow distribution throughout the stack. The modeling of the distributions of electrolyte flow and pressure within the module is needed to optimize the plumbing for electrolyte circulation and thereby to improve the performance of the module. In this work, modeling is performed to study the effect of the layout configuration of the plumbing for electrolyte circulation on the distributions of electrolyte flow and pressure within a 50 kWh Zn/Br_2 flow battery module composed of four stacks. It is seen that the plumbing layout for electrolyte circulation can have a significant effect on the flow rate distribution of electrolyte through the stacks.

443974 Observer-Based Model Predictive Control for Time-Delay Systems

Abstract:

Model predictive control is one of the most widely used advanced process control technologies in industry.[1] It has many capabilities such as dealing with input/state constrains and uncertainties. Observer-based model predictive control was developed to estimate states information when the states are not measurable directly. Many chemical processes have time delays in either states or measurements, such as polymer production and some bioprocesses. [2] In this presentation, we are proposing an unknown input observer based model predictive control algorithm for time-delay systems. Unknown input observer is an observer design to eliminate certain types of unknown inputs in the system. Stability of unknown input observer-based model predictive control without time delay was provided by our previous study. Terminal cost and terminal region are designed for a class of non-linear time delay systems. The close-loop stability of the controller is provided. [3][4]

An example of bioreactor is provided which has uncertain time delays in measurements but with a known upper bound. Two measurements are provided which are online biomass and offline ethanol measurements. Model-plant mismatch is considered as unknown inputs in the system. State estimation and control are calculated using the observer and an optimization algorithm.

Reference:

[1] Marcus Reble et al., Model predictive control of constrained non-linear time-delay systems. IMA Journal of Mathematical Control and Information, 2010, 1-19

[2] Seung Cheol Jeong and PooGyeon Park, Constrained MPC algorithm for uncertainty time-varying systems with state delay, IEEE TRANSACTIONS ON AUTOMATIC CONTROL, 50, 2005, 257-262

[3] Mayuresh Kothare et al., Robust constrained model predictive control using linear matrix inequalities. Automatica, 32, 1996, 1361-1379

[4] Wook Hyun Kwon et al., A simple receding horizon control for state delayed systems and its stability criterion, Journal of Process Control, 13, 2003, 539-551

444044 Organic Solar Cells

Abstract:

This paper describes highlights the need of next-generation “organic solar cells”. There is a continuing need for organic solar cells and arrays with very high specific power densities (1000-5000 kW/kg) for generating power in a new generation of ultra light space payloads example in NASA. An emerging technology with promise to meet these ambitious goals are organic solar cells based on very thin films of discrete layers. Theoretical efficiencies for multi junction solar cells are more ~70% as compared to ~30% for single junction devices. There is continuous development going on high mobility, band gap engineered organic donors and acceptors, as well as development of three junction devices and scaled up devices on lightweight flexible polymer substrates. The main interest in this paper is organic semi-conducting crystal: Pentacene, which has photo-conducting states. Organic solar cells manufactured today have low efficiency other than their stability issues. The efficiency of organic solar cells can be improved by either designing polymer materials that absorbs more of solar spectrum or by minimizing the charge collection efficiency by controlling the morphology of polymer films. Once fully developed, organic solar cells would revolutionary power the industry due to their flexibility and minimal weight

444055 A Refined Non-Random Two-Liquid Segment Activity Coefficient Model for Solubility Modeling

Abstract:

Accurate correlation and robust prediction for complex molecule solubility or solid-liquid equilibrium (SLE) in solvents and solvent mixtures have drawn much interest recently [Tanveer et al., *Chem. Eng. Prog.* **2014**, 110 (9), 37-47]. Solubility models such as the Non-Random Two Liquid Segment Activity Coefficient (NRTL-SAC) model [Chen and Song, *Ind. Eng. Chem. Res.* **2004**, 43, 8354-8362] are now routinely used in the industries. However, an important unfilled need is to model the solubility in oligomers/polymers/surfactants such as asphaltene nanoaggregates in petroleum fluids and drug molecules in lipid-based or solid dispersion formulation [Tung, *Org. Proc. Res. Dev.* **2013**, 17, 445-454]. This paper presents a refinement to the very successful NRTL-SAC model to better account for the contribution due to size differences of molecules. Specifically, the Flory-Huggins combinatorial entropy expression for conceptual segments in the model is substituted with the Staverman-Guggenheim expression for real molecules. We present the rationale for the refined model, the updated conceptual segment numbers for common solvents, and the newly identified conceptual segment numbers for common drug excipients. The model performance is illustrated by prediction of drug molecule solubility in various lipid solvents and solvent mixtures, compared to experimental data and previous work by other models. The refined NRTL-SAC model shows enhanced capability to correlate and predict drug molecule solubility in common solvents and excipients used in drug formulation.

443640 Preparation and Characterization of (*FURCRAEA spp*) as Green Support of Iron Nanostructured Catalyst

Abstract:

PREPARATION AND CHARACTERIZATION OF (*FURCRAEA spp*) AS "GREEN" SUPPORT OF IRON NANOSTRUCTURED CATALYST

Karen G. Bastidas G[1], Anamaria Barrera B[1], Cesar A. Sierra A[2], Hugo R. Zea R[3]

[1] Department of Chemical and Environmental Engineering. National University of Colombia (Bogota).

[2] Associate professor. Director of the Research Group of Macromolecules. National University of Colombia (Bogota).

[3] Associate professor. Director of the Research Group of Materials, Catalysis and Environment. National University of Colombia (Bogotá).

Abstract.

The availability of natural water resources has declined over time, becoming one of the biggest problems today. Environmental risk in Colombia's rivers has increase due to the widespread use of harmful chemicals in a variety of processes, ranging from illegal mineral extraction to industry wastewater, e.g. Medellin's river, due to its strategic location along a textile industrial corridor might be subject to dye contamination; the gold and platinum mining zones in Choco and Antioquia regions are being heavily affected by the use of mercury in the extraction processes. In 2009, 520 tons of mercury were imported by miners working in this area, amount that exceeds several times the mercury used by regulated industry in previous years (Ministry of Mines and Energy of Colombia, 2007). Not only the environmental risk is a concern but also the economic conditions of many low-income families whose livelihood in many cases depends exclusively of the of the mining.

Due to the issues raised above, the main goal of this study is to research, develop and implement innovative technologies in the synthesis of highly efficient, low cost and environmentally friendly iron based catalysts capable of degradate pollutants, ensuring operational stability. A native fiber (Fique, FURCRAEA spp) has been selected as a solid support for these catalysts due to its abundance, low cost and special physical and chemical characteristics.

The combination of cellulosic fibers with iron oxide nanoparticles could provide exceptional biodegradable composite materials for the treatment of organic dyes and heavy metals present in wastewater. Likewise, the use of iron particles immobilized on fique fibers allows an easy separation catalyst-treated water, turning the biocatalyst in highly recyclable and reusable material.

Fique fiber was initially characterized by determining the percentage of ashes, total lignin, soluble and insoluble lignin, cellulose, hemicellulose, total solids and humidity, obtaining 0.94, 22.58, 0.5, 22.8, 36.3, 27.5 0.701 and 2.1%, respectively. Additionally, the isoelectric point (pzc), acidic and basic sites were measured.

Then, the fiber was subjected to a pretreatment in order to expose more active sites which on time serve as nucleation points for the deposition of nanoparticles. This was done by means of an ultrasound-assisted procedure, varying the sonication time (30, 60 and 90 minutes). The effect of the sonication was followed by measuring the contents of lignin, cellulose and hemicellulose within the fiber after each time. 60 minutes of sonication was found to be most beneficial time since the amount of lignin and hemicellulose present in the fiber was reduced, while conversely the amount of crystalline cellulose exposed increases.

In order to improve the chemical ability of the nucleation sites to interact with iron and then to form stable and size controllable nanoparticles, a cationization procedure was performed on the fiber; this was achieved by the exposure of the fiber to acidic and basic conditions promoting cationization. Three cationization times (1, 2 and 3 hours) were studied and their effect on the fiber chemical surface was evaluated using FTIR-ATR and XRD analysis. The results showed 3 hours as the optimal cationizing time.

Subsequent to the cationization of the fiber surface, a procedure of ionic/cationic exchange was implemented by exposing the fiber to a solution of iron salts in order to replace the cations in the fiber surface and anchoring the iron nanoparticles. Different conditions were studied in order to synthetize zero valent iron nanoparticles (nZVI) and others with different oxidation states. Finally, to track the impregnation process on the fiber, samples from the precursor solution were taken at different times and its iron concentration was determined by atomic absorption spectrophotometry (AAS). Complementary characterization performed by X-ray diffraction, BET surface area and electronic microscopy, among others.

Key words: Natural support (fique), iron nanoparticles, bionanocatalyst.

444057 Explaining the Paradox of Asphaltene Precipitation with n-Alkanes

Abstract:

Explaining the paradox of asphaltene precipitation with n-alkanes

Md Rashedul Islam and Chau-Chyun Chen
Department of Chemical Engineering, Texas Tech University

Abstract
Paraffins, specifically alkanes, promotes asphaltene precipitation when they are blended with asphaltene-containing petroleum fluids. The alkane/oil volume ratio required for onset of asphaltene precipitation increases with the carbon number and yields a maxima. Experimental study indicates that the maxima lays at C7[1] or between C8 -- C10 [2]. However, precipitation yield decreases asymptotically with increase of the carbon number even if excessively high alkane/oil volume ratio is used [3]. For example, n-pentane be a better solvent for asphaltene than n-hexane in regard of both onset of precipitation and precipitation yield. However, high carbon number alkane beyond the maxima, n-hexadecane for instance, can be a better solvent at onset of precipitation point and a poorer solvent in regard of precipitation yield than n-pentadecane. A thermodynamic explanation for this paradox is necessarily important to find the compatibility of bitumen or crude oil with paraffinic hydrocarbons. In this work, this paradox is explained by the aggregation thermodynamics of asphaltene precipitation [4]. Both the enthalpy and entropy of mixing are investigated to explain the effect of paraffin carbon number on asphaltene precipitation. Athabasca and cold lake bitumen samples are studied.

References

[1] G. Hotier and M. Robin, "Action de divers diluants sur les produits pÃ©troliers lourds: mesure, interprÃ©tation et prÃ©vision de la floculation des asphalténes," Oil & Gas Science and Technology, vol. 38, pp. 101-120, 1983.
[2] I. A. Wiehe, H. W. Yarranton, K. Akbarzadeh, P. M. Rahimi, and A. Teclemariam, "The paradox of asphaltene precipitation with normal paraffins," Energy & fuels, vol. 19, pp. 1261-1267, 2005.
[3] D. L. Mitchell and J. G. Speight, "The solubility of asphaltenes in hydrocarbon solvents," Fuel, vol. 52, pp. 149-152, 1973.
[4] M. Wang, Y. Hao, M. R. Islam, and C. C. Chen, "Aggregation Thermodynamics for Asphaltene Precipitation," AIChE Journal, Submitted.

Using a Large Data Bank to Develop a Robust and Simple Model to Estimate the Saturation Pressure of Oil Reservoirs

By

Hassan Alkandari[1], Adel Elsharkawy[1], Adel Malallah, Osamah Alomair

Petroleum Engineering Department, College of Engineering & Petroleum
Kuwait University
Po Box 5969 Safat 13060-Kuwait

Abstract

Measurements of saturation pressure are crucial for all hydrocarbon reservoir fluids. Below the crude oil saturation pressure gas reaches a critical saturation; then, a two phase flow occurs and results in decreasing oil production and recovery. To maximize oil production and recovery, the reservoir pressure has to be maintained closer to the original saturation pressure. This pressure is normally measured using bottom-hole samples or surface recombined samples of oil and gas. Occasionally, the samples become unavailable and the pressure needs to be estimated using computational methods.

In this study, a large data bank of 231 crude oil compositions and saturation pressure measurements including literature data, unpublished data, and newly measured data were used to develop two empirical models to predict the saturation pressure of a variety of crude oils if the oil sample is unavailable or the experimental measurements are unreliable. The first proposed model utilizes the extended compositions of hydrocarbons up to the heptane plus fraction in addition to non-hydrocarbons. The second model uses the lumped compositions of light, intermediate, and heavy components in addition to non-hydrocarbon components as the input variables. The lumped model has the advantage of using fewer input parameters while maintaining the thermodynamics. The accuracy and validity of both models to calculate the saturation pressure for volatile oils, black oils, and heavy oils are presented using several compositional data. The models performance is also compared to the Peng-Robinson equation-of-state (PR-EOS), and the Soave-Redlich-Kwong equation-of-state (SRK-EOS) as well as all published methods that use compositions as input variables. The comparison indicates that the proposed models are much simpler and more accurate than the other computational methods. The proposed models treat the heptane-plus fraction as a single component, thus eliminating the splitting and characterizing the plus fractions and the binary interaction parameters needed for the EOS's calculations.

Key words: Saturation pressure (Ps), Bubble-point pressure (BPP), Equation of state (EOS), Correlations, and PVT Data

[1] Corresponding Authors: Hassan Alkandari , Hassan@hkandari.com, Adel Elsharkawy, adel.elsharkawy@ku.edu.kw

Introduction:

Reservoir fluid properties are an essential source of information for all calculations on hydrocarbon reserves. Accurate estimation of the pressure-volume-temperature (PVT) measurement is crucial. Of all the PVT properties, saturation pressure (Ps) is considered the most important parameter. It is well known that gas starts to flow once the reservoir pressure drops below the bubble point pressure and the gas saturation reaches a critical value. To minimize the gas flow and maximize the oil recovery, it is very important to maintain the reservoir pressure close to the original bubble point pressure.

Although the PVT properties are experimentally measured in laboratories and provide reliable results, this process is expensive, time consuming, and sometimes unavailable; therefore, measured field data (separator gas oil ratio, separator pressure, stock-tank oil specific gravity, and reservoir temperature) or reservoir fluid compositions are used to estimate the bubble point pressure.

There are two methods in the petroleum industry to calculate the saturation pressure of crude oil when experimental is unavailable; either using production data or compositional data.

Methods Based on Production Data (Composition is unavailable)

Numerous attempts have been made since the 1940s to develop correlations to estimate the saturation pressure from production data, such as producing gas oil ratio, oil gravity, gas gravity, reservoir pressure and temperature. Table 1 presents summary of all the methods that use production data to estimate the saturation pressure.

Methods based on Compositional data

Many authors have reported that the application of any correlation in other regions results in serious errors. These conclusions arise from the fact that regional crude oils represent a physical-chemical trend; therefore, all attempts to develop a general correlation using field measured parameters to estimate the saturation pressure from a large data bank have failed. For this reason, Elsharkawy (2003) made the first attempt to correlate the saturation pressure to the reservoir fluid composition instead of using the production data. In recent years, there have been several attempts to estimate the saturation pressure of oil reservoirs. These methods are reported in Table 1.

Elsharkawy (2003) presented a correlation of the bubble point pressure for the oil reservoir using the same input data needed for the equation-of-state (EOS) calculations, i.e., the compositional data and the reservoir temperature. Elsharkawy's model is based on the saturation pressure measurements of 60 crude oil samples from the Middle East and 75 collected from the literature. He demonstrates that the calculation of the saturation pressure depends on the type of EOS, the number of pseudo components, the method used to characterize the pseudo-components and the binary interaction number. Elsharkawy compared the performance of his model to the SRK-EOS and PR-EOS for black oils, volatile oils, heavy oils, and synthetic oils. He indicated that his model eliminates splitting the heavy plus fraction, characterizing the pseudo-components and inclusion of the binary interaction parameters needed for the EOS calculations.

Later, Tale et al. (2009) made the second attempt to model the saturation pressure of crude oil using 132 data samples published by Elsharkawy (2003). Tale's model, which is based on the alternating conditional expectation (ACE), has a set of 14 equations with 42 constants. The average absolute relative error (AARE) reported by Tale et al. indicates that the model did not perform well with the selected data. The ACE model has an AARE of 12.3% compared with AARE of 6.7% by Elsharkawy (2003) and 8.1% AARE for the SRK-EOS.

AlQuraishi (2009) attempted to model the saturation pressure using Elsharkawy's (2003) data. The model was developed using linear genetic programming. He correlated the saturation pressure to the formation temperature, methane content, and heptanes plus molecular weight. Thus, AlQuraishi's model neglected the effect of the intermediate and heptane plus content on the saturation pressure. The error analysis reported by AlQuraishi indicates that his model has a 5.8% error compared to 7.7% by Elsharkawy, 9.9% using the SRK-EOS, and 10.1% using the PR-EOS.

Bandyopadhyay and Sharma (2011) presented a model to predict the bubble point pressure, using temperature and fluid composition. They used 129 experimental and literature data originally published by Elsharkawy (2003) to develop their model. The model introduced a temperature interaction parameter that depends on compositions. The proposed model has a set of 24 coefficients and the estimated bubble point pressure is calculated as the sum of the series calculations. They reported an average absolute error (AAR) of 7.83% using their model compared to 8.3% using Elsharkawy's simple model (2003), 10.15% using SRK-EOS, and 10.58% using the PR-EOS.

Khamehchi et al. (2011) used an intelligent proxy simulator (IPS), which is based on an adaptive neural fuzzy inference system (ANFIS) and a structured artificial neural network (ANN) to predict the bubble point pressure and asphaltene onset pressure. The model uses a reservoir fluid composition and temperature to predict the Ps. It is not clear how the Ps is calculated from the proposed method, and they did not present the accuracy of the proposed method compared to the other published methods.

Farasat et al. (2013) presented a new mathematical approach to calculate the crude oil saturation pressure as a function of the temperature and the reservoir fluid composition. The model developed by Farasat et al used 130 experimental data, which were originally published by Elsharkawy (2003). Farasat used a support vector machine (SVM) to model the saturation pressure. They did not clearly demonstrate how to calculate the saturation pressure from the SVM model. Moreover, they did not indicate the overall average absolute deviation (AAD) for their model compared to other published models and both the PR-EOS and SRK-EOS for the 130 data.

Ahmadi et al. (2014) also presented an ANN technique called gene expression programming (GEP) to determine the bubble point pressure of the oil samples. Similar to Farasat et al. (2013), Ahmadi's model used most of the data previously published by Elsharkawy (2003) to construct a non-linear equation using reservoir fluid compositions and temperatures as input parameters. The proposed model is highly complicated and consists of 8 equations that have several constants compared to the one simple equation proposed by Elsharkawy (2003). They compared the performance of their proposed model to the SRK-EOS and PR-EOS; however, their paper does not mention how the heptanes plus fraction were treated and characterized for the EOS calculations. They reported an AAD of 4% using the proposed model compared to 8% using Elsharkawy's model (2003), 10.9% by PR-EOS, 10.5% by SRK-EOS, and 15% by Bandyopadhyay and Sharma (2011).

Gholami et al. (2014) estimated the bubble point pressure using a support vector regression SVR and a supervised learning algorithm SLT based on the statistical learning theory, and then the ACE was used. The model used reservoir fluid compositions and temperature to estimate the bubble point pressure. Gholami et al. used the same data originally published by Elsharkawy (2003). They presented three attempts based on genetic programming to correlate the bubble point pressure: SVR, ACE, and power-law committee machine (PLCM). The ACE-based method resulted in a set of three equations with a matrix of [8 by14] constants. Gholami et al. compared the accuracy of their model to the SRK-EOS and PR-EOS predictions reported by Elsharkawy (2003). Based on their AARD analysis, the PLCM model has the lowest errors followed by the SVR model, Elsharkawy's model, ACE model, PR-EOS, and SRK-EOS.

Lately, Jarrahian et al (2015) presented two empirical models to estimate the saturation pressure of black oils. The first model is based on production information. The second model is based on compositional information mostly from Elsharkawy (2003) and Heidaryan and Moghadasi (2011). They have not presented comparison of the accuracy of their model to Farasat (2013) model, Ahmadi (2014) model, and Gholami (2014) model.

This paper is concerned with the estimation of saturation pressure using compositional data. Therefore, the objective is to develop a robust yet simple model to estimate the saturation pressure for various types of reservoir oils using a larger data bank than that used by other published studies. The accuracy of the proposed models is compared to previously published models as well as SRK-EOS and PR-EOS.

Data bank

This study uses a large data set of 231 crude oil compositions and saturation pressure measurements. The data set consists of 130 data sets originally reported by Elsharkawy (2003); in addition to others collected from literature (Jacopy, 1958; Vogel, 1980; Coats, 1986; Droham, 1980; Dansh, 1992; Moharam, 1996; Rosenegger, 1999); 94 new data sets that have not been utilized before; as well as 7 newly measured samples at the Petroleum Fluid Research Center (PFRC) in Kuwait University. These data represents a wide range of compositions, such as volatile oils, black oils, and heavy oils. The data cover a pressure range of 537 to 5065 psia and a temperature range of 58 to 319 °F, crude oil compositions from C_1 to C_{7+}, non-hydrocarbons, such as N_2, H_2S, and CO_2, and the molecular weight and the specific gravity of the heptane plus fraction.

Data screening and classification

The compositions and the saturation pressure measurements in the data bank are subjected to a meticulous screening process to check the validity of the data. Our screening criteria uses the following steps: the sum of the molar compositions of all the components must add up to one, otherwise, the sample is eliminated. Samples with a missing molecular weight or specific gravity of the plus fraction are also eliminated.

After filtering, the data were reduced to 214 and divided into 179 measurements used to develop the new models, and 35 data sets are used for the blind test of the robustness of the proposed models. Statistical analysis of the input variables of the data set used in training and testing the extended and the lumped models are illustrated in Tables 2.

Several of the reported compositional data have extended analysis up to (C_{20+}). The following techniques are used for lumping or regrouping the extended analysis into single pseudo-fractions (C_{7+}). The mole fraction (X_i) of the individual components of the extended analysis is equal to the plus fraction (Ahmad, 2011).

$$X_{C_{7+}} = \sum_{i=7}^{n} X_i \qquad i=7, 8 \ldots n \qquad (1)$$

The average molecular weight (Mw) of the individual component of the extended analysis is equal to the molecular weight of the plus fraction.

$$Mw_{C_{7+}} = \sum_{i=7}^{n} \frac{X_i Mw_i}{X_{C_{7+}}} \qquad i=7, 8 \ldots n \qquad (2)$$

The density of the total plus-fraction can be calculated as follows:

$$\gamma_{C_{7+}} = \frac{X_{C_{7+}} Mw_{C_{7+}}}{\sum_{i=7}^{n} \frac{X_i Mw_i}{\gamma_i}} \qquad i=7, 8 \ldots n \qquad (3)$$

It is important to note the significance of the density of plus fraction because its density determines its paraffinic naphthenic aromatic (PNA) content.

Equation-of-state calculations

The EOS calculation of the saturation pressure is compared with the newly developed models as well as the other methods. The two EOS calculations considered in this paper are the ones most considered and widely used in the petroleum industry to evaluate the volumetric and phase behavior of reservoir fluid properties: the Soave-Redlich-Kwong EOS (SRK-EOS) and the Peng-Robinson EOS (PR-EOS). It should be noted that the accuracy of any EOS calculation of the saturation pressure largely depends on the number of pseudo-components and the splitting schemes used to divide the heptane plus fraction; as well as the characterization and calculation of the critical properties of the pseudo-components and the binary interaction parameters. Therefore, the two EOS calculations of the saturation pressure are not the same, even though the number of pseudo-components, characterization of the pseudo-components, and binary interaction parameters were the same.

Model development

As previously stated, the objective of this work is to provide the petroleum reservoir engineer with a simple yet accurate tool to estimate the saturation pressure for a variety of crude oils using compositional data when experimental is unavailable. Two models are presented in this study to calculate the saturation pressure of the crude oil samples. The first model uses all input parameters, such as the molar compositions of the hydrocarbon and non-hydrocarbon components similar to the EOS models. This model is referred to as the extended model. The second model uses fewer input parameters in which some of the hydrocarbon components have been lumped into a single group; this is designated as the lumped model. Both models are simpler than all published correlations, physically correct, and have an acceptable accuracy compared with the well-known EOS.

Figure 1 shows a correlation matrix of all the data available under consideration. This figure also indicates the relative importance of each of the input variables into the first proposed model. It agrees with most published correlations and the EOS that the methane content, heptane plus content and temperatures have the greatest impact on the saturation pressure followed by the intermediate content. The data understudy indicates that the non-hydrocarbon has the smallest impact. This result is simply a reflection of the molar contribution of each component comprising the crude oil composition.

Model I - Extended Model

The extended model uses 13 variables comprising the molar compositions of the pure components, heptane plus properties, and reservoir temperatures, which are similar to the input parameters for the EOS. Figure 2 shows the correlation coefficient graph of all the input parameters in the extended model. Thus, the elimination of the intermediate effect has a negative impact on the accuracy of the predictive model. Our proposed extended model has the following form:

$$Ps = 124.72N_2 + 17.57CO_2 + 22.55H_2S + 64.22C_1 - 9.80C_2 - 52.49C_3 + 6.16C_4 - 19.22C_5 - 23.63C_6 - 21.43C_{7+} + 435.31\gamma_{C7+} + 1.14Mw_{C7+} + 4.29T \quad (4)$$

Where Ps is the saturation pressure in psia, T is the temperature in °F, and the hydrocarbon and non-hydrocarbon compositions are expressed in mole.

Model II - Lumped Model

The second model is much simpler and uses only 7 -variables (the lumped model). Several lumping schemes are adapted in the petroleum industry for phase behavior modeling and reservoir simulation of the petroleum reservoir fluids (Ahmed, 2011). Hydrocarbon components are grouped into a pseudo fraction based on their molecular weight (Whitson, 1980), the similarity of their physical-chemical properties (Lee, 1979), their carbon number, their true boiling point or their number of families based on the chemical structure (Becker et al., 2015). (Emera and Samara, 2005) proposed lumping the hydrocarbon components into volatile (C_1 & N_2), intermediate (CO_2, H_2S, C_2 to C_4) and heavy categories (C_{5+}) to calculate the minimum miscibility pressure. (Duan et al., 2013) proposed lumping several components into a single carbon number (SCN) with predetermined values for properties, such as critical pressure, temperature, and eccentric factor. Others have proposed optimal lumping schemes based on the objectives (Lin et al., 2008). In lumped model, $C_{1+}N_2$, C_2, CO_2, and H_2S, the intermediate (C_3 to C_6), heptanes plus (C_{7+}) and its properties, and temperature are the input variable. The correlation matrixes of the lumped components are provided in Figure 3. Even though the non-hydrocarbon has the least effect their contribution is important, especially the H_2S in the case of sour crudes. Both models presented by AlQuraishi (2009) and Bandyopadhyay and Sharma (2011) have eliminated the effect of H_2S.

The proposed model has the following form:

$$Ps = 82.115*(C_1+N2) - 11.635*C_2 + 39.158*CO_2 + 38.244*H_2S - 1.217*(C_3—C_6) - 890.701*(C_{7+}/MW_{7+}*\gamma_{7+}) + 4.217*T - 1042 \quad (5)$$

Where Ps is the saturation pressure in psia; and T is the temperature in °F.

Result and discussion

Model accuracy

The accuracy of the two proposed models used to calculate the saturation pressures of the previously described crude oil samples is shown in Table 3. This table shows reports the average relative error, average absolute relative error standard deviation and the correlation coefficients for all the methods considered in this study. Table 3 also reports the accuracy of the PR-EOS, SRK-EOS, Elsharkawy's model, AlQuraishi's model, and Bandoypandyay and Sharma's model and Jarrahian model. The methods proposed by Gholami et al., Ahmadi et al., and Jarrahian et al are included in Table 3, even though they produce a unreasonable estimate of the saturation pressures, as explained in the appendix.

This table indicates that the extended and the lumped models have an (AARE) in the order of 8% and 8.5%, respectively. It should be noted that the complications in the Bandoypandyay and Sharma model did not improve the accuracy of the saturation pressure, because it had a large error. It is also noted that Jarrahian et al model produces an average error of 35%. Compared with the EOS, the lumped model has a few constants and eliminates the splitting, characterizing the pseudo-components and binary interaction parameters necessary for the EOS calculations. The lumped model as well as the extended model are very simple and require one-step calculations compared with all the other methods reported in Table 3.

A cross-plot of the calculated saturation pressure using the extended model based on the training data versus the measured data is illustrated in Figure 4A. This figure indicates that most of the data falls on the unit slope line, and there is no over-fit. This trend is supported by the high correlation of regression of the model (97.6%) presented in Table 3. Figure 4B shows a cross-plot of the testing data (28 from literature and 7 new measured in PFRC at Kuwait University). The extended model indicates an excellent agreement with all of the 28 data sets chosen at random, with the exception of a few data points at higher pressures. Additionally, the model has a perfect match for the 7 newly measured samples.

Figure 5A shows the performance of the lumped training-model for capturing the measured saturation pressure for most of the 179 samples. A cross-plot of the calculated saturation pressure using the lumped model based on the training data versus measured data is depicted in Figure 5B. This figure indicates that most of the data scatters falls to the unit slope line. The lumped model also has a high coefficient of regression of 97.4%, which is indicated in Table 3.

Models Validity

The accuracy and validity of the models for estimating the saturation pressure of various types of crude oil are reported in Table 4. This table illustrates the capability of the newly proposed models to predict the saturation pressure for volatile oils, black oils, and heavy oils. Table 4; provides the compositions, properties of the heptane plus fraction, temperature, and measured saturation pressures of the samples. Additionally, the calculated saturation pressure using the PR-EOS, SRK-EOS, Elsharkawy's model (2003), AlQuraishi's model (2009), and Bandoypandhyay and Sharma's model (2011), Jarrahian et al (2015); as well as the extended and lumped models are provided in Table 4. It is clear from this table that the proposed models have a wide range of applications regardless of crude oil types. We have also included information regarding the 7 newly measured data; at the PFRC, compositional, and saturation pressure. At the bottom of Table 4, we have provided the unpublished data of the compositions; and the saturation pressures at low temperatures. It is well known that the saturation pressure of a crude sample is measured at low temperatures in addition to the reservoir temperature as a method to check the quality of the sample.

Figure 6A reports the error of all the methods considered in this paper to predict the saturation pressure for volatile oils (heptane plus content below 30%). Both the extended and the lumped models have an AARE of nearly 4% followed by Bandoypandhyay and Sharma's model (2011); and Elsharkawy's model (2003); the SRK-EOS has nearly twice the AARE of the newly proposed model. AlQuraishi's model (2009) has an AARE of 12% because the model is based on methane and neglects the effect of heavy and intermediate components, whose contribution is highly significant in the reservoir and production performance of volatile crudes. Jarrhaine et al (2015) model is not suitable for volatile oils. It shows a 35% error in estimated saturation pressure.

The performance of the extended and lumped models compared to the various methods for the black oil samples ($35< C_{7+} <55\%$) is provided in Figure 6B. The newly proposed models have an AARE close to 10%; followed by Bandoypandyay and Sharma's model 14%, the SRK-EOS and PR-EOS 16%, and Elsharkawy 17%. AlQuraishi's model has an AARE of 24%. Jarrhain et al (2015) model shows a 33% error in estimated saturation pressure even though the authors claim that the model is suitable for black oils.

Figure 6C shows the error level of all the methods as well as the proposed models for the newly measured crude oil samples at Kuwait University using the PFRC. Again, Jarrhain et al (2015) model show a 40% error in estimated saturation pressure.

The performance of all published models at low temperature is illustrated in Figure 6D. The low-temperature calculation indicates that Bandoypandhyay and Sharma's model (2011) has accuracy problems and is not valid at low temperatures because the authors proposed a temperature-composition related coefficient. Also, Jarrhain et al (2015) model is not suitable at low temperature. It shows a 40% error in estimated saturation pressure.

The ability of the proposed models to capture the thermodynamic is thoroughly checked against its independent variables. The validity of the model is compared to the experimentally measured data as well as the predictions by other methods used to capture the physics of increasing the saturation pressure as a function of temperature is illustrated in Figure 7. Two main points are observed here; first, the model matches the measured saturation pressures much better than the other methods. Second, it follows the physical trend predicted by the EOS's. Figure 8 shows the model performance for capturing the effect of changing the saturation pressure as a function of the molecular weight of the heptane plus fraction compared with the SRK-EOS and PR-EOS. The extended model follows the trend of the PR-EOS in increasing the saturation pressure. Elsharkawy's model (2003) was unable to predict the correct trend as predicted by the EOS's. The model sensitivity in capturing the change in the saturation pressure as a function of increasing the methane content is shown in Figure 9. The proposed model as well predicts the correct trend of increasing the saturation pressure when increasing the methane content.

Conclusion

In this paper, two models are presented to estimate the saturation pressure of a variety of crude oils when measures are unreliable or samples are unavailable. The two models are developed using compositional data and saturation pressure measurements from a large data bank comprising literature data, unpublished data, and newly measured and reported data of 231 crude oil samples. The input parameters in the first model are the extended hydrocarbon and non-hydrocarbon molar percentages and reservoir temperatures. The second model uses lumped hydrocarbon fractions of intermediate, and heptane plus fractions in addition to its properties; and the reservoir temperature as input parameters. The accuracy of the two models has been tested and compared to that of the SRK-EOS and the PR-EOS; as well as published correlations that use compositional data as the input parameters. The comparison indicates that the two models are much more accurate than all the published methods. The validity of both models has also been tested using various types of crude oils samples and compared to experimentally measured saturation pressures and estimates using the EOS as well as the published correlations. The proposed models are physically correct and much simpler than all the published methods. The models do not require tedious computations and eliminate splitting of the plus fraction, characterization of the pseudo-components; as well as the binary interaction parameters that are required for the EOS calculations.

Acknowledgement

The authors thank the General Facility Lab at Kuwait University for their cooperation and conducting the compositional and saturation pressure measurements of the 7 new samples described in this paper using Petroleum Fluid Research Center (PFRC) project no. GE 01107.

References

Agarwal, R.K., Li, Y.K., Nghiem, L., 1990. A regression technique with dynamic parameter selection for phase – behavior matching. SPE Res. Eng. February, 115 – 120.

Ahmadi, M. A., Zendehboudi, S, James, L., Elkamel, A., Dusseault, M., Chatsiz, I., Lohi, A., 2014, New Tools to determine Bubble Point Pressure of Crude Oils: Experimental and Modeling Study, *Journal of Petroleum Science and Engineering, v.123, 204-216.*

Ahmed, T., 2011, Splitting and lumping schemes of the plus fraction, working guide to vapour-liquid phase equilibria calculations, Chapter 9, Hydrocarbon Phase Behaviour, Gulf Publishing Co. 111-126.

Al-Gathe, A., Abd-el Fattah, Kh., El-Mtwally, K., 2014, New artificial intelligent approach for bubble point pressure, Int. J innovation and Sci. Research, Vol.4, No.2, July, 121-135.

Al-Marhoun, M.A. 1985. Pressure-Volume-Temperature Correlations for Saudi Crude Oils. Presented at the SPE Middle East Oil Technical Conference and Exhibition, Bahrain, 11–14 March. SPE-13718.

Almehaideb, R.A. 1992. Improved PVT Correlations for UAE Crude Oils. Presented at the Middle East Oil Show and Conference, Bahrain, 15-18 March. SPE-37691.

Al-Najjar, H.S., Al-Soof, N.B.A., and Al-Khalisy, K.M. 1988. Correlations for Bubble-Point Pressures, Gas Oil Ratios and Formation Volume Factors for Iraqi Crude Oils. *Journal of Petroleum Research* (June 1988): 13.

AlQuraishi, A. A., 2009, Determination of crude oil saturation pressure using linear genetic programing, Energy & Fuel, 23, 884-887.

Al-Shammasi, A.A. 2001. A Review of Bubblepoint Pressure and Oil Formation Volume Factor Correlations. *SPE Res. Eval. & Eng.* 4 (2): 146-160. SPE71302.

Asgarpour, S., McLauchlin, L., Wong, D. et al. 1989. Pressure-Volume-Temperature Correlations for Western Canadian Gases and Oils. *J Can Pet Tech.* 28 (4): 103.

Asoodeh, M., Bagheripour, P.,2012, Estimation of bubble point pressure from *PVT* data using a power-law committee with intelligent systems, *J. Pet. Sci. and Eng., Vol. 90–91, July 2012, Pages 1-11.*

Bandyopadhyay, P., Sharma, A., 2011, Development of a new semi analytical model for prediction of bubble point pressure of crude oils, *Journal of Petroleum Science and Engineering, Volume 78, Issues 3–4, September 2011, Pages 719-731.*

Becker, P. J., Celse, B., Guillaume, D., Dulot, H., Costa, V., 2015, Hydrotreatment modelling for variety of VGD feedstocks: a continuous lumping approach, Fuel, 139, 133-143.

Coats, K.H., and Smart, G.T., 1986, Application of regression base EOS PVT Program to laboratory data, SPERE, May 277-299.

Dansh, A. Xu, D. H., Todd, A. C., 1992 A grouping method to optimize reservoir description for compositional simulation of gas injection process, SPERE, Aug. 343-348.

De Ghetto, G. and Villa, M. 1994. Reliability Analysis on PVT Correlations. SPE-28904, Presented at the European Petroleum Conference, London, United Kingdom, 25-27 October.

Dindoruk, B. and Christman, P.G. 2001. PVT Properties and Viscosity Correlations for Gulf of Mexico Oils. Presented at the SPE Annual Technical Conference and Exhibition, New Orleans, 30 September-3 October. SPE-71633

Dokla, M.E. and Osman, M.E. 1992. Correlation of PVT Properties for UAE (United Arab Emirates) Crudes. *SPE Form Eval.* (March 1992): 41.

Droham, J.K., 1988, Enhancing the evaluation of PVT data, OSEA paper presented the 7th Offshore South East Asia, Feb. 2-5, Singapore.

Duan, J., Wang, W., Liu, H., Cong, J., 2013, modelling the characterization of the plus fraction by using continuous distribution function, Fluid Phase Equilibria, 345, 1-10.

Elam, F.M. 1957. *Prediction of Bubble Point Pressures and Formation Volume Factors from Field Data.* MS thesis, University of Texas at Austin, Austin, Texas.

Elmabrouk, S., Zekri, A., Shirif E., 2014, The prediction of bubble point pressure and bubble point oil formation volume factor in absence of PVT analysis, PET. Sci. & Tech., 32, 1168-1174.

Elsharkawy, A. M., 2003, An empirical model for estimating the saturation pressures of crude oils, Journal of Petroleum Science and Engineering 38 (2003) 55-77.

Elsharkawy, A.M. and Alikhan, A.A. 1997. Correlations for predicting solution gas/oil ratio, oil formation volume factor, and undersaturated oil compressibility. *J. Pet. Sci. Eng.* 17 (3–4): 291-302.

Emera, M., Samara, H., 2005, Use of genetic algorithm to estimate CO2-oil minimum miscibility pressure- a key parameter in design of CO2 miscible flood, J. Pet. Sci. Eng., Vol. 46, issue 1-2, 37-52.

Farasat, A., Shokrollahi, A., Arabloo, M., Gharagheizi, F., Mohammadi, A., 2013, toward an intelligent approach for determination of saturation pressure of crude oil, Fuel Processing Technology 115 (2013) 201-214.

Frashad, F., LeBlanc, J.L., Garber, J.D. et al. 1996. Empirical PVT Correlations for Colombian Crude Oils. SPE-36105 Presented at the SPE Latin American and Caribbean Petroleum Engineering Conference, Port of Spain, Trinidad and Tobago, 23–26 April. -MS.

Gharbi, R.B.C., Elsharkawy, A.M., 1999, neural network model for estimating the PVT properties of Meddle East crude oils, SPE Reservoir Eval. Eng. 2 (3), 255-265.

Gholami, A., Asoodeh, M., Bagheripour, P., 2014 How committee machine with SVR and ACE estimates bubble point pressure of crudes, *Fluid Phase Equilibria, Volume 382, 25 November 2014, Pages 139-149.*

Glasø, Ø. 1980, Generalized Pressure-Volume-Temperature Correlations. *J Pet Tech.* 32 (5): 785-795.

Heidaryan, E., Moghadasi, J., 2011, Crude saturation pressure through imperial correlation, Thermodynamics 2011, Athens, 31 August- 3 September 2011.

Hasan, M., 1993, PVT Correlations for Indonesian Crude Oils. *Proc.* 22nd Annual Convention of the Indonesian Petroleum Association, Indonesia, Vol.1, IPA93-2.1-070.

Jarrhian, A., Moghadsi, J., Heidaryan, E., 2015, Empirical estimating of black oil bubblepoint (saturation) pressure, J. Petroleum Science and engineering, (126), 69-77.

Jacopy R.H., Berry V.J.,1958, A method for predicting the pressure maintenance for reservoir producing volatile oil, AIME 213, 59-65.

Kartoatmodjo, T., Schmidt, Z. 1994, Large Data Bank Improves Crude Physical Property Correlations. *Oil Gas J.* 92 (27): 51–55.

Kartoatmodjo, T.R.S. and Schmidt, Z. 1991, *New Correlations for Crude Oil Physical Properties*, SPE 23556.

Khairy, M., El-Tayeb, S., Hamdallah, M., 1998. PVT Correlations Developed for Egyptian Crudes. *Oil Gas J.* 96 (18): 114.

Khamekhcki, E., Behvandi, R., Rashid, F., 2011, Prediction of bubble point pressure & asphaltene onset pressure during CO2 injection using ANN & ANFIS models, Pet. Sci. & Tech. Vol. 1 No.2, 35-45.

Labedi, R.M. 1982, *PVT Correlations of the African Crudes*, PhD thesis, CSM.

Labedi, R.M. 1990, Use of Production Data to Estimate the Saturation Pressure, Solution GOR, and Chemical Composition of Reservoir Fluids, SPE-21164-MS. Presented at the SPE Latin America Petroleum Engineering Conference, Rio de Janeiro, Brazil, 14-19 October,

Lasater, J.A. 1958, Bubble Point Pressure Correlations. *J Pet. Tech.*10 (5): 65–67. SPE-957-G.

Levitan, L.L. and Murtha, M. 1999. New Correlations Estimate Pb, FVF. *Oil Gas J.* 97 (10): 70.

Lee, S., 1979, Experimental and theoretical studies on the fluid properties required for simulation of thermal process, Paper SPE 8393.

Li, Y.K., Nghiem, L.X., Siu, A., 1985. Phase behavior computations for reservoir fluids: effect of pseudo-components on the phase diagrams and simulation results. J. Can. Pet. Technol. Nov. – Dec., 29 – 36

Lin, B., Leibovici, C. F., Jorgensen, S. B., 2008, optimal component lumping: problem formulation and solution techniques, Computer & Chem. Eng., 32, 1167-2272.

Macary, S.M. and El-Batanoney, M.H. 1992, Derivation of PVT Correlations for the Gulf of Suez Crude Oils. *Proc. of the* 11th EGPC Pet. Exp. & Prod. Conference, Vol. 1, 374, Cairo, Egypt,

Moharam, H.M., 1995. Prediction of viscosity of heavy petroleum fractions and crude oil using a corresponding states method. Ind. Eng. Chem. Res. 34, 4140 – 4144.

Moradi, B., Malekzadeh, E., Amani, M., Boukadi, F.H., Kharat, R., 2101, Bubblepoint pressure imperial correlation, SPE 132756.

Obomanu, D.A. and Okpobiri, G.A. 1987, Correlating the PVT Properties of Nigerian Crudes. *J. Energy Res. Tech.* 109 (4): 214-217.

Okoduwa, I. G., Ikiensikimama, S.S., 2010, Bubble point pressure correlation for Niger Delta Crude Oils, SPE 136968.

Omar, M.I. and Todd, A.C. 1993. Development of New Modified Black Oil Correlations for Malaysian Crudes. SPE-25338 Presented at the SPE Asia Pacific Oil and Gas Conference, Singapore, 8-10 February.

Owolabi, O.O. 1984. *Reservoir Fluid Properties of Alaskan Crudes*. MS thesis, University of Alaska, Fairbanks, Alaska (May 1984).

Pedersen, K.S., Thomassen, P., Fredenslund, Aa., 1988a. Characterization of gas condensate mixtures. AIChE Spring National Meeting, March 6 – 10, New Orleans, LA.

Peng, D. y., Robinson. D. B., 1976. A new two-constant equation of state , Ind. Eng. Chem. Fund. 15, 59-63.

Petrosky, G.E. Jr. 1990. *PVT Correlations for Gulf of Mexico Crude Oils*. MS thesis, University of Southwestern Louisiana, Lafayette, Louisiana.

Petrosky, G.E. Jr. and Farshad, F. 1998. Pressure-Volume-Temperature Correlations for Gulf of Mexico Crude Oils. *SPE Res Eval & Eng.*1 (5): 416-420. SPE-51395.

Riemens, W.G., Schutle, A.M., Jong, L.N.J., 1988. Birba field PVT variations along the hydrocarbon column and confirmatory field tests. J. Pet. Technol. Jan., 83 – 88

Rosengger, L., Wu., R.1990, Integrated PVT data characterization- lessons from four case histories, J. Can. Per. Tech., 38, (13), (paper 97-05)

Shojaei, M., Bahrami, F., Barati, P., Riahi, S., 2014, Adaptive neuro-fuzzy approach for reservoir oil bubble point pressure estimation, *J. of Natural Gas Sci. and Eng., Volume 20, September 2014, Pages 214-220.*

Soave. G. 1972. Equilibrium constants from a modified Redlich-Kwong equation of state, Chem. Eng. Sci. 27, 1197-1203.

Standing, M.B. 1947. A Pressure-Volume-Temperature Correlation for Mixtures of California Oils and Gases. *API Drilling and Production Practice* (1947): 275-287.

Tale, M., Alquraishi, A., Al Mushaigeh, E. 2009, New model for crude oil saturation pressure prediction using alternation conditional expectation, Offshore Mediterranean Conf. & Exhibit. Italy 25-27, 2009.

Talebi, R., Ghiasi, M. Talebi, H., Mohammadyian, M., Zendehboudi,S., Arabloo, M., Bahadori, A., 2014, Application of soft computing approaches for modelling saturation pressure of reservoir oil, *J. Nat. Gas Sci. and Eng., Vol. 20, Sep. 2014, 8-15.*

Vazquez, M., Beggs, H. D. 1980, Correlations for Fluid Physical Property Prediction. *J Pet. Tech.* 32 (6): 968-970. Vazquez, M.E. 1976. *Correlations for Fluid Physical Property Prediction,* MS thesis, University of Tulsa.

Velarde, J., Blasingame, T.A., and McCain Jr., W.D. 1997. Correlation of Black Oil Properties at Pressures below Bubble Point Pressure - A New Approach. Presented at the Annual Technical Meeting of CIM, Calgary, Alberta, 8–11 June. PETSOC 97-93.

Vogel, J.R., Yarborough, L. 1980, the effect of nitrogen on phase behavior and phase properties of reservoir fluids, SPE 8815.

Whitson, C. H., 1980, Characterizing hydrocarbon plus fraction, SPEJ, 24 (4), 683-694.

Williams, C.A., Zana, E.N., 1980. Use of the Peng-Robinson equation of state to predict hydrocarbon phase behavior and miscibility for fluid displacement. SPE 8817, SPE/DOE on Enhanced Oil Recovery, April 20–23, Tulsa, OK.

Appendix

In this appendix, sample calculations are presented to show the deficiency of some of the published methods.

Input parameters:

N_2	CO_2	H_2S	C_1	C_2	C_3	C_4	C_5	C_6	C_{7+}	GR.C7+	MWC7+	T, °F	Ps, psia
0.65	0.02	0	45.02	12.45	8.93	6.03	3.02	1.44	22.44	0.81	184	140	3002

Ahmadi et al (2014)

Ahmadi et al. proposed an 8 steps calculation (copied from their paper) to estimate the BPP as follows:

$$BPP = A \times Mw_{C7+} - \frac{B}{C_{7+}} + C \times Vol./Inter. + D \times SG_{C7} + \frac{E}{SG_{C7+}} + \frac{F}{Mw_{C7+}} + T \times 505.967 + \frac{2.45789 \times 10^6}{T} + \frac{G}{Vol./Inter.} \quad (5)$$

$$A = 1488.46 - SG_{C7+} \times 1087.64 - \frac{601.512}{SG_{C7+}} + 1670.58 + T \times 0.40771 + \frac{6345.73}{T} + \frac{8.72863}{Vol./Inter.} \quad (6)$$

$$B = \frac{4132.53}{Vol./Inter.} + \frac{6.1977 \times 10^7}{Mw_{C7+}} + T \times 318139 + \frac{1.06393 \times 10^8}{T} \quad (7)$$

$$C = \frac{20964.7}{C_{7+}} - \frac{1.76162 \times 10^6}{Mw_{C7+}} - T \times 19.1963 - Mw_{C7+} \times 21.5427 + \frac{12851.5}{SG_{C7+}} - \frac{712613}{T} - 3477.49 + SG_{C7+} \times 9532.73 \quad (8)$$

$$D = 1.05345 \times 10^{+06} - \frac{1.50748 \times 10^8}{Mw_{C7+}} - T \times 676.048 - \frac{1.20252 \times 10^7}{T} - \frac{12178.2}{Vol./Inter.} \quad (9)$$

$$E = 641357 - \frac{1.02891 \times 10^8}{Mw_{C7+}} - T \times 301.034 - \frac{1.91446 \times 10^6}{T} - \frac{3873.62}{Vol./Inter.} \quad (10)$$

$$F = 2.51471 \times 10^{+08} + T \times 25332.5 + \frac{3.47893 \times 10^8}{T} - \frac{244573}{Vol./Inter.} \quad (11)$$

$$G = 20125.6 - T \times 23.9271 - \frac{703193}{T} \quad (12)$$

Ahmadi et al. define the Vol. /Inter. as the mole percentage ratio of the volatile (C_1& N_2) to the intermediate components (C_2 to C_6, CO_2 and H_2S). Application of the above equations from their paper using the input parameters results in the following:

Vol./Inter.	A	B	C	D	E	F	G	BPP
1.43211	1643.9	1545062	-271.127	45122.71	23642.47	2.57E+08	11753	1,794,123

The calculated saturation pressure (BBP) from Eq.5 is 1,794,123 psia which is unreasonable. This indicates a deficiency in the model.

Gholami et al (2014)

Gholami et al. presented 3 steps calculation (copied from their paper) to estimate the BPP as follows:

$$Tr(M) = \sum_{i=0}^{5} B_i M^i \quad (15)$$

Application of Eq. 15 resulted in the following

N_2	0.00584
CO_2	0.060917
H_2S	0.050636
C_1	-0.47658
C_2	-0.41547
C_3	-0.45083
C_4	-0.02003
C_5	0.068769
C_6	0.273921
C_{7+}	1.814523
SG.C7+	-0.0197
MWC7+	37.03525
T., F	-0.11139

Application of Eq. 16

$$Tr(P_b) = Tr(N_2) + Tr(CO_2) + Tr(H_2S) + Tr(C_1) + Tr(C_2) + Tr(C_3) + Tr(C_4) + Tr(C_5) + Tr(C_6) + Tr(C_{7+}) + Tr(SG_{C7+}) + Tr(MW_{C7+}) + Tr(T) \quad (16)$$

Resulted that Tr(Pb) = 37.8159

$$P_b = (57.0840387741 \times (Tr(P_b))^2) + (1166.9102304199 \times (Tr(P_b))) + 2207.3573204358 \quad (17)$$

Application of Eq. 17 resulted in saturation pressure (P_b) = 127,967 psia, which is unreasonable, compared to the measured one is 3002 psia. This indicates that there is a deficiency in the model.

Jarrahian et al (2015)

Jarrahian et al. proposed the following Eq. (copied from their paper) to estimate the saturation pressure:

$$P_{sat} = A_0 \exp\left(A_1 z_{N_2} + A_2 z_{CO_2} + A_3 z_{H_2S} + A_4 z_{C_1}^{A_5} + A_6 \sum_2^6 z_{C_i} + A_7 z_{C_{7+}}^{A_8} \gamma_{C_{7+}}^{A_9} Mw_{C_{7+}}^{A_{10}} + A_{11}(T - 459.67)\right) \quad (2)$$

Application of Eq. 2 resulted in saturation pressure (Ps) = 1812 psia, compared to the measured one is 3002 psia. This indicates that model has inaccuracy problem. This has been the pattern in all the crude oil samples shown in Table 4 and figures 6A through 6D.

Table 1- Summary of attempts to calculate the saturation pressure of crude oils

Method based on production information	Method based on compositional data
Standing, 1947;	Elsharkawy , 2003
Elam, 1957;	Tale et al., 2009
Lasater, 1958;	AlQuraishi, 2009
Vasquez, 1976;	Bandyopadhyay and Sharma, 2011
Vasquez and Beggs, 1980;	Khamehchi et al., 2011
GlasØ, 1980;	Farasat et al., 2013
Labedi, 1982&1992;	Ahmadi et al., 2014
Owolabi, 1984;	Gholami et al. 2014
Al-Marhoun, 1985;	Jarrahian et al 2015
Obomanu and Okpobiri, 1987;	
Asgarpour et al., 1989;	
Al-Najar et al., 1988;	
Dokla and Osman, 1990;	
Petrosky, 1990;	
Kartoatmodjo and Schmidt, 1991&1994;	
Macary and El-Batanoney, 1992;	
Omar and Todd, 1993;	
Hasan, 1993;	
De Ghetto and Villa, 1994;	
Farashad et al., 1996;	
Almehaidaeb, 1997;	
Elsharkawy, 1997;	
Velarde et al., 1997;	
Kairy et al., 1998;	
Petrosky and Farshad, 1988;	
Al-shammasi, 2001;	
Gharbi & Elsharkawy 1999	
Levitan and Murtha, 1999;	
Dindrouk and Christman, 2001	
Moradi et al., 2010	
Okoduwa and Ikiensikimama 2010	
Asoodeh and Bagheripour, 2012	
Shojaei et al., 2014	
Talebi et al., 2014	
Elmabrouk et al., 2014	
Kanu and Ikiensikimama, 2014	
Jarrahian et al 2015	

Table 2- Description of the data used to develop and test the extended model

Input	Training model (179 Data)			Testing (28 Data)			Testing (7 new experimental)		
	Min	Ave	Max	Min	Ave	Max	Min	Ave	Max
N_2	0.00	0.35	1.67	0.00	0.27	1.67	0.10	0.26	0.35
CO_2	0.00	1.27	11.37	0.00	1.17	11.37	0.14	0.41	0.83
H_2S	0.00	0.15	3.68	0.00	0.15	3.22	0.00	0.00	0.00
C_1	13.16	33.86	74.18	19.50	34.79	73.36	20.64	28.25	42.79
C_2	3.36	8.07	13.71	4.64	8.26	11.63	6.69	7.88	10.82
C_3	0.89	6.59	11.87	2.46	6.76	12.03	6.05	6.59	7.10
C_4	0.95	4.46	8.4	1.66	4.41	6.58	4.03	5.05	5.93
C_5	0.40	3.21	5.95	1.00	3.11	5.23	3.36	4.33	5.52
C_6	0.00	3.14	6.37	0.99	3.15	5.46	2.88	3.74	4.33
C_{7+}	10.72	38.89	57.73	11.18	37.88	51.22	27.61	43.49	51.31
$SG.C_{7+}$	0.74	0.87	0.959	0.70	0.86	0.93	0.85	0.89	0.92
MWC_{7+}	134	240	368.9	105	223	292	216	262	298
T, ^{o}F	58	156	319	58	165	261	121	157	230
Ps, psia	537	2198	5065	1025	2307	5000	1227	1779	3344

Table 3: Error analysis of the proposed models in comparison with all the investigated methods

Method	ARE	AARE	SD	R^2%
PR-EOS	6.24%	10.19%	1028	97.9
SRK-EOS	-3.12%	10.36%	1024	96.7
Elsharkawy (2003)	0.85%	8.57%	1015	96.4
Alquraishi (2009)	-6.06%	10.16%	1088	96.7
Bandyopadhyay (2011)	8.05%	12.28%	1054	96.8
Farasat (2013)	NA	NA	NA	NA
Gholami et al. (2014)	?	?	?	?
Ahmadi et al. (2014)	?	?	?	?
Jarrahian et al (2015)	35.52	35.56	983	95.2
New Extended Model	-0.37%	8.20%	1030	97.6
New Lumped Model	-0.50%	8.46%	1036	97.4

ARE = Average Relative Error

AARE = Average Absolute Relative Error

SD = Standard Deviation

R^2 = Correlation Coefficient

NA = Calculation was not possible, because the weight and bias were not provided.

? = The method produces unreasonable results as indicated in the appendix.

Table 4- Comparing model accuracy for calculations of saturation pressures for various crude oils

Volatile-Oil-Samples (C7+<30)																							
Fluid Source	index	N2	CO2	H2S	C1	C2	C3	C4	C5	C6	C7+	GR.C7+	MWC7+	T, °F	Ps, psi	SRK	PR	Elsh	Alq	Bandy	Jarr	Extd	Lump
Rosenegger	7	0.03	8.39	0	47.4	10.3	6.12	4.2	2.89	2.05	18.6	0.83	180	295	4000	3933	3879	4460	4420	4165	3317	4139	4178
Vogel-80	13	0.65	0.02	0	45	12.5	8.93	6.03	3.02	1.44	22.4	0.81	184	140	3002	2624	2548	2742	2895	2643	1812	3019	2996
Dansh-92	18	0	0	0	46.8	8.77	7.44	4.01	2.56	4.02	26.4	0.77	158	212	2941	2550	2594	3248	3617	3269	1944	3244	3374
	31	0.44	0.38	0	49.1	7.6	6.13	3.84	2.52	2	28	0.84	231	199	3739	3522	3507	3418	3951	3508	2309	3625	3644
Middle East	45	0.03	0.97	0	41.6	10.8	7.52	4.5	3	2.43	29.1	0.85	208	241	3069	3109	2979	3080	3513	3138	1962	3127	3140
Coats-86	51	0.55	1.03	0	36.5	9.93	8.85	6	3.78	3.56	30.4	0.84	200	234	2746	2537	2472	2714	2945	2643	1682	2777	2720

Black-Oil -Samples (35<C7+<55)																							
Fluid Source	index	N2	CO2	H2S	C1	C2	C3	C4	C5	C6	C7+	GR.C7+	MWC7+	T, °F	Ps, psi	SRK	PR	Elsh	Alq	Bandy	Jarr	Extd	Lump
Coats-86	73	1.64	0.08	0	28.4	7.16	10.5	8.4	3.82	4.05	36	0.84	252	131	1694	1674	1618	1615	1738	1644	1084	1713	1713
Middle East	88	0.08	1.37	0.69	36	8.67	5.94	2.97	1.5	3.1	39.7	0.9	274	168	2505	2710	2318	2247	2608	2321	1476	2441	2445
Middle East	114	0.29	0.48	0	28.4	8.29	7.38	5.06	3.42	4.41	42.3	0.88	252	133	1632	1672	1477	1544	1748	1517	984	1587	1598
Middle East	122	0	1.25	0	33.4	9.24	6.07	2.4	1.62	2.78	43.3	0.85	252	230	2110	2313	2278	2424	2737	2438	1410	2388	2412
Middle East	134	0	0.17	0	28.6	8.63	7.43	3.81	2.61	4.11	44.7	0.86	249	134	1548	1551	1429	1477	1764	1433	937	1503	1567
Middle East	146	0.44	0.83	0	27.8	7.54	6.9	4.61	3.1	3.11	45.7	0.9	272	133	1565	1799	1505	1558	1730	1528	1002	1602	1590
moharam-96	149	0.21	0.34	0	20	7.93	8	6.6	5.87	5.08	45.9	0.86	230	235	900	1353	1290	1531	1514	1225	773	1303	1294
Middle East	162	0.35	0.47	0	26.5	7.71	6.05	4.03	4.08	4.22	46.6	0.89	253	135	1500	1613	1380	1427	1631	1391	923	1478	1456
Middle East	176	0.12	0.51	0	28.8	7.91	6.46	3.17	2.07	3.25	47.7	0.88	250	134	1615	1665	1476	1574	1783	1502	981	1571	1614
Middle East	194	0.21	0.15	0	27.8	7.68	6.19	3.41	1.81	2.84	49.9	0.86	228	134	1590	1532	1377	1545	1680	1426	923	1466	1493
Rosenegger	207	0.37	0.02	0	17.3	6.11	8.23	5.47	3.76	3.96	54.8	0.85	242	156	537	938	888	914	1064	836	558	661	731

Newly-Measured Kuwait University Laboratory																						
Fluid Source	index	N2	CO2	H2S	C1	C2	C3	C4	C5	C6	C7+	GR.C7+	MWC7+	T, °F	Ps, psi	SRK	PR	Elsh	Alq	Bandy	Jarr	Extd	Lump
KU-1	209	0.33	0.22	0	25.6	6.87	6.39	5.61	4.68	4.33	46	0.879	222	133	1595	1360	1530	1442	1514	1394	860	1357	1337
KU-2	210	0.35	0.47	0	26.5	7.71	6.05	4.03	4.08	4.22	46.6	0.893	253	135	1500	1400	1620	1427	1631	1391	924	1478	1456
KU-3	211	0.27	0.74	0	23.1	7.2	6.45	5.93	4.66	4.16	47.5	0.906	290	138	1283	1140	1360	1284	1448	1296	800	1287	1219
KU-4	212	0.22	0.26	0	20.6	6.69	6.65	5.11	5.52	3.6	51.3	0.919	298	171	1227	1120	1300	1188	1415	1170	742	1176	1131
KU-5	213	0.19	0.14	0	27.8	8.37	6.88	5.28	4.33	3.85	43.2	0.89	278	121	1524	1300	1560	1402	1659	1409	928	1534	1489
KU-6	214	0.33	0.24	0	31.3	7.48	6.58	4.86	3.7	3.17	42.3	0.9	274	171	1978	1800	2200	1975	2242	1985	1224	2060	2026
KU-7	215	0.1	0.83	0	42.8	10.8	7.1	4.51	3.36	2.88	27.6	0.85	216	230	3344	3070	3400	3083	3568	3139	2003	3213	3200

Low-Temperature Oil Samples (59°F<T<71°F)																						
Fluid Source	index	N2	CO2	H2S	C1	C2	C3	C4	C5	C6	C7+	GR.C7+	MWC7+	T, °F	Ps, psi	SRK	PR	Elsh	Alq	Bandy	Jarr	Extd	Lump
Middle East	101	0	1.12	0.21	27	10.8	9.23	4.97	2.62	3.33	40.8	0.88	249	71	1247	1090	1033	1040	1235	564	808	1148	1205
Middle East	117	0.36	0.17	0	27.2	8.93	8.6	6.07	3.71	2.55	42.4	0.88	271	68	1130	1028	1245	1094	1248	766	820	1239	1228
Middle East	135	0	0.17	0	28.6	8.63	7.43	3.81	2.61	4.11	44.7	0.86	249	70	1270	1055	1255	1130	1311	783	826	1229	1297
Middle East	138	0.14	0.54	0	29.4	7.96	6.19	3.96	3.25	3.84	44.7	0.9	290	59	1265	1065	1350	1191	1301	880	894	1408	1390
Middle East	147	0.44	0.83	0	27.8	7.54	6.9	4.61	3.1	3.11	45.7	0.9	272	67	1320	1125	1400	1201	1268	921	879	1319	1312
Middle East	173	0.06	0.81	0	31.3	7.42	5.68	2.82	1.63	2.64	47.6	0.89	270	68	1460	1215	1535	1441	1463	1026	975	1548	1576
Middle East	193	0.16	0.3	0	24.7	7.51	6.92	4.76	2.65	3.43	49.6	0.88	239	71	1050	975	1156	1008	1107	701	714	958	986

Elsh: Elsharkawy (2003)
Extd: Extended model

Alq: AlQuraishi (2009)
Lump: Lumped model

Band: Bandyopadhyay and Sharma (2011)

Jarr: Jarrahian et. al (2015)

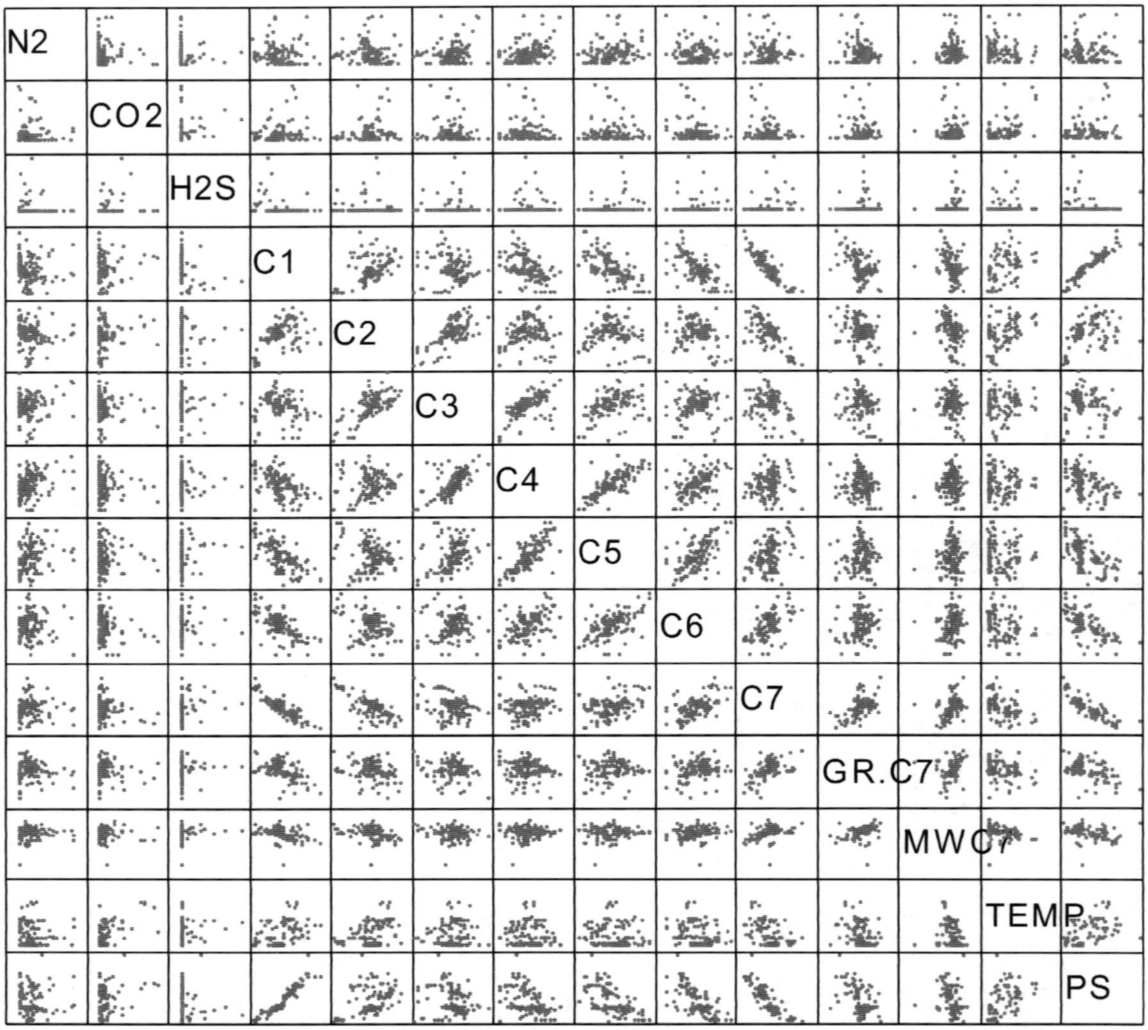

Figure 1: Correlation matrix for all the variables in the oil data bank

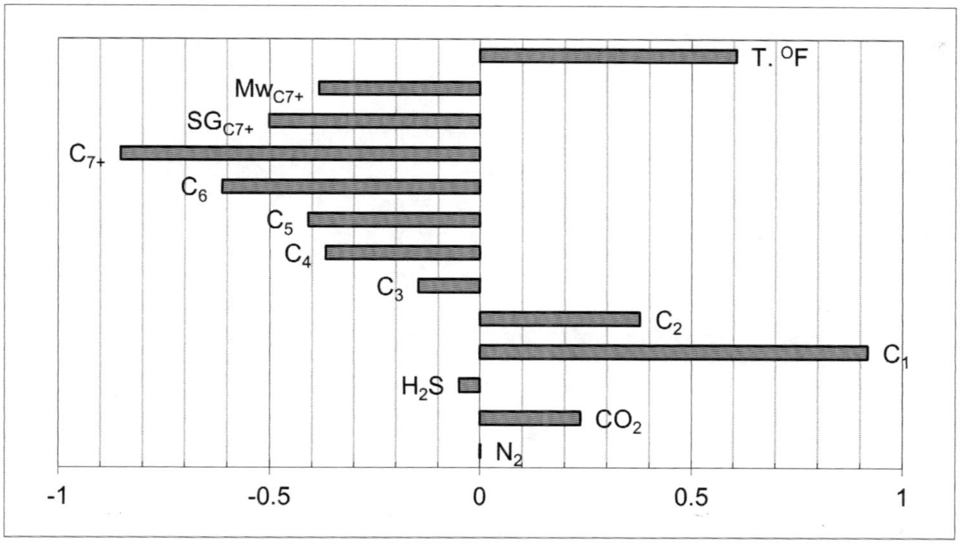

Figure 2: Correlation coefficient for all the variables in the training model (179 data)

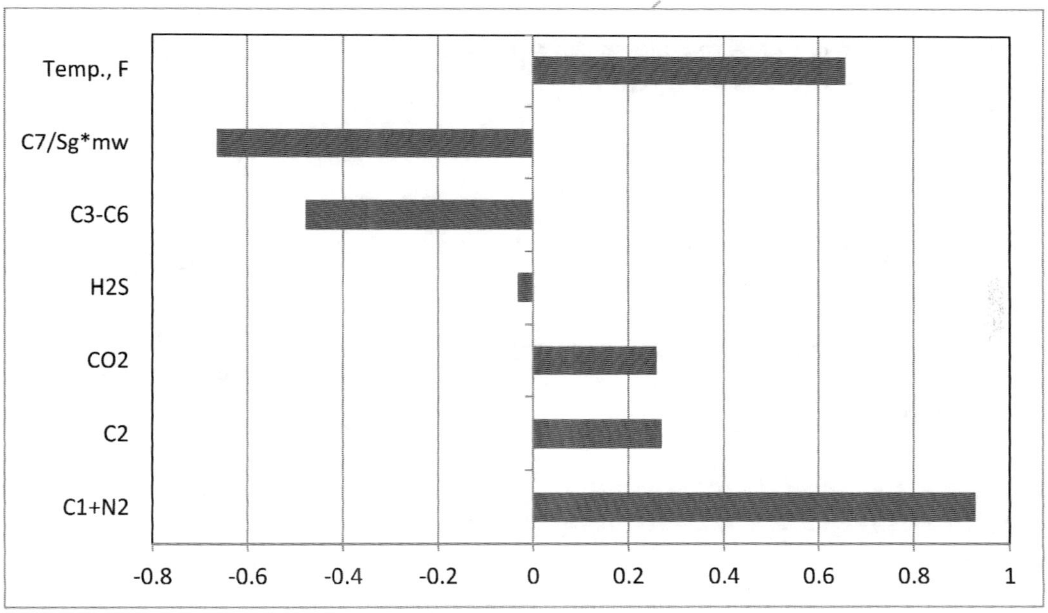

Figure 3: Correlation matrix for the lumped model (179 data)

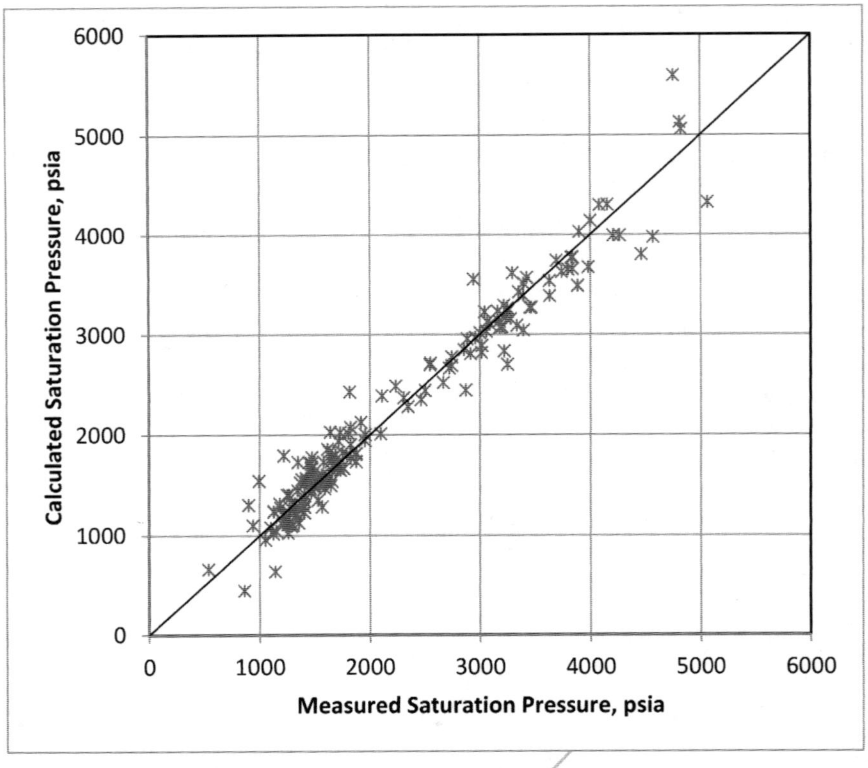

Figure 4A: Cross plot of the extended model using 179 training data

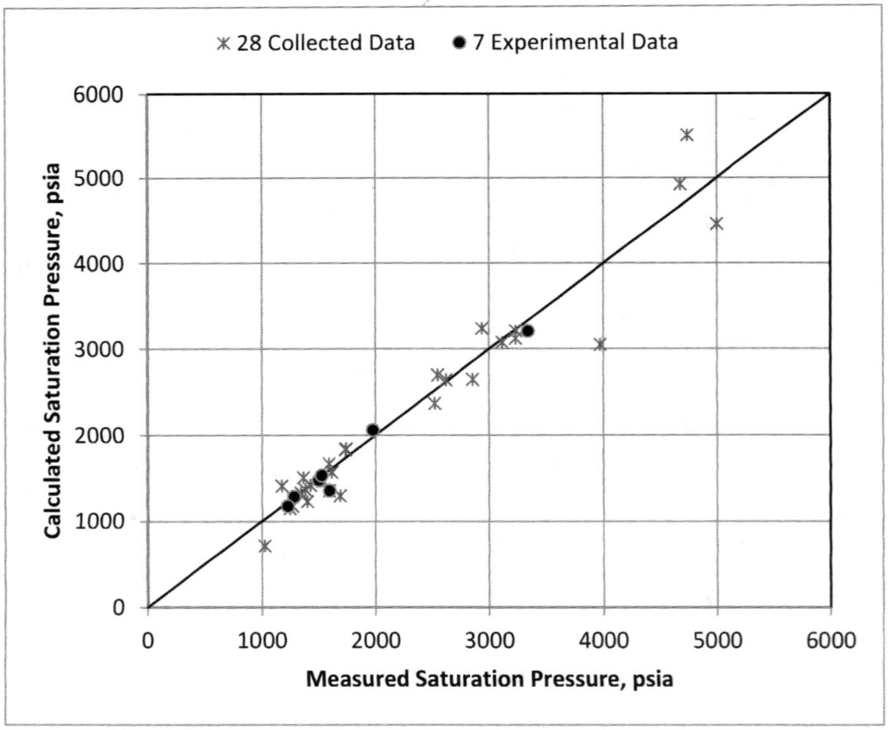

Figure 4B: Cross-plot of the extended model using 35-testing data

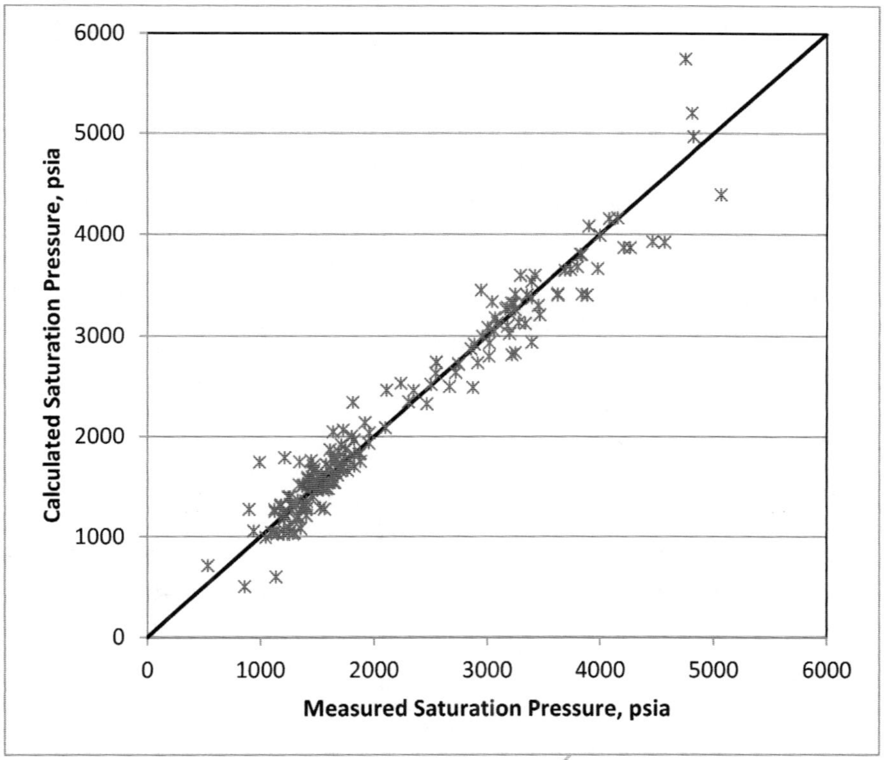

Figure 5A: Cross-plot of the lumped model using 179 training data

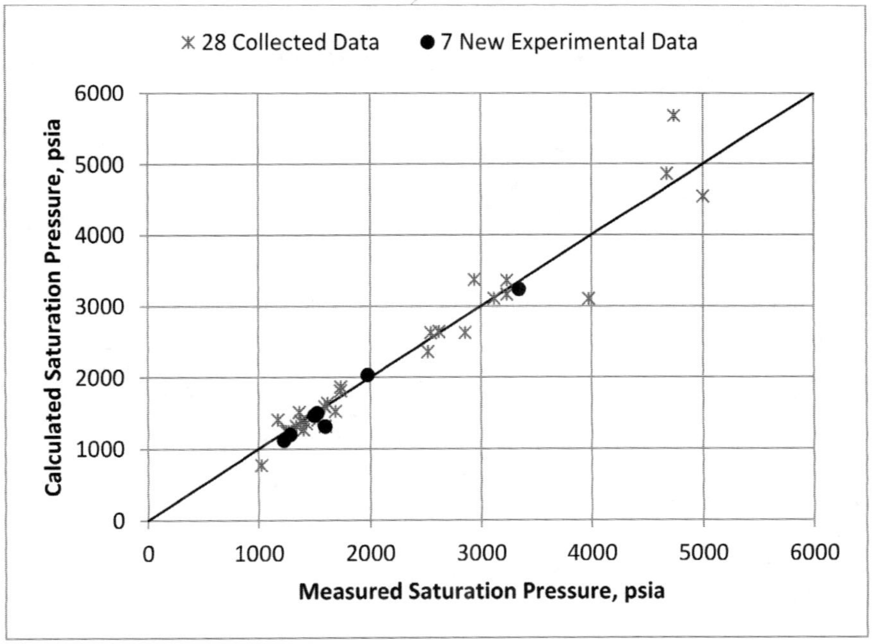

Figure 5B: Cross-plot of testing data for the lumped model

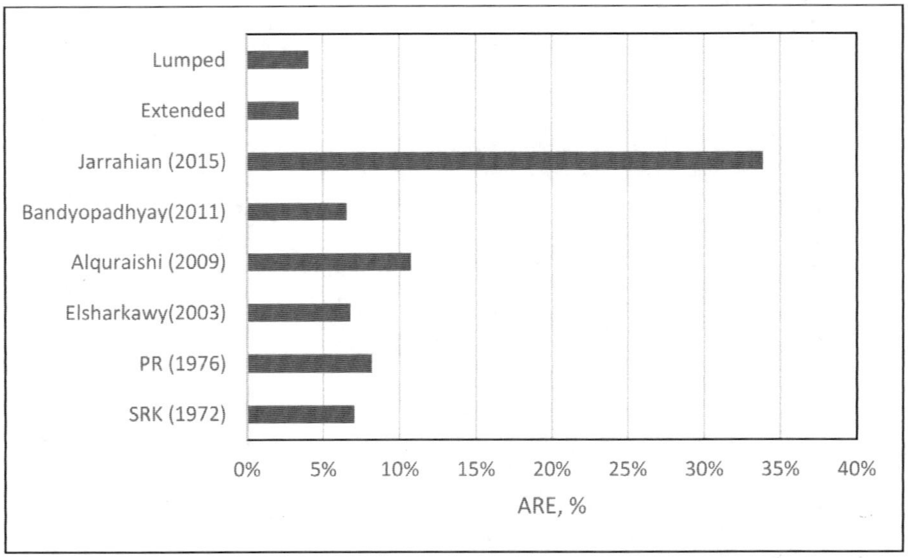

Figure 6A: Comparison of all the models' accuracy for the volatile oil samples

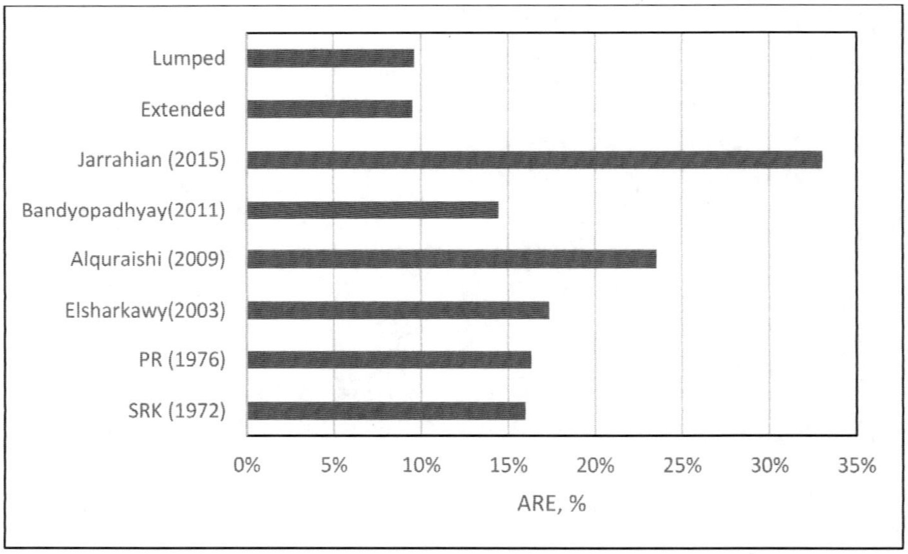

Figure 6B: Comparison of all the models' accuracy for the black oil samples

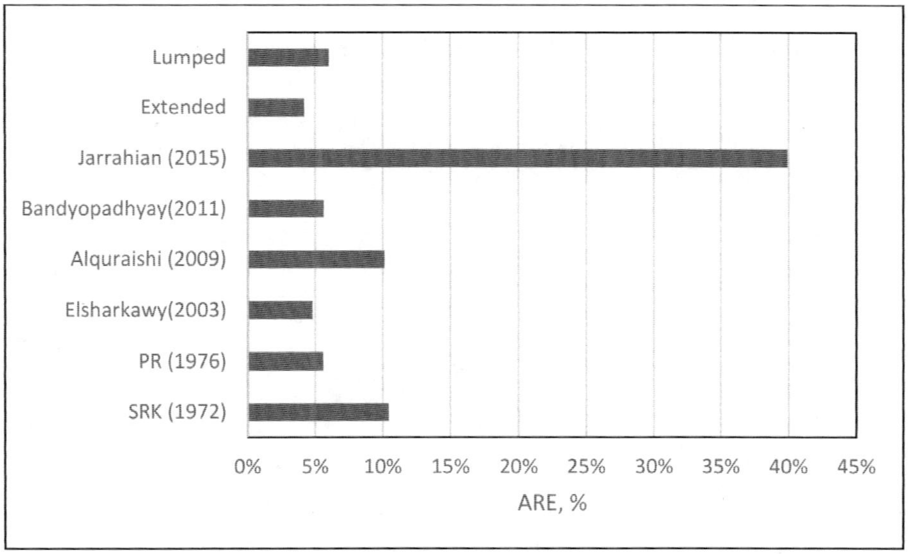

Figure 6C: Comparison of all the models' accuracy for the newly measured samples at KU

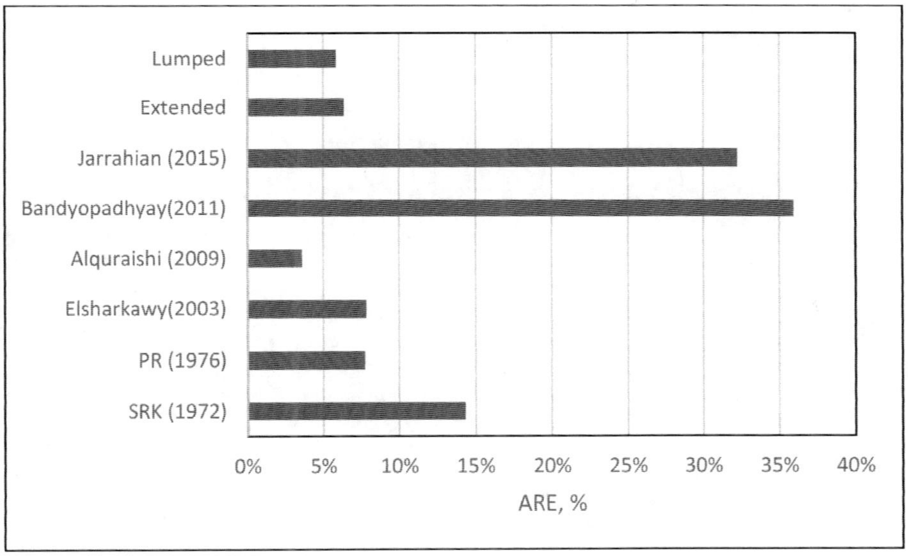

Figure 6D: Comparison of all the models' accuracy for low-temperature oil samples

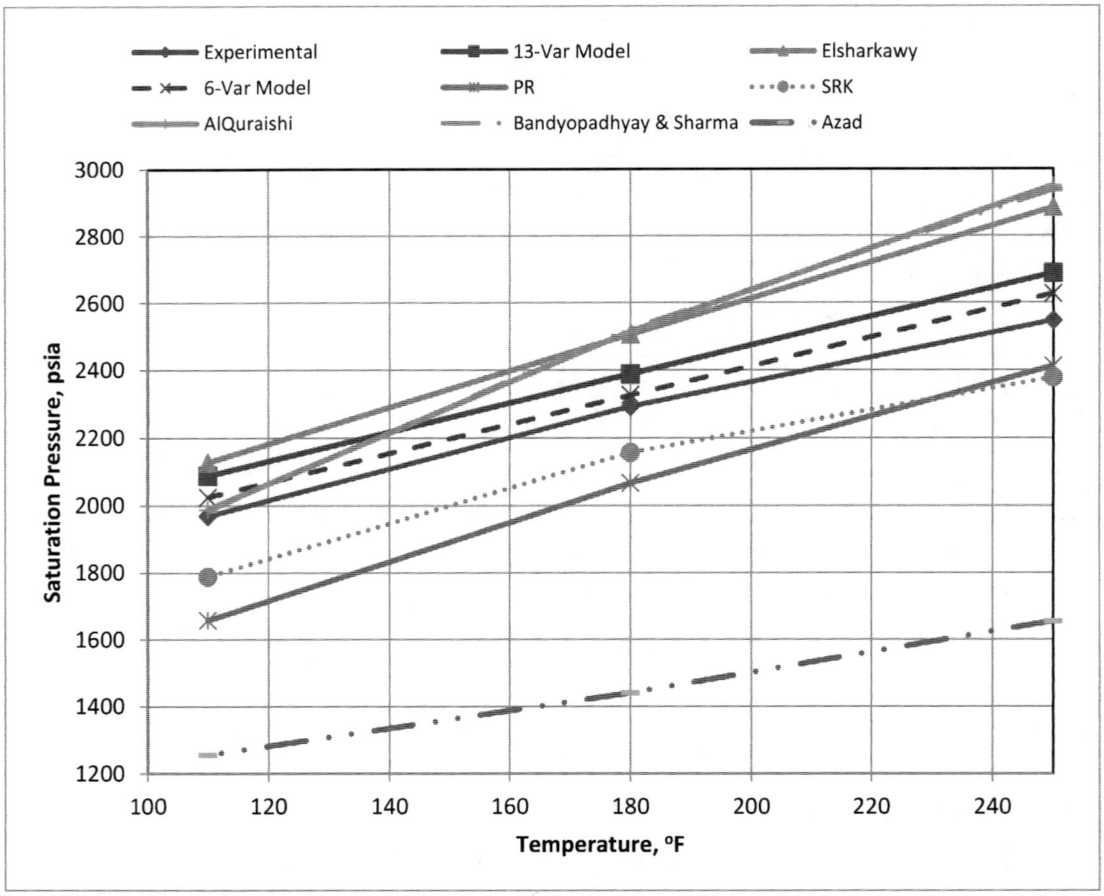

Figure 7: Changes in saturation pressure as a function of temperature for the different models (Sample #70)

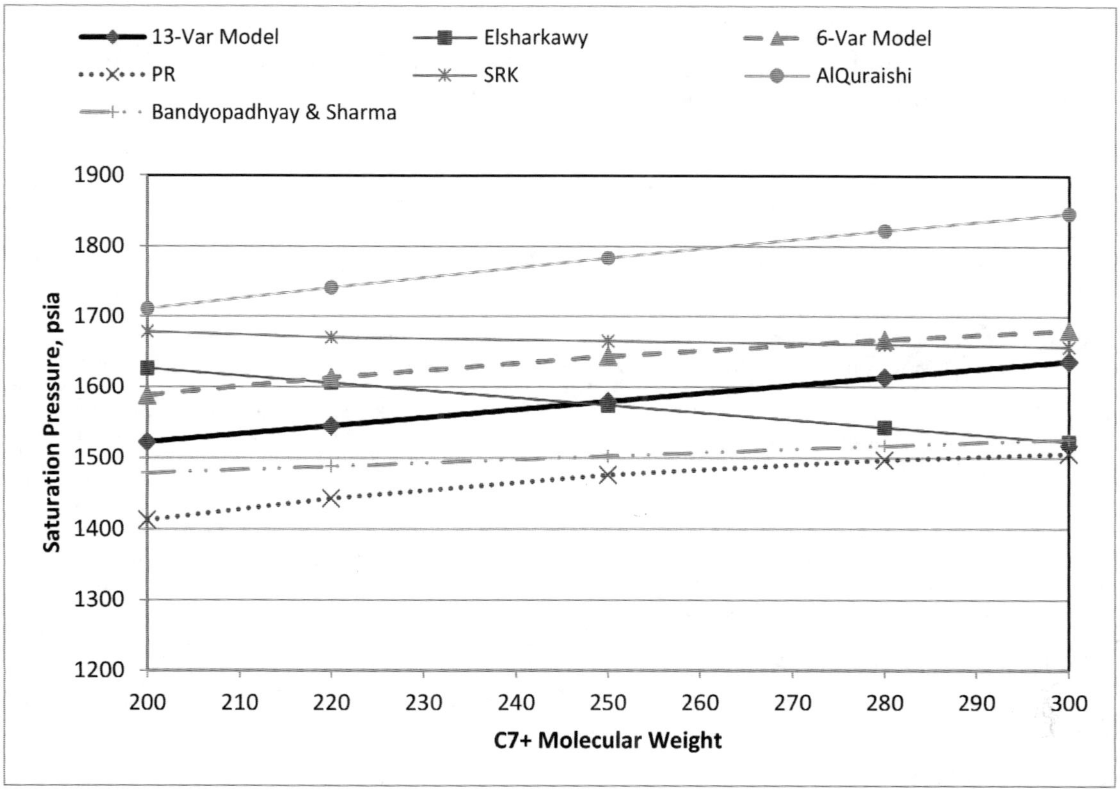

Figure 8: Comparing model performances: Changes in BPP as a function of C_{7+} molecular weight at 134°F (Sample #176)

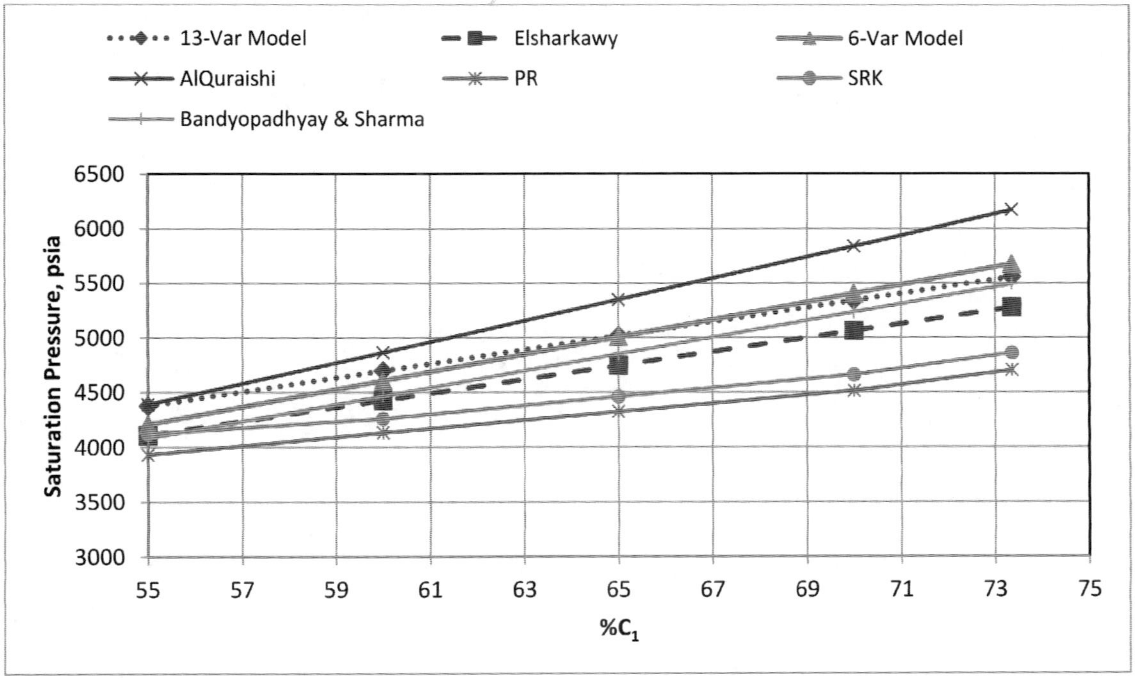

Figure 9: Sensitivity of saturation pressure to methane content at 212°F (sample #2)

444834 Stability and Rheology of Emulsified Acid in Waste Oil: Effect of Emulsifier Concentration, Mixing Speed, Temperature and Droplet Size

Abstract:

Acid in oil emulsion is used in the industry for stimulation purposes. In particular, HCl as the acid phase and diesel as the oil. Emulsifying HCl has numerous benefits over regular HCl. Perhaps the main reason is due to deeper penetration. Apart from that, less corrosion damages are caused since the external phase is hydrocarbon. Several studies showed the success of replacing diesel as an external phase with other hydrocarbon oil, such as crude oil and xylene.

This work utilizes the extra hydrocarbon left unused –or sometimes dumped- from refineries. This unused hydrocarbon is referred to as waste oil. First the chemical composition of waste oil was studied using HPLC, SARA fractionation as well as FTIR. This helped in selecting the proper emulsifier agent for the waste oil. The HCl-waste-oil emulsion was prepared using 15wt% HCl with a ratio of 70:30 of acid-to-oil. Extensive work was done to ensure reproducibility of data. This study was done to experimentally investigate the effects of distinct important variables on thermal stability and apparent viscosity of the HCl-waste-oil emulsion. These variables are; emulsifier concentration, mixing speed and the addition of corrosion inhibitor. All experiments were conducted at a high temperature 120˚C. To assess the quality of the HCl-waste-oil emulsion, droplet size was measured as a function of temperature (up to 100 ˚C).

The results showed that the HCl-waste-oil emulsion is a shear-thinning fluid. Power-law model was applicable seamlessly to all of the apparent viscosity data for all of measured temperatures. The optimum conditions were found to be; 0.5vol% and 600 rpm for emulsifier concentration and mixing speed, respectively. Apart from that, the emulsions prepared with corrosion inhibitor were less stable compared to the ones without it. However, the addition of corrosion inhibitor tended to generate monodispersed emulsion but bigger size droplets for temperatures 25˚C and 50˚C and polydispersed droplets for higher temperatures. Droplet size measurements showed an average size of 1.47 to 3.09 µm. Overall, the lab results showed a promising potential for the HCl-waste-oil emulsion to be used in the field.

This work takes into account; reducing the budget of acidizing job and at the same time environmental concerns. It makes use of the unwanted hydrocarbon from refiners. And, it saves the environment by using the waste oil instead of dumping it.

443836 Oxidative Dehydrogenation of n-Butane on Carbon Catalysts

Abstract:

Butenes and butadiene, which are one of the important industrial chemical products to produce synthetic rubber , are sythesized traditionally by oxidative dehydrogenation (ODH) of n-butane over various mixed metal oxides. Metal-free carbon nanomaterials has recently received much attention by researchers as a catalyst for the ODH reaction due to its environmental acceptability, corrosion resistance, and unique surface properties. For the reason, the reaction system was modelled in this study for the sythesis of butene and butadiene. For the modelling of the reaction system, the catalytic performance of several carbon catalysts, such as graphite and activated carbon, were initially tested. Then, the reaction mechanisms and kinetics were developed based on experimental results. Lastly, the mathematical modeling of the reaction system was optimized to maximize an overall yield of the main products. This approach allowed us to develop the optimized reaction system more efficiently.

This research was respectfully supported by Engineering Development Research Center (EDRC) funded by the Ministry of Trade, Industry & Energy (MOTIE). (No.N0000990)

443837 Blowdown Dynamic Modeling and Analysis of Pressure and Temperature

Abstract:

Vessel blowdown is one of the safety operation procedures that discharges flammable gas or liquid to flare system during emergency situation such as fire. When blowdown occurs, the liquefaction and freezing effect cause a pressurized vessel to be brittle, exposing it to possible hazard situation. Also the blowdown systems are recommended to be accordance with API(American Petroleum Institute) 521 code. API 521 requires pressure in vessel should be half of the initial pressure or be lower than 100 psig in 15 minutes after commencing blowdown in fire case. Therefore, accurate estimation over the change of temperature and pressure of the vessel during blowdown period is crucial for the design of blowdown system. For the reason, researchers have developed a few dynamic blowdown models so far. However, the model was found to be rather inaccurate under certain conditions. Therefore a new dynamic blowdown model was developed in this work. For the accurate analysis, the model algorithm considers both vapour and liquid phase in non-equilibrium condition that the temperatures of vapour and liquid phases are different. Futhermore, heat transfer from the vessel wall to discharge gas at either laminal or turbulent flow in the vessel was considered in the model to increase the accuracy of blowdown estimation. For thermodynamics, both Peng-Robinson and Soave-Redlich-Kwong equations of state were applied to the model, and the results were compared each other. Finally, the simulation results were compared with Haque et al's experimental and steady-state simulation results.

444364 Molecular Simulation of Controlling Ions and Their Role in Low Salinity IOR for Carbonate Reservoirs

Abstract:

Carbonate geology represents around 50% of hydrocarbon-bearing reservoirs in the world and more than 70% in some areas such as the Middle East. Water flooding has not been very efficient in carbonate reservoirs due mainly to the low water wetness. Modifying the ionic composition and salinity of the injected seawater demonstrated that the oil recovery from these carbonate reservoirs could be improved. Although the underlying mechanisms are still a subject of debate, the improved oil recovery has been mostly attributed to the increased water-wetness of the rock through the interactions of specific ions like Ca^{2+}, Mg^{2+}, SO_4^{2-}, and Cl^- , which are present in seawater with the rock surface under favorable conditions

The study of hydrated gas-phase metal ions provides a link between intrinsic chemistry of the isolated ion and its chemistry in solution. The structures and binding enthalpies of a variety of gas-phase and solution mono- and divalent ion hydrates containing up to 12 water molecules have been studied computationally. A DFT and molecular dynamics simulations were employed to study hydrated structures of the form $[\mathbf{M}(H_2O)_n]^{2+}.mH_2O$ for calcium and magnesium and $[\mathbf{M}(H_2O)_n]^{+}.mH_2O$ for sodium, in which there are n water molecules in the first coordination shell and m water molecules in the second coordination shell. The hydration, dehydration, and interactions of metal ions with Cl^{-1} and SO^{-2}_4 have been studied to better understand the mechanism of low salinity IOR. Density functional (DFT) theory using the B3LYP functional, and 6-311++g** basis set has been used. The charge of the metal ion decreases monotonically as the number of water molecules increases in the complex. Extensive calculations of the energetic contribution of individual structural characteristics as well as the energetic trends in binding energy are developed, which will lead to better understanding of how low salinity water IOR works.

Keywords: Hydration, carbonate reservoir, DFT, binding energy, Low Salinity

447119 Development R&D of Manoir Industries and Emphasis on Coating

Abstract:

The Senior VP Technical Support & Projects will present the development R&D of Manoir Industries and especially the coating (it is our lattest development) as well the services we can provide such as metallurgy expertise, advice, convector cleaning, site welding, welding supervision in order to increase the furnace productivity. So it is 60% R&D and 40% services. Manoir Industries manufactures cast components for the petrochemical market and develops high temperature alloys such as the Manaurites.

442925 Identifying Gas Liquid Contactor Performances for Optimized Natural Gas Desacidification Amines Units - from Process Calculations to Experimental Strategy

Abstract:

Amine units are widely used for gas sweetening. The absorber performances of these reactive absorption processes are typically mass transfer limited. The proper selection of gas liquid contactor is thus critical to optimize the size of towers, the solvent flowrate and the duty of the reboiler. This work proposes an approach to identify the most suitable packings for a given application.

The case-study selected here to illustrate the approach deals specifically with a CO_2removal application where the sour gas is at 7 % vol, and the sweet gas specification is 2% vol. The gas flowrate is fixed at 0,18 MMSCFD with a solvent flowrate fixed at the minimum of the thermodynamic constraints (via equilibrium curves). An in-house rate-based simulator with accessible forms of mass transfer correlations, based on a double film model, is used to calculate the absorber height and diameter which are needed to reach the sweet gas specifications. A reference structured packing is compared with standard valve trays.

The calculations show that the bed height required with the first contactor is considerably higher than with the second one while the column diameter is reduced. In this case the use of standard valve trays is more attractive for cost optimization. Interestingly, because mass transfer is accelerated, the absorption rate with the structured packing is found to be thermodynamically limited due to a more severe temperature increase which reduces the driving force in the bottom section of the absorber.

The study determined that the optimum design for the investigated case is a compromise between a full trayed tower, that generally imposes a larger diameter, and a packed tower where the higher mass transfer forces result in a temperature rise in the bottom of the tower which in turn imposes a height increase and a much higher solvent flowrate. . To determine the mass transfer characteristics of a gas-liquid contactor providing this optimum, a sensitivity analysis has been conducted. Mass transfer parameters, i.e., liquid and gas side mass transfer coefficients and interfacial area, were varied by a factor 2 relative to the ones obtained with the reference structured packing. Calculations show that one should use a contactor that reduces liquid phase resistance, specifically, the mass transfer conductance $k_L a_e$ should be increased by at least 50%. This target could be reduced to +30% if one could simultaneously reduce the gas phase conductance $k_G a_e$ by 20%. This result can be explained by the link between mass and heat transfer (Chilton-Colburn analogy) whereby reducing $k_G a_e$decreases the temperature level which in turn thermodynamically increases the mass transfer driving force.

The next step will be to review the literature more extensively to make a short list of pre-selected random and structured packings that best meet the targets. Experimental tests will then be run on IFPEN facilities to measure flooding limits and mass transfer parameters in a small 150 mm diameter column. This step is required to confirm packing's behavior, considering the disparity of the values reported by different authors in the literature. The last step is to consolidate the existing correlations for the most promising packing, and mitigate differences between various authors by performing characterization tests on larger diameter test columns. The final correlations will be implemented in our in-house rate based simulator for industrial use.

The present approach can be applied to determine the required mass transfer characteristics for optimizing other gas treatment applications. However, the conclusions reached here as regards to mass transfer targets and the problematic temperature levels, are valid only to the present case –study, and are not directly applicable to other gas sweetening applications (selectivity, LNG, …). For example, in a case where H_2S and CO_2 are treated simultaneously the mass transfer limitations could be very different.

443784 Measurement and Correlation of the Mutual Solubility of Diisopropylamine + Water and Triethylamine + Water Systems at High Pressure

Abstract:

The mutual solubility data of diisopropylamine + water and triethylamine + water systems were measured at the pressure up to 100 MPa and the temperature range was 280.2~316.2 K. Experiments were performed in a high-pressure autoclave with a moveable piston. The experimental data were correlated satisfactorily with NRTL activity coefficient model and the binary parameters were obtained.

444161 Novel Microwave Synthesis Carbon Dots from PINE Bark and Its Application for Heavy Metal Sensing

Monday, April 11, 2016
Exhibit Hall E (George R. Brown)
Saheed Sanni, Chemistry, Vaal University of Technology, Vanderbijlpark, South Africa, Puthalapattu ReddyPrasad, Chemistry, Vaal University of Technology, Vanderbiljpark, South Africa and Augustine E. Ofomaja, Chemistry, Vaal university of Technology, Vanderbiljpark, WV, South Africa

NOVEL MICROWAVE SYNTHESIS CARBON DOTS FROM PINE BARK AND ITS APPLICATION FOR HEAVY METAL SENSING

Sanni S.O., ReddyPrasad P., Ofomaja A.E.

Biosorption and Waste water Treatment Research Laboratory, *Department of Chemistry, Faculty of Applied and Computer Sciences, Vaal University of Technology, P. Bag X021, Vanderbijlpark-1900, South Africa.*

INTRODUCTION

Microwave pyrolysis for the production of carbon has the advantages of being cheap, produces high yield of carbon and is absolutely a green synthetic method [1]. In recent times, the use of microwave pyrolysis have been extended to the preparation of carbon dots. Carbon nanodots (C-dots) falls under the class of carbon nanoparticles with size below 10 nm[2]. C-dots are carbon materials which processes high water solubility and strong fluorescence depending on their size, edge, shape, surface ligands, and defects. Since C-dots preparation can be achieved from difference carbonaceous sources and by varying a range of parameters during their synthesis, the properties of the final product may vary depending on these factors. C-dots have sp^2 characteristics and possess several functional groups on their surfaces that makes them highly hydrophilic and easily functionalized with various organic, polymeric, inorganic, or biological spices[3]. Several synthesis methods that have been explored includes electrochemical synthesis, combustion/thermal/hydrothermal/acidic oxidation, supported synthesis, microwave/ultrasonic, arc discharge, laser ablation/passivation, and plasma treatment[4]. C-dots processes excellent such as water dispersibility, chemical stability and photostability, ease of modification, excitation-dependent multicolor emission, low toxicity, and good cell permeability, great efforts have been paid to their potential applications in biosensing, bioimaging, drug delivery, and other biological related aspects.

METHODS

The carbon dots (C-dots) with green luminescence was synthesized from microwave assisted pyrolysis of pine bark waste our carbon source at 800, 1000 1200 W for 1 Hr. The dots produced were characterized using FTIR, TGA, XRD, UV-Vis and Raman spectroscopy. The dots were applied for heavy metal sensing.

RESULTS AND DISCUSSION

The functional groups present on the C-dots (**Fig.1**) includes -OH stretching band mode at 3413 cm^{-1}, C-H stretching at 2934 and 2845 cm^{-1}, stretching peaks of C=O, C=C at 1731, 1604 and 1437 cm^{-1}. The XRD pattern (**Fig 2a &b**) of produced C-dots depicts sharp peak at 20.9° and 26.8°, which is highly crystalline cores and possess amorphous nature of C-dots. This peculiar peak is ascribed to 002 of the graphitic carbon, which in line with inter-planar spacing of 3.77 Å with higher spacing than bulk graphite at 3.34 Å. The thermogravimetric and differential thermal analysis display (**Fig 3**) that the C-dots possess thermal stability more than 100 °C and mass losses occurs afterwards. The thermal analysis depicts 3 endothermic peaks at 331, 542 and 735 °C, which are ascribed to oxygenated groups such as carboxyl, hydroxyl, carbonyl and phenols. The optical properties of the synthesize C-dots were explored using UV-vis absorption and photoluminescence (PL) spectroscopy. The UV-vis spectra display (**Fig 4**)intense absorption peak in the range of 260 to 290 nm, which is ascribed to π-π^* transition of the aromatic sp2 domain and also presence of multiple polyaromatic chromophores. A shoulder peak at 330 nm which is ascribed to n-π^* transitions of C=O and structural defects of the C-dots. PL spectrum of the C-dots (**Fig 5a**)exhibits an intense maximum emission peak at 432 nm upon excitation at 330 nm, which also shows characteristic yellowish green color under sunlight and intense blue coloration UV lamp (365 nm). The PL properties of as prepared C-dots was varied at (**Fig5b**) different excitation wavelength from 300 to 405 nm. There was red-shifting in emission peak as the excitation peak was increased from 300 to 405 nm and also simultaneously decrease in fluorescent intensity, which is ascribed to quantum effect and surface defects. This behavior of the prepared C-dots is dependent on excitation-fluorescence, which is accordance with other report [13]. The quantum yield of synthesized C-dots was measured to be 17% at maximum excitation at 330 nm

Fig.6 shows the Raman spectrum of the PB-GQDs, which has been usually used to confirm the quality of the prepared PB-GQDs. There is two major Raman peaks at 1370 cm^{-1} (D band), 1580 cm-1 (G band). The peak of G band corresponds to the Sp^2 bonded carbon atoms in a two-dimensional (2D) hexagonal lattice and the peak of G band is associated with the vibrations of carbon atoms with dangling bonds in the termination plane of tangled graphite or glassy carbon. The D/G ratio band ratio is ID/IG: 0.68 is confirm the formation of graphene like structure from synthesized pine bark.

Fluorescence detection of Cu^{2+} ion

The detection of Cu^{2+} was performed at room temperature in phosphate buffer (PBS) solution pH 4.3. 1.0 mL of PB-CQDs solution was added into 0.5 mL of PBS buffer, followed by the addition of 1.0 mL different concentrations of Cu^{2+}. The fluorescence emission spectra were recorded after reaction for 5 min at room temperature (~28 °C).

Selectivity and sensitivity studies

Moreover sensitivity and selectivity is additional significant parameter to assess the performance of the sensing system. Consequently, we examined the fluorescence intensity changes in the presence of representative metal ions such as including Cu^{2+}, Hg^{2+}, Ca^{2+}, Al^{3+}, Ba^{2+}, Mn^{2+}, Ni^{2+}, Pb^{2+}, Co^{2+}, Zn^{2+}, Fe^{3+}, $Cs^{2+}$and Mg^{2+}, of each at a concentration of 15.0 μg mL^{-1} were reacted with a PB-CQDs solution. Fig. 7 A displays the relative change in PL intensity of PB-CQDs in the presence of various metal ions. Fe^{2+}, Mn^{2+} and Hg^{2+} ions cause the slight PL changes (defined as the relative change of PL intensity in 60–80% as compared to blank) which can be attributed to the nonspecific interactions between the carboxylic and/or amine groups and the metal ions. Among these metal ions, Cu^{2+} displays the strongest PL quenching effect on PB-CQDs, assigning to the special coordination between the Cu^{2+} ions and the phenolic hydroxyl and/or amine groups of PB-CQDs which has been widely used for the detection of Cu^{2+} ions or colored reactions in traditional organic chemistry. Such a specific fluorescence quenching effect may originate from the strong interactions between Cu^{2+} ions and the surface groups of PB-CQDs which transfer the photoelectrons from PB-CQDs to Cu^{2+} ions. These results clearly demonstrate that the PB-CQDs-based Cu^{2+} sensor is highly selective to Cu^{2+} over the other metal ions. Fig.8. Fluorescence responses of CCDs to the addition of different metal ions under acetate buffer (pH 4.3). The concentration of each metal ion is 15 μg $mL^{-1.}$ F_0 and F correspond to the fluorescence intensities of PB-CQDs at emission wavelength of 440 nm excited at 320 nm in the absence and presence of metal ions, respectively.

As shown in Fig. 3, the fluorescence intensity of PB-CQDs gradually decreased with the Cu^{2+} concentration. The fluorescence intensity displayed liner responses with the Cu^{2+} concentrations, with two linear ranges of 0.0□14.0 μg mL^{-1} and 15.0□25.0 μg mL^{-1} (Fig.3B). The linear equations are $F/F0$ =0.9827□0.8485C_{Cu2+} (R = 0.993) and $F/F0$ = 0.7565–0.0033 C_{u2+} (R = 0.996), respectively.

REFERENCES

[1] W. B. Lu, X. Y. Qin, S. Liu, G. H. Chang, Y. W. Zhang, Y. L. Luo, A. M. Asiri, A. O. Al-Youbi, X. P. Sun, Economical, green synthesis of fluorescent carbon nanoparticles and their use as probes for sensitive and selective detection of mercury(II) ions, Anal. Chem. **2012**, 84, 5351-5357.

[2] J. Jiang, Y. He, S. Li, H. Cui, Amino acids as the source for producing carbon nanodots: microwave assisted one-step synthesis, intrinsic photoluminescence property and intense chemiluminescence enhancement. Chem Commun **2012**, 48, 9634–6.

[3] Y. Wang, A. Hu, Carbon quantum dots: synthesis, properties and applications, J. Mater. Chem. C **2014,** 2, 6921–6939.

[4] C. Liu, P. Zhang, X. Zhai, F. Tian, W. Li, J. Yang, et al., Nano-carrier for gene delivery and bioimaging based on carbon dots with PEI-passivation enhanced fluorescence, Biomater. **2012,** 33, 3604-13.

Figures:

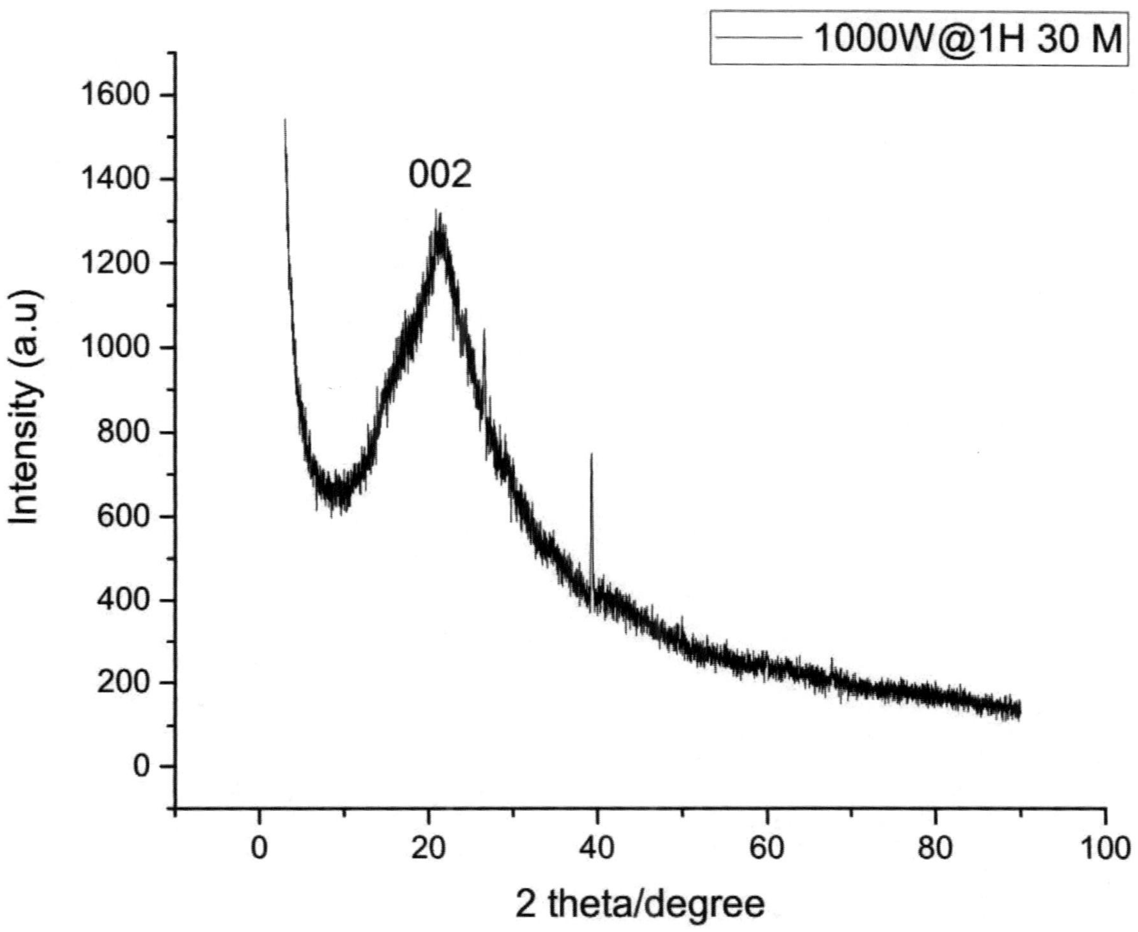
1000W@1H 30 M
002
Intensity (a.u)
2 theta/degree
1600
1400
1200
1000
800
600
400
200
0
0
20
40
60
80
100

Fig 1. FTIR spectra of as-prepared C-dots from Pine Bark

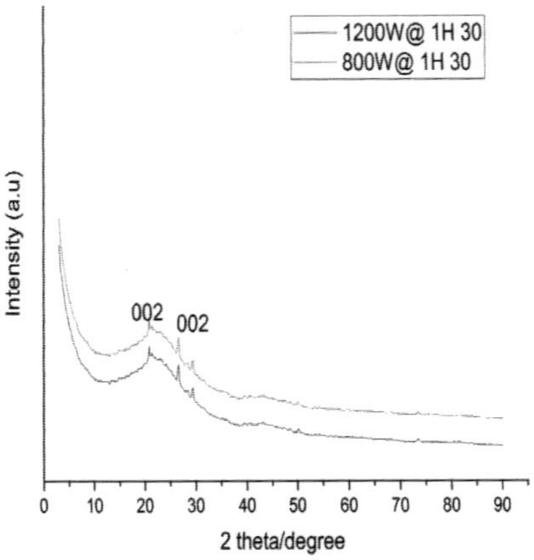

(a)

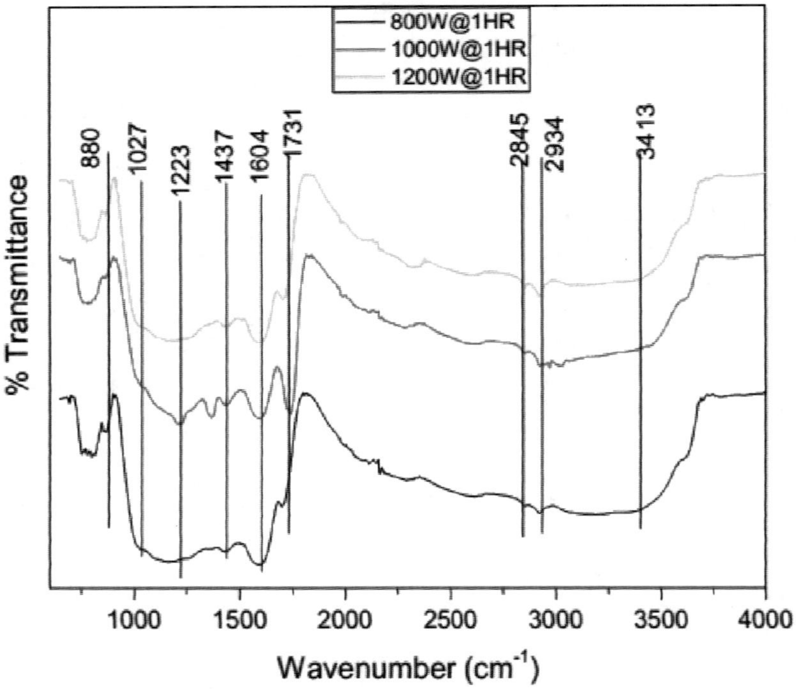

(b)

Fig. 2a & b. XRD pattern of C-dots.

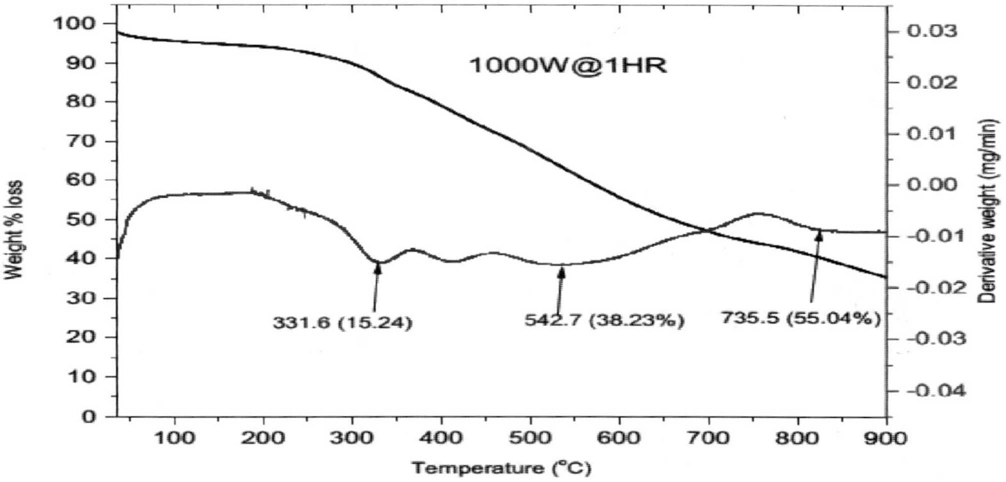

Fig 3. Thermal curves for C-dots under N_2 atmosphere.

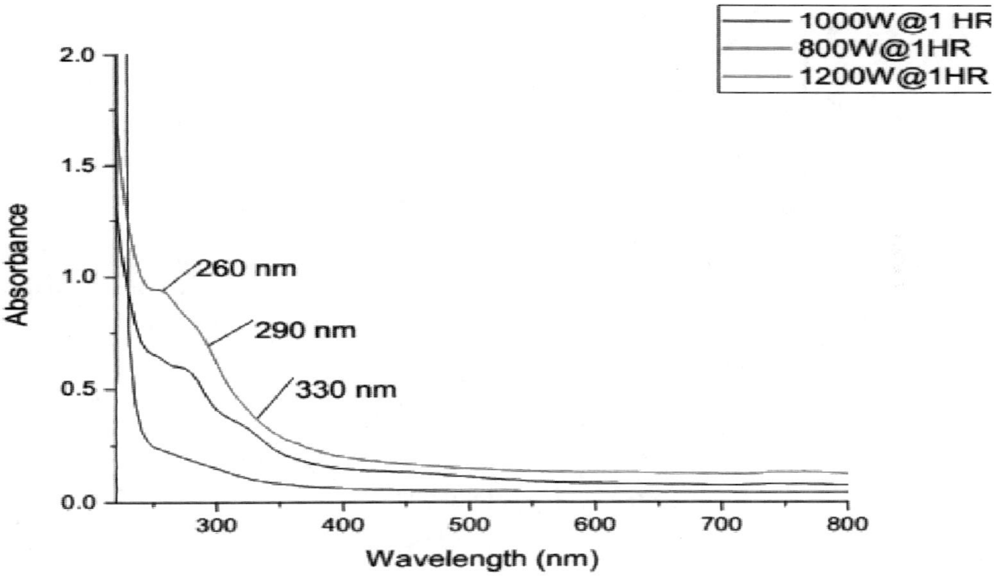

Fig 4. UV-vis absorption of C-dots.

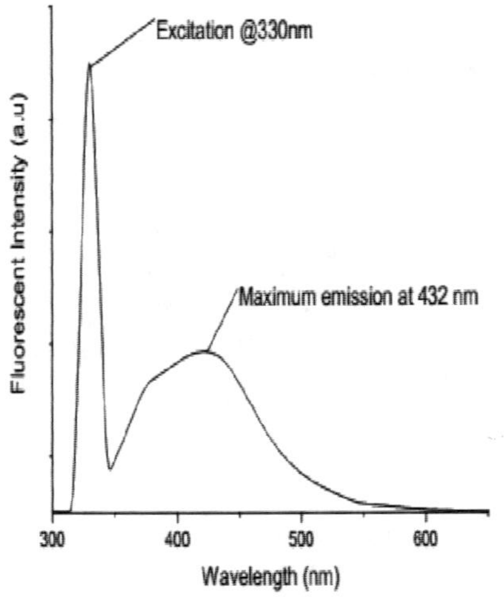

(a)

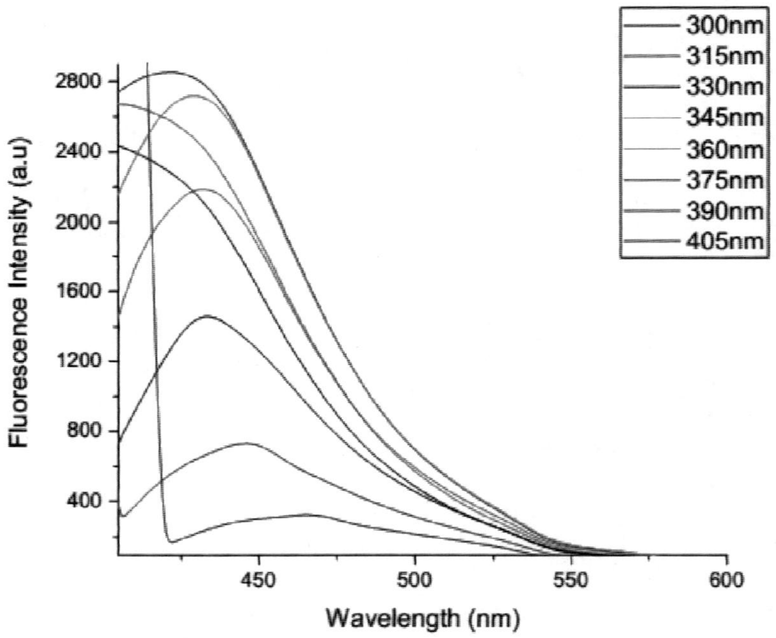

(b)

Fig 5a. PL excitation and emission wavelength (b) PL emission spectra at different excitation wavelength from 300 to 405 nm in 15 nm increments.

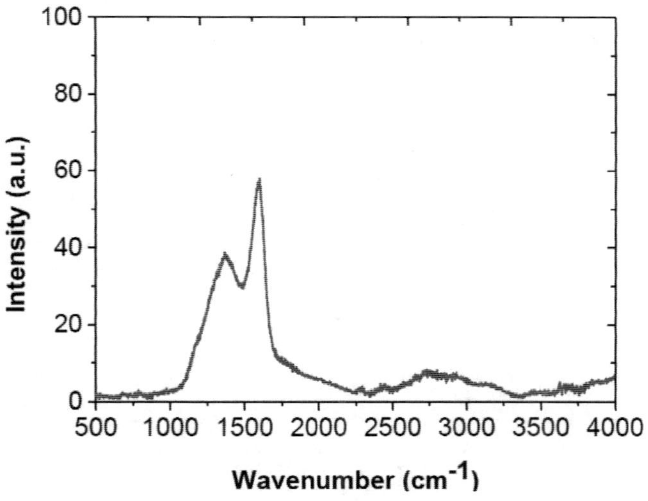

Fig. 6. Raman Spectra of PB-CQDs

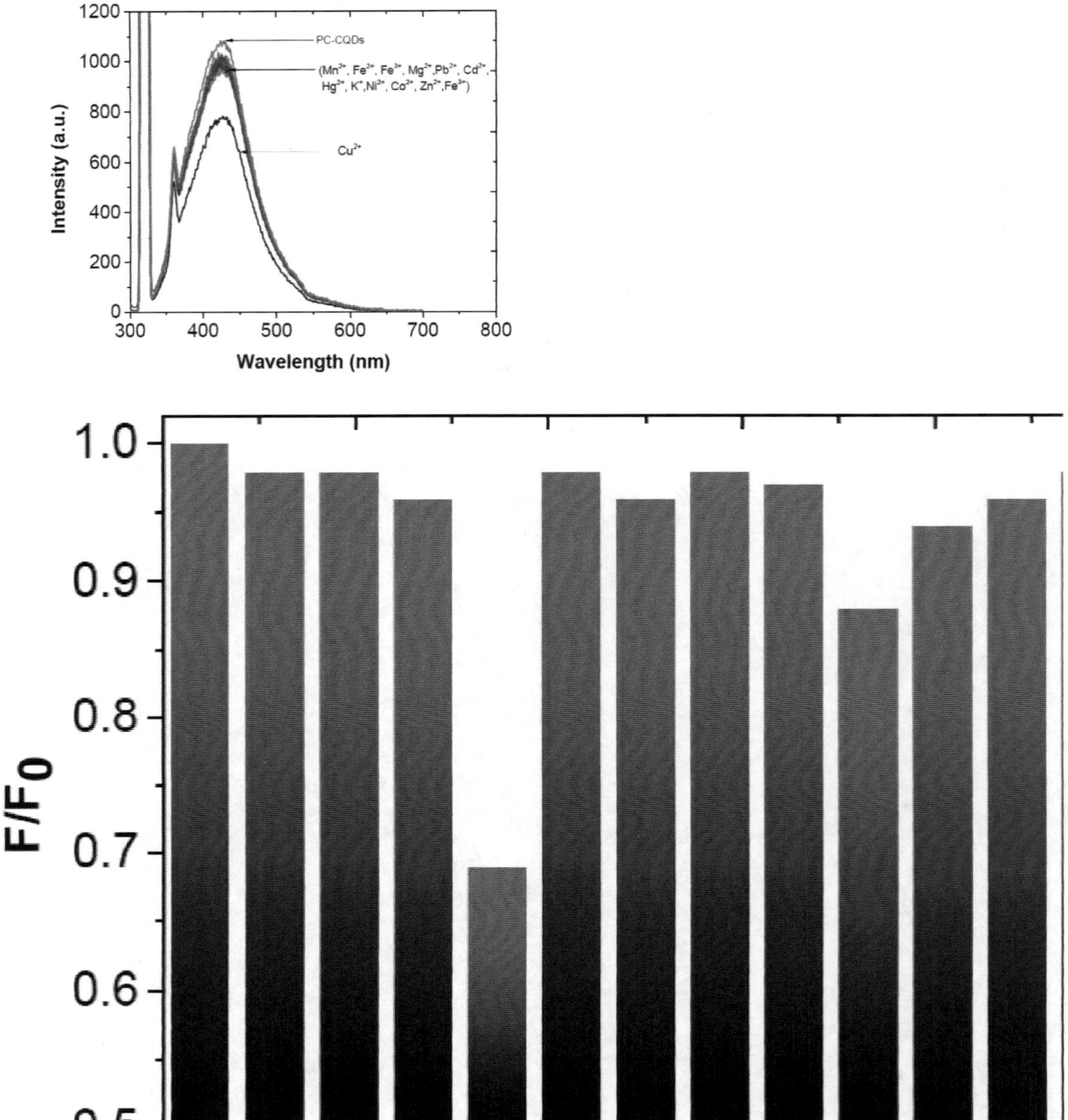

Fig.7. Fluorescence responses of CCDs to the addition of different metal ions under acetate buffer (pH 4.3).

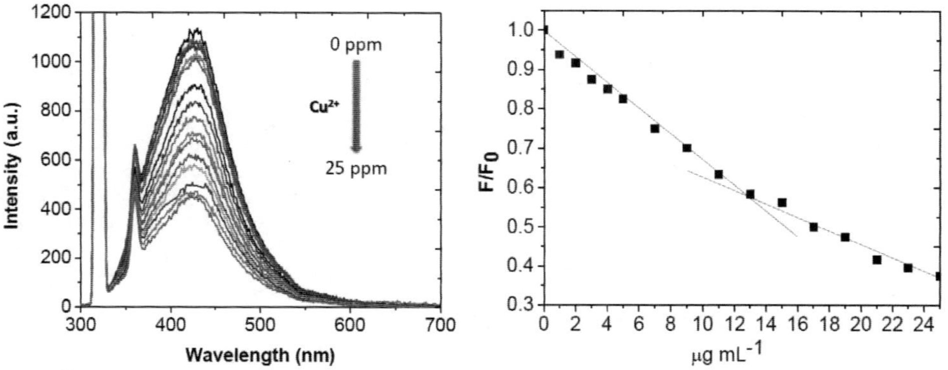

Figure 8. Fluorescence titrations of the y-CDs with Cu^{2+} from 0 to 25 μg mL^{-1}. (Influence of Cu^{2+} concentration on the fluorescence intensity of PB-CQDs (top to bottom: 0, 1, 2, 3, 4, 5, 7, 9, 11, 13, 15, 17, 19, 21, 23 and 25.0 μg mL^{-1})

443095 Experimental and Numerical Study of Transport Phenomenon of Falling Film Flow with Different Heating Methods

Abstract:

Falling film flow is a common phenomenon in chemical industry. It is widely used in chemical instruments, such as structured packed columns and falling film evaporators. An experiment was implemented to study the mass transfer phenomenon of binary-mixture (ethanol-water) falling film flows with different heating methods (uniform or concentrated heating). In order to study the detailed flow and mass transfer behaviors, a 2-D multiphase computational fluid dynamics (CFD) model based on volume of fluid (VOF) was developed. This model divides mass transfer into two parts: one is caused by the vapor-liquid equilibrium (VLE) and the other is caused by the heat flux. Results from simulation are inconsistent with the experimental data. Both experiments and simulations show that the ethanol concentration of the vapor with concentrated heating is 1.5% higher than the one with uniform heating.

443774 Thermodynamic Anaysis of Dehydration of n-Propanol Using Heterogeneous Azeotropic Distillation

Abstract:

Heterogeneous azeotropic distillation is a widely popular process in chemical process industries for the separation of binary azeotropic mixtures into their constituent pure components. It involves the addition of an entrainer which is partially miscible and forms a heterogeneous azeotrope preferably with one of the original components. This is a powerful technique which combines several physical phenomena such as presence of azeotropes and liquid-liquid immiscibilities, and is used widely in industries to enhance difficult or intractable distillation separations. For the design of heterogeneous azeotropic distillation columns, a detailed study of the heterogeneous region is crucial to plan the distillation sequence, which requires experimental vapor-liquid-liquid equilibrium (VLLE) data.

The present study focuses on the investigation of the heterogeneous region of the well known industrially available azeotropic mixture n-Propanol-Water system using Propyl acetate as entrainer. The VLE data of the system was generated by means of modified Othmer still and the still is modified using a probe type ultrasonic homogenizer to account for the heterogeneity. Isobaric vapor- liquid- equilibria and vapor - liquid - liquid - equilibria data at atmospheric pressure have been obtained for the ternary system n-Propanol-water-propyl acetate and the constituent binary systems. The equilibrium data has been correlated using NRTL, UNIQUAC and UNIFAC models and compared with the experimental data.

444269 Functionalization of OH-Terminated Nanoparticles By Cyclic Azasilanes

Abstract:

Colloidal suspensions of functionalized nanoparticles have been suggested for enhanced oil recovery, formation damage remediation, flow assurance, and water treatment operations. Surface modified nanoparticles can also be used as tracers in reservoir studies, active media in thermal management fluids, and fillers in high performance elastomers. Conventional functionalization of nanoparticles with alkoxysilanes and chlorosilanes is limited by needs for catalytic amount of water for the reaction to proceed, requirement of multiple hours to fully saturate the surface, and release of by-products upon reaction completion. Monolayer deposition with a high density of functional groups represents a particularly difficult challenge.

Our ongoing studies indicate that application of cyclic azasilanes that have a strained ring structure with adjacent Si and N centers provides highly efficient functionalization of hydroxylated silica, alumina, magnetite, and diamond nanoparticles which is catalyst-free and by-product-free. The data on synthesis and characterization of functionalized nanoparticles as well as possible reaction mechanism of functionalization will be discussed.

443776 Study of Ionic Liquids As Entrainers for the Extractive Distillation of *N*-Propanol and *N*-Propyl Acetate Binary Azeotropic Mixture and the Isobaric VLE Studies of the System

Abstract:

Distillation is a widely accepted technique for the separation of binary liquid mixtures. Separation by simple distillation is not possible for binary mixtures that form azeotropes. One of the commonly used methods to separate azeotropes is extractive distillation. It is the process of adding an entrainer that can alter the relative volatility of the mixture and thereby breaking the azeotropic point. Ionic liquids on account of their physiochemical properties as well as the recyclability have been used as a preferred replacement for the conventional entrainers. Extractive distillation with ionic liquids has the advantages of both liquid solvent (easy operation) and solid salt(high separation ability). The corrosion and dissolution problems by using inorganic salts can be rectified by these non-volatile, chemically stable, high electrical conductive molten salts.

The azeotropic system under consideration in the present study is *n*-propanol - *n*-propyl acetate mixture. These compounds which constitute the products/reactants of esterification reaction of n-propanol form a minimum boiling azeotrope at 94.7°C. *n*-propyl acetate has been included in a list of volatile acetates as an ecologically benign CFC-free solvent for aerosol sprays.

Vapour liquid equilibrium(VLE) measurements of *n*-propanol and *n*-propyl acetate are performed at atmospheric pressure in a modified Othmer still which is a continuous distillation still in which vapour phase is recirculated. The most suitable ionic liquids for shifting of azeotropic composition of the system are selected using COSMO-RS model with suitable combinations of anions and cations. The isobaric VLE of the system with [EMIM][HSO_4] is obtained and compared with the existing data. The influence of alkyl chain length on the performance of ionic liquids is also analysed with regard to the previous studies. The effect of polarity is the most important factor on the behaviour of ionic liquid solvation in alcohols. Experimental VLE data were correlated using modified UNIFAC model.

444280 Copper-Phyllosilicate Core-Shell Nanocatalysts with Balanced Active Sites for Carbon-Oxygen Hydrogenolysis Reactions

Abstract:

Copper-Phyllosilicate Core-shell Nanocatalysts with Balanced Active Sites for Carbon-Oxygen Hydrogenolysis Reactions

Hairong Yue, Weimao Fang, Changmin Liu, Jiankang Ying

College of Chemical Engineering, Sichuan University, Chengdu, China

Email: hryue@scu.edu.cn

Abstract

Hydrogenolysis of carbon-oxygen (C¨CO) bonds (e.g., esters, ethers, furfural, and CO_2) has emerged as a versatile synthetic tool in organic methodology, as it could produce a variety of products (e.g., chemicals, fuels, and polymers). Copper-based catalysts have been intensively explored for hydrogenation reactions as the copper sites account for the selective hydrogenation of carbon-oxygen bonds and relatively inactive for the hydrogenolysis of carbon-carbon bonds. Work on understanding the active sites of copper catalysts for hydrogenation reactions indicted that both Cu^0 and Cu^+ species were crucial to the activity of Cu-based catalysts. However, researchers were not able to establish a relationship between activity and Cu^0/Cu^+ active species since the conventional Cu-based catalysts is the deactivation by metal particle growth and unstable surface Cu^0 and Cu^+ active species in the strong reductive H_2 and oxidizing carbon-oxygen atmosphere.

We renctly present on efficient approaches for fabrication of a series of copper-phyllosilicate core-shell nanocatalysts and nanoreactors with balanced and stable Cu^0 and Cu^+ active species. We chose the hydrogenation of dimethyl oxalate, ethylene cyclicarbonate and as probe reactions for a better understanding of properties of active sites and the structure-activity relationship. The results indicated a synergy between copper and the oxide components and the balanced surface Cu^0 and Cu^+ species can greatly improve the catalytic C-O hydrogenolysis performance of the catalysts. These nanocatalysts with balanced and stable Cu^0 and Cu^+ active species, confinement effect, intrinsic high surface of Cu^0 and Cu^+ and unique tunable tubular morphology, has potential applications in the high-temperature hydrogenation reactions.

Keywords: core-shell nanocatalysts, Hydrogenation reaction; copper catalyst

References

1. Yue, H.; Zhao, Y.; Ma, X.; Gong, J.: Ethylene glycol: properties, synthesis, and applications. *Chem Soc Rev* 2012, 41, 4218-4244.

2. Gong, J.; Yue, H.; Zhao, Y.; Zhao, S.; Zhao, L.; Lv, J.; Wang, S.; Ma, X.: Synthesis of Ethanol via Syngas on Cu/SiO2 Catalysts with Balanced Cu^0-Cu^+ Sites. *J Am Chem Soc* 2012, 134, 13922-13925.

3. Zhao, S.; Yue, H.; Zhao, Y.; Wang, B.; Geng, Y.; Lv, J.; Wang, S.; Gong, J.; Ma, X.: Chemoselective synthesis of ethanol via hydrogenation of dimethyl oxalate on Cu/SiO_2: Enhanced stability with boron dopant. *J Catal* 2013, 297, 142-150.

4. Yue, H.; Ma, X.; Gong, J.: An Alternative Synthetic Approach for Efficient Catalytic Conversion of Syngas to Ethanol. *Acc Chem Res* 2014, 47(5):1483-1492

5. Yue, H.; Zhao, Y.; Zhao, L.; Lv, J.; Wang, S.; Gong, J.; Ma, X.: Hydrogenation of dimethyl oxalate to ethylene glycol on a Cu/SiO2/cordierite monolithic catalyst: Enhanced internal mass transfer and stability. *AIChE J* 2012, 58, 2798-2809.

6. Yue, H.; Zhao, Y.; Zhao, S.; Wang, B.; Ma, X.; Gong, J.: A copper-phyllosilicate core-sheath nanoreactor for carbon¨Coxygen hydrogenolysis reactions. *Nat Commu*n 2013, 4, 2339.

443819 Improving High Shear Rate Friction Reduction Behavior of Polyacrylamide-Based Fracturing Fluids Using Xanthan Gum

Abstract:

The main component of slickwater in fracturing low permeability formations is a high molar mass polymer such as polyacrylamide (PAM), which acts as a drag reduction agent in turbulent flow. Polyacrylamide does not degrade in the environment and also suffers from mechanical degradation at high turbulent flow rates. Five different concentrations of PAM as well as mixtures of polyacrylamide/xanthan were prepared to investigate the possibility of improving shear stability of PAM. Drag reduction measurements were performed in a flow loop and friction factors at different flow rates were calculated using rheological parameters. Based on the observations, Xanthan (XG) is not as good a friction reducer as PAM, even at three-fold higher concentration. XG slightly improved the shear stability in mixed polymer solutions at low concentrations. But at high concentrations, XG significantly improves the friction reduction behavior of PAM at nonionic and ionic environments.

Stimuli-responsive Rubber in Oil and Gas Applications

Rostyslav Dolog, Sankaran Murugesan, and Valery Khabashesku

Baker Hughes, 14990 Yorktown Plaza Dr, Houston, Texas, 77040, USA

ABSTRACT

A problem that exists today in the oil & gas industry is the loss of mechanical properties by reactive rubber seals due to swelling, and the limited space in downhole well applications to accommodate larger swellable packing elements.

This paper presents cost-effective, reinforcing, stimuli-responsive fillers, such as certain classes of fly ash, for use in rubber composite materials in various oil and gas applications. Apart from cost effectiveness, these fillers can also enable stimuli-responsive effects, as well as improved mechanical properties (e.g. modulus) for rubber composites when swelled in water. Fly ash (Class C) is a common component of Portland cement which was used in the literature to achieve a stimuli-responsive effect when added to Hydrogenated Nitrile Butadiene Rubber (HNBR) as filler. Exposure to water brines can cause a filler to promote an increase in pH which enables the hydrolysis of HNBR nitrile groups to carboxyl groups, thus creating an ionomer. This process is followed by neutralization of the new ionomer with metal ions, and formation of ionic crosslinks, further improving mechanical properties of the rubber.

This paper demonstrates that incorporation of fly ash as a reinforcing filler allows development of swellable seals that can retain, or even increase, their modulus and stiffness upon swelling. These characteristics allow avoiding the loss of mechanical properties by seals due to swelling and also eliminating the need for large swellable packing elements that occupy additional space downhole.

KEYWORDS: Elastomers, stimuli-responsive, fly ash, composites

INTRODUCTION

There is a growing demand for high performance elastomers in the oil & gas industry. Rubber compounds are used for various sealing applications, packer elements, stators for pumps, etc. In general, rubber formulations are comprised of elastomer matrix, curing system, antioxidants, plasticizers, along with reinforcing filler, which provides the compounds with their strength. The reinforcing filler is usually represented by carbon black and, to a lesser extent, silica. Along with other factors, filler mechanical properties and concentration generally define hardness, modulus and tensile strength of the elastomer compound. Throughout the years, many different materials have been tried as alternative fillers for rubber composites, including nanomaterials (CNTs, graphene, nanofibers), recycled rubber particles, and ash/fly ash. In the case of ordinary carbon black, the filler serves not only for reinforcement, but also as a cost reducer, because raw elastomer is often a more expensive component. Mechanical properties of the composites are affected by the aspect ratio of the filler, quality of the dispersion, and interface between blend components. The latter determines the effectiveness of load transfer from polymer matrix to the reinforcing filler. High performance compounds can be designed by incorporating well dispersed fine particles with a high aspect ratio and good compatibility with the polymer matrix.

In this context, the fly ash is a cheaper, low reinforcing filler alternative to carbon black, which will help reduce cost of the compounds for "non-demanding" sealing applications requiring large amounts of raw materials. Fly ash is produced in large quantities as a byproduct of coal burned at power plants, and can be considered as a recycled material. Fly ash costs several cents per pound, while carbon black, depending on the grade, is usually one order of magnitude more expensive. Hemmings et al[1] demonstrated that fly ash can be used as filler for polymeric composites with enhanced mechanical properties. Adoption of this technology in the oil and gas industry could change the design approach for reactive rubber compounds. Further functionalization of fly ash through a cost-efficient process could ensure compatibility with the rubber matrix, while promoting filler dispersion and ensuring an efficient load transfer. Such functionalization's include, but are not limited to, surface modification with fatty acids, dicarboxylic acids or phosphoric acid. The reasoning for surface modification is to attach functional groups with aliphatic chains to fly ash particles to make them compatible with aliphatic rubber chains, or to attach acidic groups on their surface that can interact with nitrile groups of HNBR. Initial attempts towards chemical functionalization of fly ash have been made by Seal et al for treatment of offshore oil spills.[2]

Apart from cost effectiveness, composites with fly ash can show stimuli-responsive properties. It has been shown that adding Portland cement to the HNBR matrix improves its mechanical properties in the swollen state by water uptake.[3] Exposure to water causes Portland cement to grow in volume and mass, and promotes an increase in pH which enables the hydrolysis of HNBR nitrile groups to carboxyl groups, thus creating an ionomer. This process is followed by neutralization of the new ionomer with metal ions and formation of ionic crosslinks, further improving mechanical properties of the rubber. Fly ash (Class C) is a common component of Portland cement, which contains a large variety of metal oxides, and is capable of increasing the

[1] Filler Comprising Fly Ash For Use In Polymer Composites. Hemmings et.al., US 6,916,863 B2, July 2005.

[2] Hydrophobic High Surface Area Zeolites Derived from Fly Ash for Oil Spill Remediation, Seal et.al., Environ. Sci. Technol., 2013, 47 (11), pp 5843–5850

[3] Stimuli-Responsive Cement-Reinforced Rubber, Musso et. al., ACS Appl Mater Interfaces. 2014 May 14;6(9):6962-8.

pH of the water solution by hydrolysis, which has a potential to achieve a stimuli-responsive effect when added to HNBR as filler.

Degradation of tensile properties is a common problem for swellable packers and other sealing elements. By selectively using certain classes of fly ash as effective reinforcing filler, one can develop swellable seals that can retain or even increase their modulus and stiffness upon swelling. Therefore, there is a potential to minimize or even avoid the loss of mechanical properties by water swellable reactive packing elements due to swelling. This provides an opportunity to decrease the size of these elements, which are often represented by rather long sections, and hence occupy a significant amount of space downhole. In this paper we have studied the reinforcing effects of fly ash fillers with HNBR compounds as well as stimuli responsive properties of these compounds.

EXPERIMENTAL SECTION

Rubber compounds T1, T2, T3, and T4 are elastomers based on a custom-built peroxide curing system. Formulation F2 was prepared by adding 15 wt. % of fly ash class C to compound F1, which is a commercial swellable NBR sulfur-cured compound.

T series compounds were prepared using a C.W. Brabender® internal mixer operated at 60 RPM for 20 minutes at ~160°F and then molded into sheets and cured at 340°F for 30 minutes in a hot press. Compound F2 was prepared by mixing fly ash with compound F1 in the Brabender mixer for 6 minutes at 60 RPM at ~100°F. Compound F1 is a commercial water swellable formulation that was used as received. F series compounds were cured at 300°F for 60 minutes in a hot press.

Tensile properties of rubber samples were tested using dog bone specimens, according to the ISO37 standard, at room temperature and at 350°F using an MTS® testing system.

Swelling tests in 3% KCl brine and LVT 200 oil were performed to assess the effect of fillers on the chemical resistance of elastomers. Standard one-inch rubber buttons were tested at 200˚F at ambient pressure for 6 days. Thermal properties were studied using TGA Q50 and DSC Q2000 instruments.

To analyze the stimuli-responsive effect introduced by fly ash fillers, rubber buttons were aged for 10 days in 3% KCl in the oven at 200°F and then removed from the solution, dried for 30 minutes in the oven and then for 3 hours at room temperature. Samples were later tested in the MTS testing system in compression mode.

Viscoelastic properties of stimuli-responsive samples, based on water swellable compounds, were accessed using the RSA G2 dynamic mechanical analysis (DMA) instrument. This experiment was conducted in a temperature ramp mode of 2˚C/minute at 1Hz frequency and 0.2% amplitude strain. A Nicolet 6700 Fourier Transform Infrared (FTIR) spectrometer was used to confirm neutralization of carboxylic groups by metal cations.

RESULTS AND DISCUSSION

The reinforcing properties of fly ash were evaluated at room temperature (Table 1) and at 350°F (Table 2) and compared with N550 carbon black. Sample T3, reinforced with fly ash, demonstrated inferior tensile properties and hardness compared to a compound filled with carbon black. However, at room temperature, fly ash was capable of delivering a reinforcing effect to the rubber compound without sacrificing its elasticity, e.g. the ultimate elongation of elastomer compound remained virtually unchanged after incorporation of 50 parts per hundred rubber (PHR) of fly ash, while its modulus and tensile strength have increased. Therefore, potentially, one can substitute carbon black with high loadings of fly ash in certain applications, thus further decreasing the cost of the rubber product. The summary of mechanical properties is provided in Table 1, E_x denotes modulus of a compound at certain strain (x), i.e. stress at strain.

Sample	N550 CB (PHR)	Fly ash Class F (PHR)	Tensile strength (psi)	E_{25} (psi)	E_{50} (psi)	E_{100} (psi)	Elongation (%)	Hardness Shore A
T1	0	0	829	71	97	121	472	48
T2	50	0	3402	165	273	744	290	68
T3	0	50	1176	94	127	166	467	54

Table 1. Room temperature tensile properties and hardness of elastomer samples

The relatively low increase in modulus and hardness of T3 compared to T1 at room temperature can be explained by a low aspect ratio of fly ash and weak interactions between blend components. The latter problem can be solved by functionalization of fly ash particles.

The general data trends from the high temperature tensile tests closely resemble the results of the tensile test conducted at room temperature. However, at high temperature, rubber specimens break at much lower elongation, which leads to a great decrease in the measured tensile strength. This behavior is typical for elastomers and limits their application to a certain maximum service temperature and is also due to the fact that high temperature tests are conducted at conditions that push the limits of the material. The modulus values at 15% (E_{15}) and 20% (E_{20}) strain are not provided herein, because half of the tested specimens were broken before they reached such strains.

The superiority of carbon black as an efficient filler became even more pronounced at high temperature. However, incorporation of fly ash doubled the tensile strength and increased the elongation at break by one-third when compared to the unfilled elastomer.

Sample	E_5 (psi)	E_{10} (psi)	E_{15} (psi)	E_{20} (psi)	Tensile strength (psi)	Elongation (%)
T1	31	46			91	23
T2	61	130	194	261	832	39
T3	42	73	98	119	173	30

Table 2. Summary of high temperature (350˚F) tensile data for elastomer samples

Thermal properties of samples T1-T3 were assessed with Differential Scanning Calorimetry (DSC) and Thermogravimetric analysis (TGA). TGA allows evaluating thermal stability of samples and to verify the actual composition of a sample from the ash content. The results of the DSC and TGA analysis are summarized in Table 3. During TGA, samples were heated to 600°C at 10°C/minute rate in a nitrogen (N_2) environment, and subsequently heated at 10°C/min to 600 °C in air to burn off carbon-based fillers. This allows decomposing of organic matter first by applying heat, and then burn the remaining carbon-based material in air atmosphere.

The initial degradation of the polymer started at 200°C and progressed to complete decomposition of aliphatic chains in the temperature range of 420-500°C with a highest rate of decomposition at ~460°C. Sample T2 incurred the second rapid mass loss after 600°C due to burning of the carbon black, which comprised most of the ash content at 600°C in a N_2 environment. The DSC experiment demonstrated the glass transition temperature of the elastomer remained largely independent of the type of filler - the respective DSC heating scan was conducted with a 10°C/minute temperature ramp after precooling the sample to -50°C.

Sample	W_{600} (%)	W_{800} (%)	$T_{deg}1$, °C	$T_{deg}2$, °C	T_g, °C
T1	2.4	2.4	452.8	-	-29
T2	29.8	2.7	458.7	616.5	-29
T3	32.8	32.7	447.3	-	-28

Table 3. TGA/ DSC results for elastomer samples

Table 3 contains ash contents (W) at 600°C and 800°C, degradation temperatures (T_{deg}) which represent the temperature of the highest rate of decomposition during TGA experiment, and glass transition temperature (T_g) of the samples, as determined by DSC scan.

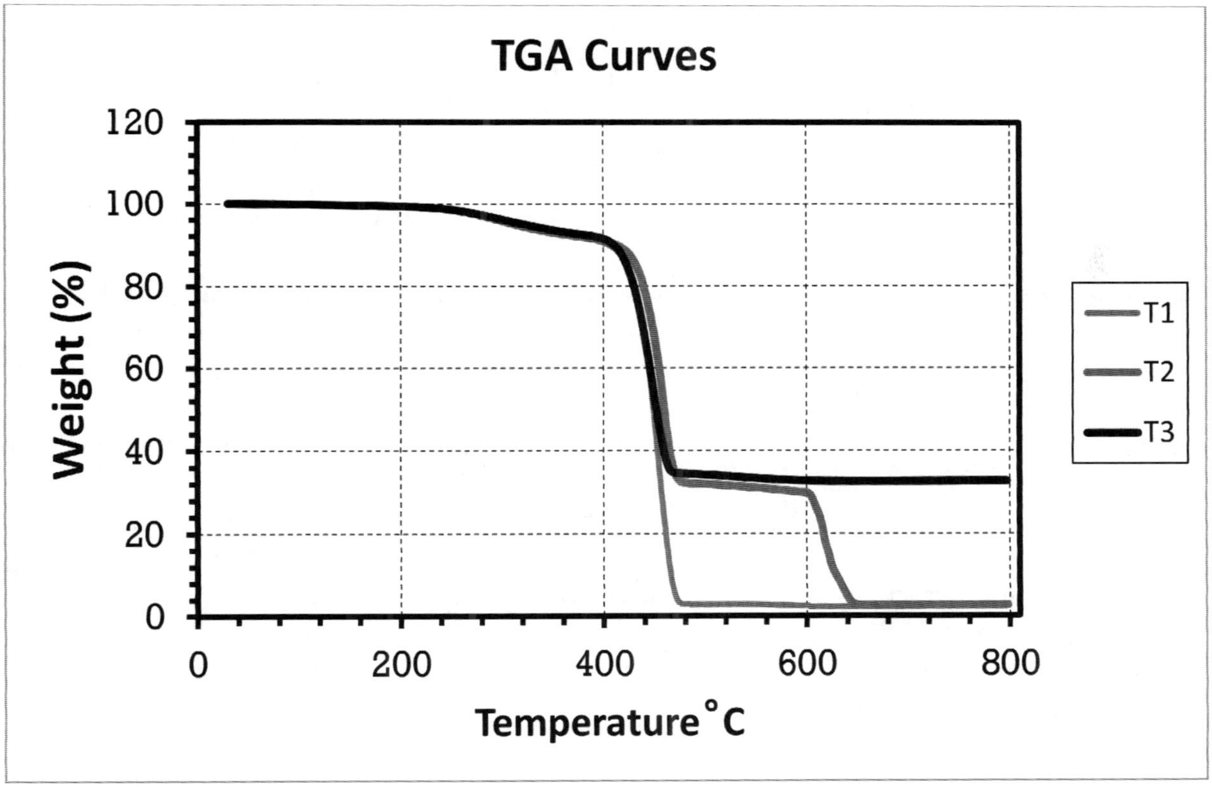

Figure 1. TGA Curves for samples T1, T2, and T3

Sample \ Testing fluid	3% KCl brine	LVT 200	Specific gravity
T1	0.6	13.0	0.96
T2	0.3	11.6	1.13
T3	0.8	9.1	1.18

Table 4. Volume increase (%) of elastomer samples due to swelling for 6 days at 200°F

Musso et. al. developed a stimuli-responsive HNBR composite reinforced with Portland cement that was increasing its stiffness and modulus when swelled in water[3]. Fly ash (Class C) is a common component of Portland cement, which contains the components necessary to achieve a similar effect. Therefore, one can avoid the loss of mechanical properties by water swellable packing element seals upon activation. To test this hypothesis, we studied the effect of aging in water brines on the properties of elastomers with fly ash fillers.

Rubber samples were molded and cured into standard size rubber buttons (cylinders of ~0.25" height and ~1" diameter). These buttons were aged for 10 days in 3% KCl brine in the oven at 200°F, removed from the solution, dried for 30 minutes in the oven and for 3 hours at room temperature. The swelling of both compounds was close to ~1% by volume (Table 4). Samples were later tested in an MTS machine in compression mode. Each button was compressed to 70% strain, buttons remained intact during the experiment, and visually all appeared to have a certain level of compression set at the end of the experiment. The results of the experiments are summarized in Table 5 and Figure 2. Only one button was tested for each sample. The provided compression modulus values (C_n) represent values of stress at a percentage strain (n), e.g. C_5 is stress at 5% compression strain.

Compound T4 was reinforced by fly ash Class C, which contains larger weight fraction of CaO than fly ash Class F (~25% vs ~10-15% respectively) and is expected to produce a higher pH and closely resemble cement that was used to prepare stimuli-responsive rubber in the original paper by Musso et.al.[3] The results demonstrate that exposure to water (3% KCl) at high temperatures can improve the modulus for the compounds filled with fly ash Class C (T4), while the control sample (T1) showed the opposite behavior. After aging, the compressive modulus at 50% strain (e.g. stress at 50% strain) of control sample T1 decreased by 19%, as compared to the unaged sample, while compressive strength of sample T4, reinforced by fly ash, increased by 15%. The difference in modulus values is even more drastic at 70% compression strain (-35% and 31%, respectively). This is in agreement with the hypothesis stated earlier.

Sample	Filler	Compressive modulus (psi)				
		C_5	C_{10}	C_{25}	C_{50}	C_{70}
T1	None	19	42	127	468	1810
T1 swollen	None	16	36	109	380	1185
T4	Fly Ash class C (50 PHR)	21	46	134	434	1628
T4 swollen	Fly Ash class C (50 PHR)	29	57	153	501	2125

Table 5. Summary of compressive modulus properties after 3% KCl soaking and 200°F drying

To further investigate the stimuli-responsive property of fly ash filler, the latter was incorporated into a water swellable compound, which provided for significantly more interaction of filler with water to occur. Sample F2 comprises unmodified commercial water swellable NBR formulation F1 mixed with 15 wt.% of fly ash Cass C. Sample F2 was processed into a 0.04” slab, then a rectangular sample was cut from the slab and swollen in 3% KCl solution until 100% volume increase. The F2 sample was then dried overnight in the oven, which lead to de-swelling of the sample due to solvent evaporation. Mechanical properties of the F2 sample were assessed with DMA prior to the swelling test, in a swollen state, and after drying - the data summary is shown on Figure 3.

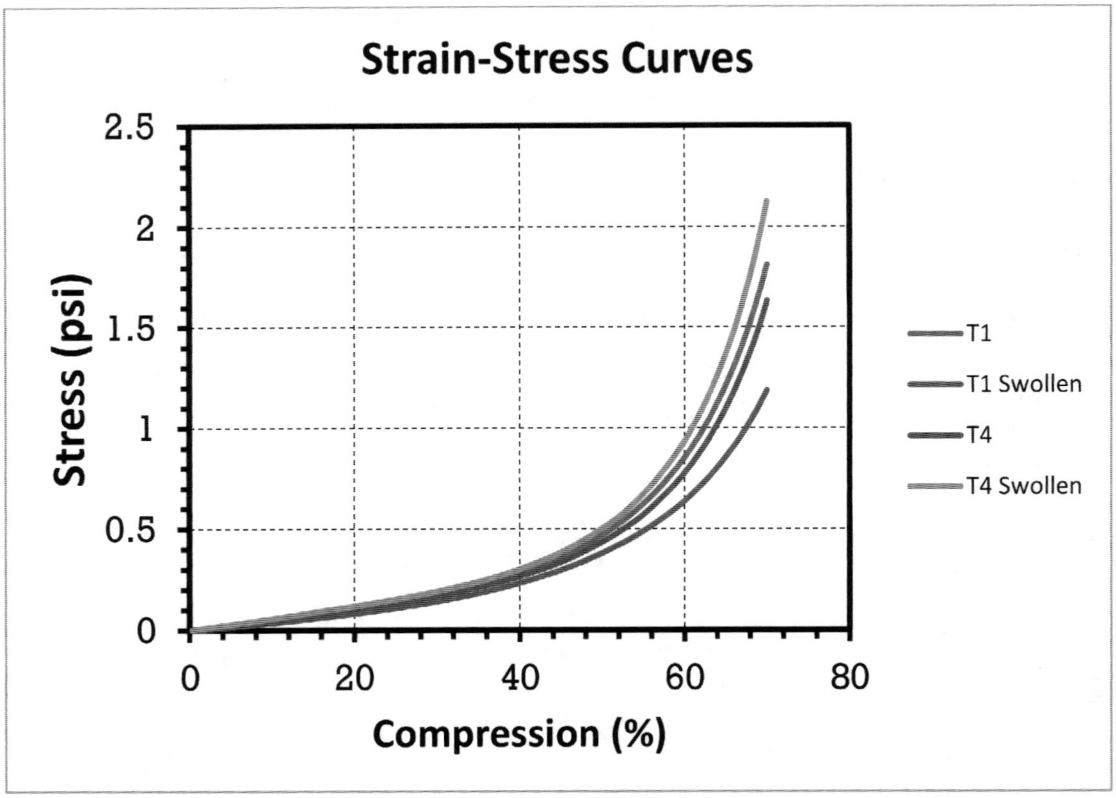

Figure 2. Stress-strain curves

The storage modulus of sample F2 after swelling and subsequent drying was about an order of magnitude higher than that of original sample F2 measured prior to swelling experiment. The storage modulus of the dried F2 sample was also less sensitive to the increasing temperature, contrary to the weakening original sample F2. The increase in the storage modulus of swollen F2 with temperature is simply due to the drying of the sample. The obtained result demonstrated that fly ash can essentially act as a stimuli-responsive material.

The significant improvement in the modulus is believed to be due to the neutralization of the swellable component of the commercial F1, and potential hydrolysis with subsequent neutralization of NBR elastomer. This is confirmed by the shift in T_g to a lower temperature range after drying as illustrated in Figure 3.

Carboxylic groups were detected in the original sample F2 prior to swelling, hence, they cannot be attributed solely to hydrolyzed HNBR, but rather to the super-absorbent polymer present in water swellable rubber compound (see Figure 4). FTIR confirms neutralization of carboxylic groups because a significant increase of the peak near ~1540 cm^{-1} and a lowering of peak intensity at ~1720 cm^{-1} have been observed. This neutralization is believed to promote the formation of ionic clusters in the compound with fly ash considered to be the source of cations. The neutralization of carboxylic groups could as well be attributed to K^+ cations present in the KCl brine solution, however, in that case, the respective peak in FTIR spectra would appear at higher wave numbers, approaching ~1600 cm^{-1}.

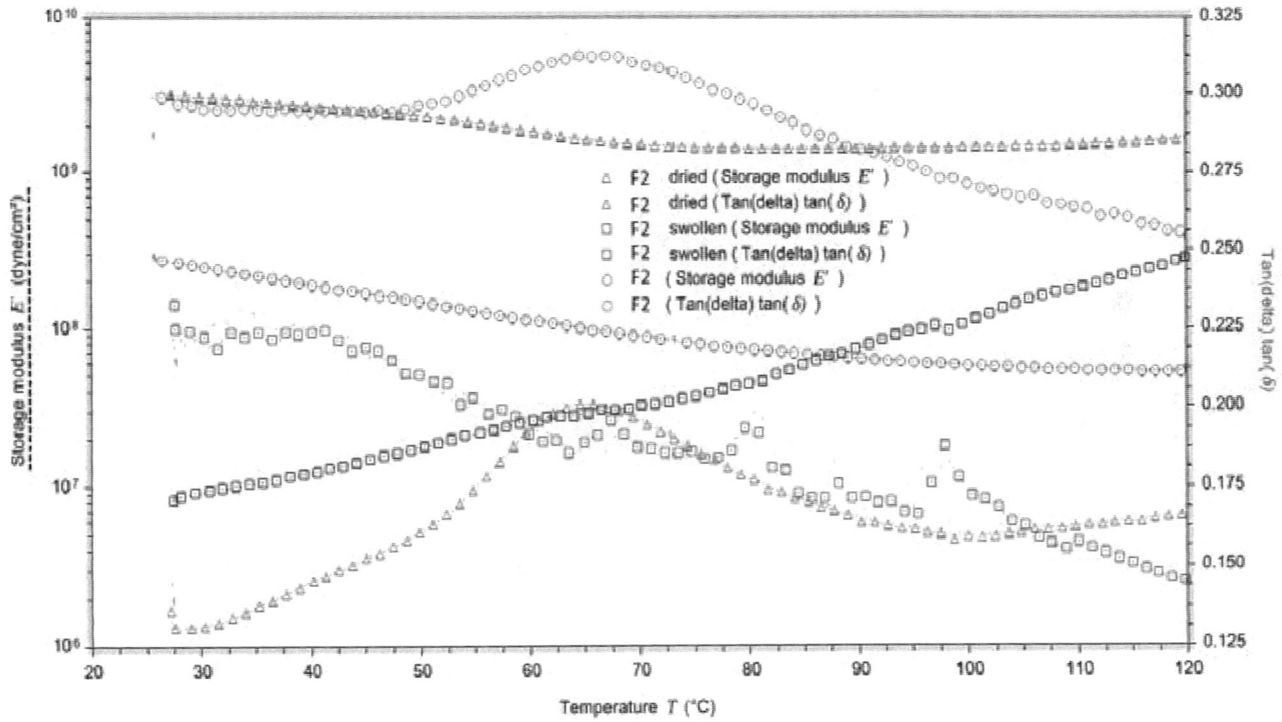

Figure 3. DMA data on F2 sample

It is important to mention that sample F2 exhibited a certain degree of brittleness after drying, which is undesired in rubbers, but this can be managed by incorporation of common ionomer plasticizers.

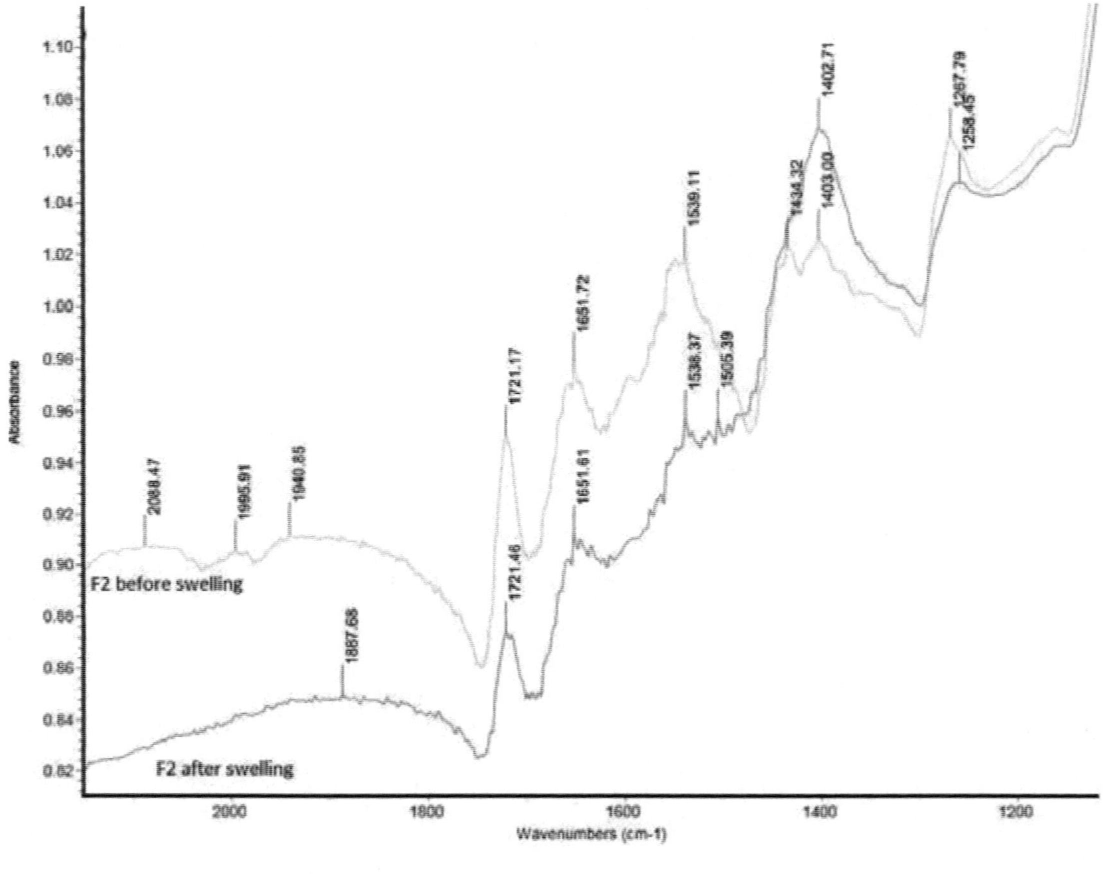

Figure 4. FTIR spectra of F2 sample taken before and after swelling

Conclusions

Fly ash demonstrated a great potential for consideration as cost effective, low reinforcing filler for downhole oil and gas applications. Although the reinforcing properties of fly ash are inferior to carbon black, incorporation of fly ash can improve the mechanical properties of HNBR elastomer compared to the unfilled raw crosslinked elastomer at both room temperature and at 325°F.

Furthermore, composites of HNBR with fly ash Class C demonstrated a stimuli-responsive effect by significantly increasing their modulus upon swelling in water brine. This effect was even more pronounced in composites of fly ash with swellable water compounds, which provided an easy access of filler to water upon swelling. A ten-fold increase in modulus was achieved in such fly ash reinforced sample, after swelling in 3% KCl water brine and drying in the oven overnight. The FTIR confirmed neutralization of carboxylic groups present in the investigated system, which contributed to overall improvement of mechanical properties of the compound upon swelling.

444093 New Development on Hoist Scheduling with the Relaxation of Cyclic Precondition

Abstract:

Multi-Stage Material Handling (MSMH) processes are designed to manufacture materials requiring a series of treatments and in industry, they are broadly used due to their capability to handle large work orders and compatibility with materials with different recipes. A typical example is electroplating process. A hoist handles the transportation of each job among units (stages) according to its processing recipe. The hoist movement plan is predetermined through optimization and this plan is repeated to perform continuity. Obviously, the ultimate goal is to maximize the productivity within each movement plan, which is equivalent to minimize the cycle time to finish one work order. The optimization process is generally called Cyclic Hoist Scheduling (CHS), since every movement plan is a cycle due to its repeatability. This precondition in term of hoist operation and unit usage is general for most cases. However, it could block the really optimal results.

In this study, this precondition has been relaxed for multi-capacity units, thus the unit usage can be more flexible. Thus, the real optimality can be found through CHS. To deal with the cycle connection issue from the relaxation of the precondition, a new module Cycle Connection Identification (CCI) is proposed. Obtained hoist schedule will be checked. Once a schedule with cycle connection problems is identified, a new cycle will be scheduled forcing the smooth connection with the first cycle. The objective of scheduling a new cycle is to match the beginning of the first cycle and the ending of the second cycle. If the two-cycle group is still with cycle connection issue, more cycles will be scheduled till the cycle group is free from cycle connection problems. Each cycle is called a sub-cycle and the constructed cycle group without cycle connection problems is called a full cycle. Note that in each sub-cycle, the work order and all transportation operations are completed.

In this paper, hoist scheduling models with/without such a precondition have been developed and proposed as mixed integer linear programming. The performance of the two models is tested through case studies and the difference is compared and demonstrated in detail.

444310 Formal Modeling of Accidents and Automatic Generation of Tutoring Scenarios for Intelligent Safety Training of Process Operators

Abstract:

With the emphasis of accident-free operations and prevention of catastrophic events, safety-training demands for the chemical plants have been on the increase globally. However, the research results on modeling of accidents and involving human operators and the advanced theory and technology for (safety) operator training have been shared limitedly among domain experts and commercial developers. Even with the recent trend of simultaneously delivering the operator training system with the construction of a new process plant, in commercial operator training systems (OTS), safety training used to implement quite limited and obvious scenarios (and the burden was almost up to the trainer), partially due to the difficulty in modeling propagations of probable accidents. By the regulatory requirements or proactive investment by companies, effective scenarios for training of safety and abnormal situation management are necessary to be improved in the aspects of the size of volume as well as the training depth. Thus, considering the emergence of enhanced virtual reality technologies and inexpensive gadgets like HMD, supporting the immersive training environment, the future efficiency of the training heavily relies on the quantity and quality of safety scenarios, which would be essential in guiding the entire training process.

In this research we first overview a government-sponsored research project developing an enhanced safety training system for chemical process operators, using haptic devices in addition to enhanced audio and visual equipment. The initial version of the system runs with a fixed set of training scenarios developed for predetermined accident scenarios. However, with a limited set of training scenarios, the trainees simply memorize and rote learn the steps, required to be mastered to proactively solve the problems with an advent of possible accidents. In addition to that, without formal modeling, the system cannot give timely feedbacks and the performance of the trainee cannot be measured in real-time. Then to enhance the contents and usefulness of the training, the knowledge of process operations is collected and systematically stored as a form of knowledge base, in addition to the accident scenarios obtained by executing Fault Tree Analysis and HAZOP study.

By using a formal modeling language like UML and Petri nets, the accident scenarios and operators' problem-solving tasks are modeled hierarchically and classified into distinguishable subtasks. Then a new training scenario is generated on-line through automatic synthesis based on planning using the knowledge base. Safety training scenarios are generated intelligently for the tutoring of a single operator's job function as well as effective collaboration among a group of operators for real-scale malfunction diagnosis and intervention at the initial propagation stage of an accident. As case study, safety trainings for city gas governing stations are executed based on systematic scenario generation and analyzed for improved effectiveness and field applicability of the system. Reduction of time and cost in the training scenario production and real-time customized training based on new operator's level of knowledge and skill are also discussed in the framework of intelligent tutoring systems.

Experimental Study on Najas Minor of Producing Bio-Oil

Abstract: Due to the diminishing of petroleum reserves and the environmental degradation caused by consumption of fossil energy , many attentions have been paid to the application of biomass energy which is acceptable as an abundant renewable energy resource. Najas minor grows in dense clusters with highly branched stems of which a small fragment could propagate into a thrive one.In this study, we researched the properties of Najas minor to make biofuels .Firstly,the najas minor after heavy metal ions degradation were fed into pyrolyser to prepare biofuels. Significant differences of the product was observed compared with only Najas minor pyrolysed.

KEY WORDS: Najas Minor; catalytic pyrolysis; bio-oil

1.Introduction

Energy demands and environmental problems caused by fossil energy have been a matter of growing concern.Bio-oil has been considered to be a promising way for producing renewable bio-fuels or chemicals[1-3].In recent years,many countries pay attention to the importance of biomass that is an abundant renewable energy resource.Biomass can be defined as any organic material mainly consisting of carbon, hydrogen,oxygen, nitrogen, and some other components in small proportions[4].Although the biomass has been studied for nearly a century,most studies mainly concentrated in Lemna minor, Lemna gibba, Spirodela polyrhiza and Landoltia punctata.We still know little about the biological characteristics and potential application value of other varieties of biomass[5].Najas minor,a submersed herb,grows in dense clusters with highly branched stems of which a small fragment could propagate into a thrive one.Najas Minor is native to Europe, Asia and North Africa from the Netherlands to Morocco east to Japan and the Philippines,including China,Siberia, Central Asia, Iran, Turkey, Ukraine, Germany, France Italy and a host of other Countries.

Catalysts used for catalytic cracking are basically divided into the following categories: molecular sieve, alkali metal salt, metal oxide, etc.Different catalysts have different catalytic effects.The use of metal nanoparticles has shown to be very important in recent industrial applications[6].The application of precious metal nanometer materials in catalysis,which has caused the extensive concern of researchers in recent 30 years,has become a hot research topic in the field of catalysis.Gold nanoparticles have received a great deal of attentions due to their special optical and electrical performances, catalytic properties and remarkable quantum size effect, surface effect etc[7]. The synthesis of metal nanoparticles can be classified broadly into wet and dry methods[8]. The cost of production is relatively high both materially and environmentally. There is an unequivocal need to develop more cost-effective and environmental benign methods to alternate these existing methods[9].Researchers in this field turn to use organisms to produce inorganic nanostructures extracellularly[10].

Although there has been a large volume of work on the synthesis of gold nanoparticles[11-13], and obtainning bi-oil by abundant biomass, there are relatively fewer attempts to combine these two methods.In this work, gold nanoparticles were obtained by aqueous chloroauric acid solution with the of Najas Minor at room temperature. The direct pyrolysis and catalytic pyrolysis experiments of Najas Minor were carried out to observe the effects of catalysts on pyrolysis process of Najas Minor.

2. Materials and Methods

2.1. Experimental samples

The Najas Minor collected from Tianjin city and the solutions of Au(III) were used in this study.The stock solutions of Au(III) were prepared by dissolving analytical grade chloroauric acid

($AuCl_3 \cdot HCl \cdot 3H_2O$, Kewei Reagent, China) in deionized water.

2.2.Experimental apparatus and procedure

Pyrolyzer with GCMS,shown in Fig. 1,is mainly used for analyzing the composition of polymer.Pyrolyzer with GCMS includes three parts :quartz pyrolysis tube,pyrolyzer and GCMS.

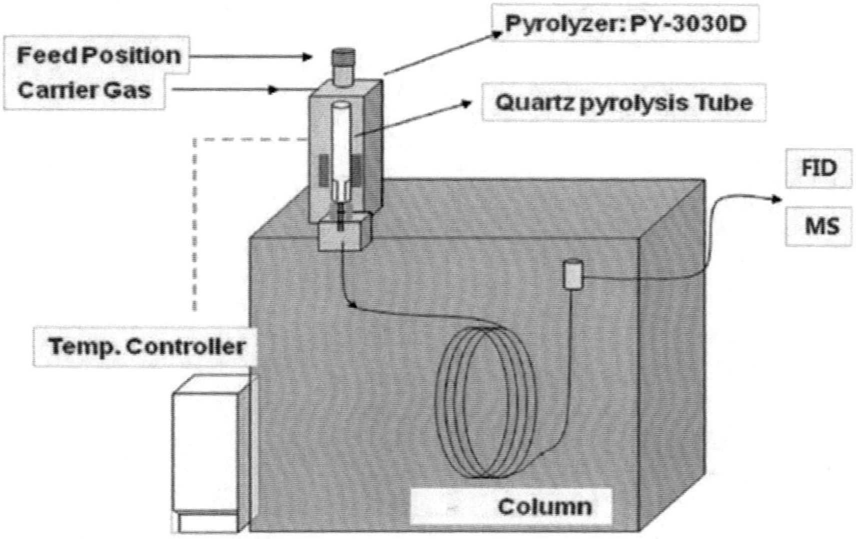

Fig. 1 Diagram of Pyrolyzer with GCMS

The quartz pyrolysis tube (shown in Fig.2) length is 12cm.The sample end diameter is 5 mm,and the pyrolysis products end diameter is 3mm.This special structure is beneficial to the fixation of the reactant in the tube.The top and bottom of the reactants are both supported by quartz wool to prevent reactants blowed by the carrier gas ,or fine particles into the chromatographic column.All pyrolysis experiments were carried out on the EGA/PY-3030D pyrolyzer(Frontier Lab,Japan) .The samples were filled in the special quartz pyrolysis tube to reduce the amount of valuable catalyst and reduce the cost of research work.

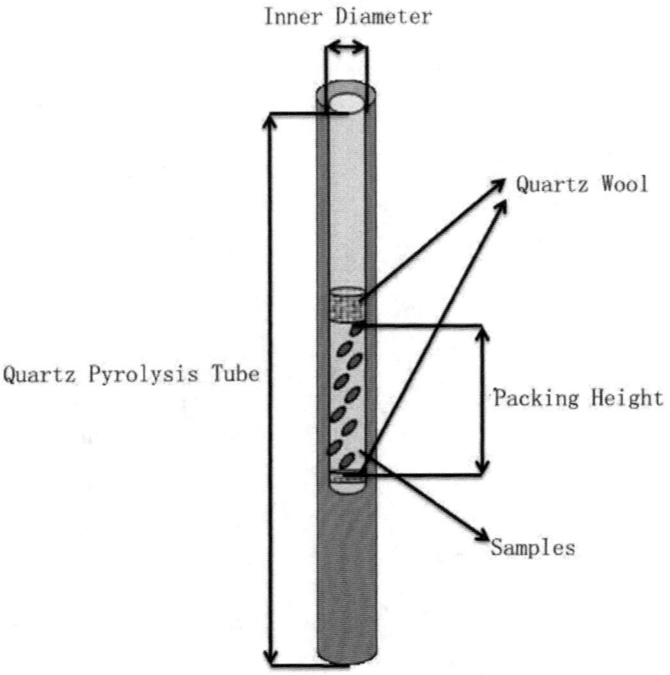

Fig. 2 Diagram of Quartz Pyrolysis Tube

Approximately 0.5g Najas Minors were used in each experiment.The samples were impregnated in 50ml deionized water and 50ml solution of chloroauric acid(0.0793mmol/L) for a night,followed by drying at 80℃ until the weight didn't change.Each experiment used 2.5mg Najas Minors as reactants.Each sample was heated, under the control of heating controller, from ambient temperature to 600 ℃ repeatly and kept at 600 ℃ for 2 h.

3. Results and discussion

3.1.MS Analysis of the Products during the Pyrolysis

Under the same pyrolysis conditions, bio-oil obtained by direct pyrolysis and catalytic pyrolysis were analyzed by GCMS,and the main chemical compounds were shown in Table1 and Table2.

Table 1 Indentification and quantification of main chemical compounds by direct pyrolysis

NO.	peak area%	Formula	Compound
1	12.67	$C_3H_6O_2$	2-Propanone, 1-hydroxy-
2	3.67	C_7H_8	Toluene
3	3.28	C_7H_8O	p-Cresol
4	3.81	$C_5H_{10}O$	Pentanal
5	3.53	C_8H_8O	Benzofuran, 2,3-dihydro-
6	2.17	C_8H_8N	Indole
7	5.93	$C_{20}H_{40}O$	3,7,11,15-Tetramethyl-2-hexadecen-1-ol
8	2.52	$C_{22}H_{42}O_2$	Phytol, acetate
9	5.25	$C_{16}H_{32}O_2$	n-Hexadecanoic acid
10	8.99	$C_{18}H_{30}O_2$	9,12,15-Octadecatrienoic acid, (Z,Z,Z)-

Table 2Indentification and quantification of main chemical compounds by catalytic pyrolysis

NO.	peak area%	Formula	Compound
1	6.01	C_7H_8O	p-Cresol
2	1.75	$C_7H_7NO_2$	3-Hydroxypyridine monoacetate
3	4.58	C_8H_7N	Indole
4	3.05	$C_{10}H9N$	1-Phenyl-1-cyclopropanecarbonitrile
5	3.90	$C_{12}H_{22}O_{11}$	.alpha.-D-Glucopyranose, 4-O-.beta.-D-galactopyranosyl-
6	8.04	$C_{20}H_{40}0$	3,7,11,15-Tetramethyl-2-hexadecen-1-ol
7	2.82	$C_{20}H_{40}$	2-Hexadecene, 3,7,11,15-tetramethyl-, [R-[R*,R*-(E)]]-
8	3.99	$C_{15}H_{28}$	1-Pentadecyne
9	2.12	$C_{15}H_{28}$	3,7,11-Trimethyl-2,4-dodecadiene
10	8.75	$C_{16}H_{32}\ O_2$	n-Hexadecanoic acid

3.2 discussing

According to the chemical category and common characteristics, the products were divided into hydrocarbons, alcohols, acids, ketones, nitrogen compounds and other compounds.Nitrogen compounds detected by GC/MS analysis ,mainly come from the pyrolysis of amino acids in Najas Minor.Table 3 showed the distribution of pyrolysis products.As shown in the table,the hydrocarbons ,alcohols and nitrogen compounds in biological oil of catalytic pyrolysis were significantly increased compared with direct pyrolysis.The content of hydrocarbons increased from 12.4% to 34.6%.The content of alcohols increased from 9.1% to 16.6%.The gold nanoparticles loaded in surface of Najas Minor, was conducive to the generation of hydrocarbon and alcohol in pyrolysis.By analyzing the GC/MS results, the main hydrocarbons in bio-oil by catalytic pyrolysis were long chain alkanes,and olefins.The main components in gasoline and diesel is hydrocarbons. As a result, whether the biomass can be effective as future power fuel oil, depends largely on the hydrocarbon content of bio-oil after further treatment.

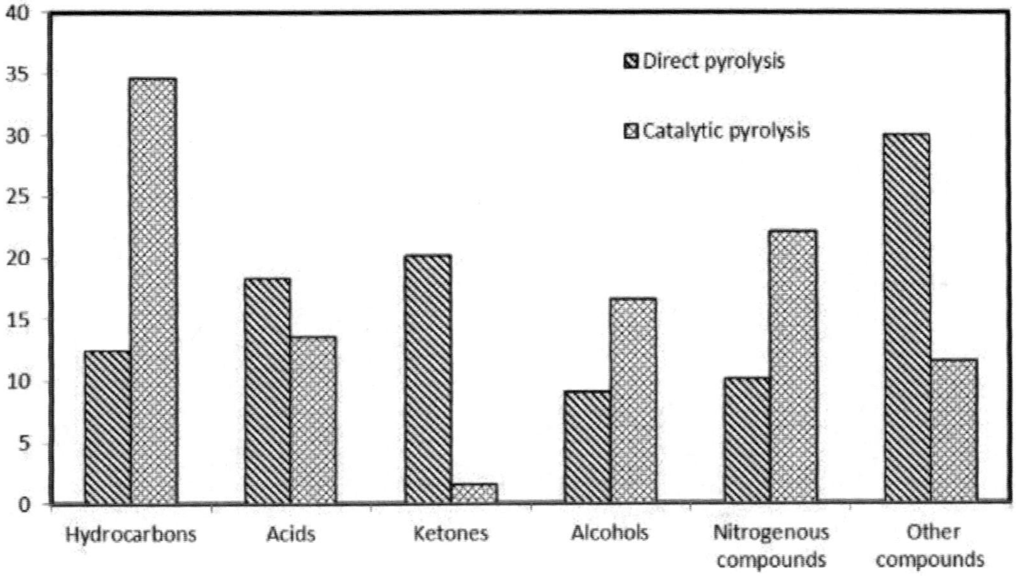

Fig.3 Distribution of pyrolysis products

4.Conclusion

Biomass was explored as a potential biomass energy resource. The experimental study on the direct and catalytic pyrolysis of Najas Minor had been presented, using Pyrolyzer with GCMS,to observe the catalytic effects of gold nanoparticles.In this study,remarkable catalytic effects for pyrolysis were observed.The pyrolysis of Najas Minor with gold nanoparticles,provided more hydrocarbons, alcohols and nitrogen compounds.

参考文献:

[1]Gong F, Yang Z, Hong C, et al. Selective conversion of bio-oil to light olefins: Controlling catalytic cracking for maximum olefins[J]. Bioresource Technology, 2011, 102(19):9247-54.

[2]Zhang Z, Bi P, Jiang P, et al. Production of gasoline fraction from bio-oil under atmospheric conditions by an integrated catalytic transformation process[J]. Energy, 2015, 90:1922-1930.

[3]Hui S, Yuqi D, Jinzhao D, et al. Transesterification of sunflower oil to biodiesel on ZrO2 supported La2O3 catalyst.[J]. Bioresource Technology, 2010, 101(3):953–958.

[4]Yaman S. Pyrolysis of biomass to produce fuels and chemical feedstocks[J]. Cheminform, 2004, 35(31):651-671.

[5]Hejný S. Ellias Landolt The family of Lemnaceae—A monographic study—Vol.1 biosystematic investigations in the family of Duckweeds (Lemnaceae) (vol. 2)[J]. Folia Geobotanica Et Phytotaxonomica, 1993, 28(1):50-50.

[6]Gardea-Torresdey J L, Tiemann K J, Gamez G, et al. Gold Nanoparticles Obtained by Bio-precipitation from Gold(III) Solutions[J]. Journal of Nanoparticle Research, 1999, 1(3):397-404.

[7]Marie-Christine D, Didier A. Gold nanoparticles: assembly, supramolecular chemistry, quantum-size-related properties, and applications toward biology, catalysis, and nanotechnology.[J]. Chemical Reviews, 2004, 35(16):293-346.

[8]Liu B ,, Xie J ,, Lee J Y, et al. Optimization of high-yield biological synthesis of single-crystalline gold nanoplates.[J]. Journal of Physical Chemistry B, 2005, 109(45):15256-63.

[9]Raveendran P, Fu J, Wallen S L. Completely "Green" Synthesis and Stabilization of Metal Nanoparticles[J]. Journal of the American Chemical Society Jacs, 2003, 125(46):págs. 13940-13941.

[10]Absar Ahmad , Satyajyoti Senapati , MIK, et al. Extracellular Biosynthesis of Monodisperse Gold Nanoparticles by a Novel Extremophilic Actinomycete, Thermomonospora sp.[J]. Langmuir, 2003, 19(8):3550-3553.

[11]Selvakannan P, Mandal S, Phadtare S, et al. Water-dispersible tryptophan-protected gold nanoparticles prepared by the spontaneous reduction of aqueous chloroaurate ions by the amino acid[J]. Journal of Colloid & Interface Science, 2004, 269(1):97-102.

[12]Gittins D I, Caruso F. Biological and Physical Applications of Water-Based Metal Nanoparticles Synthesised in Organic Solution[J]. Chemphyschem, 2002, 3(1):110 - 113.

[13]And A G, Murphy C J. Seed-Mediated Synthesis of Gold Nanorods: Role of the Size and Nature of the Seed[J]. Chemistry of Materials, 2004, 16(19):3633-3640.

COMPARATIVE ANALYSIS OF MOLECULAR STRUCTURE IDENTIFIABILTY BASED ON SIGNATURES AND DESCRIPTORS

Zelimir F. Kurtanjek
University of Zagreb, Croatia

Abstract

This work is focused on molecule computer representation aimed for molecule predictive model (QSAR) and inverse modelling. Two different approaches are compared, i.e. the molecule structure coding based on graph theory (Faulon et al.[1] extended valence) and the chemical based descriptors[2]. Applied are 1-3rd level molecule graph fragments and 630 2D molecular descriptors. The molecule test sets are alkanes, alkenes, acetones, aromatics, organic acids and halogenated hydrocarbons in the range of C_1-C_{12}, selection of biomolecules (including aminoacids) and a set of binary ionic liquids (cations: imidazole, pyridinium, quinolinium, ammonium, phosphonium). The comparative comparison of molecule description based prediction potential are statistically evaluated by octanol-water (PWA) partition coefficients (for hydrocarbons), viscosity of ionic liquids and biological impact on acetycholinesterase inhibition. The evaluation is based on two distinct levels. On the first level the comparative analysis based on internal and external correlation between "scores" in multidimensional spaces of the separate first, second and third order valance spaces, and the molecular signature space. Firstly, the potential of information content applicable for inverse modeling is tested on ability for molecule classification by use of linear discrimination (LDA) based on the singular value decomposition (SVD) of corresponding molecular fragments and descriptors. For chemical-biological property inverse modelling proposed is application of constrained genetic algorithm (GA). Tested are linear structure multivariate (chemometric) models such as principal component regression and partial least squares and nonlinear decision trees (random forest). Applied are singular value decompositions for reduction of the molecule signature and the descriptor spaces up to the significance level of p=0.05. The selected level of significance is presumed to be in coherence with experimental data of molecule physical and chemical properties. The main focus is on extraction of the main molecular fragments (partial graphs extracted from molecular signatures) and specific molecule descriptors. Importance of application of open source software modules provided by Bioclipse[3] and R software packeges[4] for computational-statistical numerical algorithms is emphasized.

Key words: Molecule signatures, descriptors, inverse modelling, decision tree

Introduction

Advancement of high-throughput experiments have lead to „big data" databases applied in pharmaceuticals, medicine, new material science, catalysis, and new industrial processes. The role of chemoinformatics as a science for computer encoding of molecule properties and algorithms for data mining and inference are the essential steps in the development. Computer encoding of molecular properties has been a part of chemical engineering from the beginning of process design. It is mostly based on molecule group contributions, but has been extended by application of molecular descriptors (1-2-3 D), molecule signatures, and application of molecular quantum mechanics. Modeling is faced with the problem of „high dimension curse". For example, about 5 000 various molecular descriptors can be determined for each molecule, although most of them are highly correlated. It results in an ill-conditioned problem of modelling with large uncertainties in parameter and poor model prediction accuracy outside modelling training sets. Great improvement has been achieved by application by use of model ensembles, like random forest of many individual decision trees. These models are highly flexible, can be trained on discontinuous and nonlinear functional relations, can encompass various data formats like numerical and categorical variables, and due to large number of randomly selected individual predictors in principle do not have tendency to overfitting. However, when the object of modelling is extraction of important molecular features, due to autocorrelation between descriptors random forest, produced lists of importance are large sets of poorly distinguishable variables. The focus of this work is to apply feature regularization for optimization and reduction of list of variable importance. Compared are two methods for structure encoding, molecular descriptors and molecular signatures of first and second order.

Modeling and Factor Regularization

In this work are selected two distinct data sets of organic molecules from open source data bases (J. Sangster, 1989) and ionic liquids ILs (The UFT/ Merck and NIST, 2015), and lists of the selected molecules are given in Table 1. As target properties for organic molecules are selected the corresponding molecular classes and partition octanol-water (POW) coefficients, and dynamic viscosity in the range of 30-80 ^{0}C of ILs is selected The molecules are coded into SMILES and used as input lists into software for evaluation of molecular signatures. Most popular open source software tools are DRAGON (R. Todeschini, et al., 2009) and PADEL (Chun Wei Yap, 2010). Here are used 1-2 D descriptors generated by PADEL. For the same list of molecules are determined first and second order molecular signatures or extended valances (J.P. Faulon, 2003). The signatures are two dimensional graph based structural molecular fragments. The vertices are atoms and the edges are corresponding bonds (lengths or heights), and are here only used of the first and second level. A molecule is characterized by a complete set of signatures as input data for modelling. Since the models are based on large number of variables (molecular features) possibly lager than number of samples (molecules), usually denoted as p>>n models. They are intrinsically „ill conditioned" leading to good predictions on train data but very poor predictions on test data. Also, produced list of variable importance are very dense "clusters are not distinguished" with large uncertainties (broad confidence intervals). This makes the problem of inverse modelling aimed to predict a set of candidate molecules for given set of target properties, such as in a case of high dimension combinatorial task for selection of ILs cations and anions, very unreliable. In order to reduce model dimensionality and improve robustness and parsimonious of important molecular features here is applied modelling with feature regularization. The model is based on ensemble of random trees (CART models, L. Breiman et al. 1998) with included feature regularization.

Data sets of organic molecules		Range of number of atoms			
Molecular classes	C	H	O	Cl	F – I - Br
Alkanes	5-12	12-30			
Alkenes	4-9	6-18			
Acetones	6-24	6-26			
Halogenated hydrocarbons	3-10	6-22	1		
Organic acids	1-16	2-32	2		
Esters	4-11	6-22	2		
Aromatics	1-12	1-21		3	2 – 2 - 1
Data sets of ionic liquids (ILs)					
Cations		**Anions**			
1,3-Dimethylimidazolium (+)		PF_6 (-) Cl (-) CH_3COO (-)			
1-Methylpyridinium (+)		BF_4 (-) $MeSO_4$ (-) $EtSO_4$ (-)			
1,1-Dimethylpyrrolidinium (+)		Tf_2N (-) CF_3SO_3 (-)			

Table 1. The molecular groups of the organic and ionic liquid molecules selected for random forest models and extraction of the most important molecular features. The properties for the organics are obtained from J. Sangster (1989) and for the ionic liquids from Merck and NIST data bases (2015). Molecular descriptors and signatures of ILs ions are recorded separately and adjoined as a single record.

The modelling is formulated as an optimization (functional and multivariable minimization) problem with two parts of the objective function (T. Hastie et al., 2009).

$$\min F_{object.}(X,T,\omega) = SSE(y,\hat{y}) + \phi(T,\omega) \tag{1}$$

The first part is the sum of square errors (SSE) based on difference between observed y and prediction $\hat{y}$ based on input data by an ensemble T of interrelated CART models (L. Breiman et al. 1998), The second part is a selected metrics ϕ for the model complexity depending the sum of leaf probabilities for population of individual CART models and their total number N_T.

$$\phi(T,\omega) = \gamma \times N_T + \frac{1}{2} \times \lambda \times \sum_i^{N_T} \omega_i^2 \tag{2}$$

The ensemble output prediction is a sum of predictions of individual T_i tree CART models:

$$\hat{y} = \sum_i^{N_T} T_i(X) \tag{3}$$

A very efficient open source code Extreme Gradient Boosting XGB (T. Chen, 2016) is available in R language (R Developm-ent Core Team, v. 3.2.3, 2015). The optimization algorithm is consecutive, starting with first CART model which is appended with a next tree boosted by errors in predictions of previous tree. At each step optimization is solved by linearization of the objective function to quadratic terms. Number and structure of trees as well as the parameters for regularization need to be found by trial and error and minimal number of trees by folding and cross validation error on the training set. The critical step is validation on untrained set to have acceptable model error and avoid model overfitting with minimal model complexity.

Results and Discussion

Tested are models for prediction of the seven molecular classes (Table 1.) based on molecular descriptors and molecular signatures (first and second level). Results for molecular descriptor feature importance are shown on Fig. 1. The effect of regularization is very pronounced. Out of hundreds of descriptors for the molecular classification needed are only few, the most important being BCUTw.1h. On contrary, when an ensemble of independent CART trees is applied, i.e. random forest model, level of accuracy of prediction is the same, but the list of variable importance is densely populated with numerous descriptors.

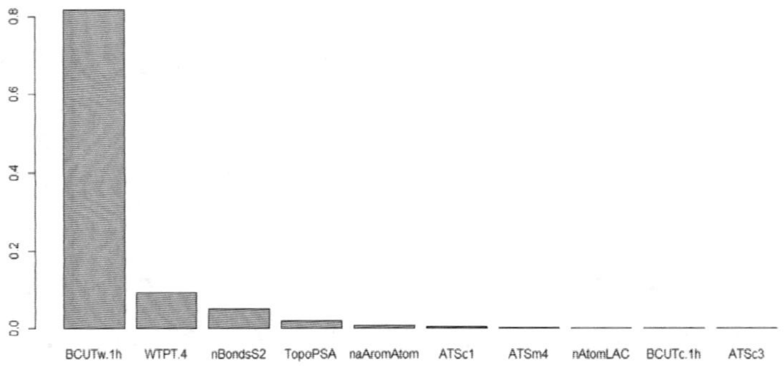

Fig 1. Variable importance molecular descriptor factors for the inverse modeling of the selected molecular classes. The first five descriptors are: BCUTw.1h (based on a weighted version of the Burden matrix which takes into account both the connectivity as well as atomic properties of a molecule), WTPT.4 (Randić's weighted path), nBondS2 (bond count), and naAromAtom (number of aromatic atoms).

The same classification problem was tested by the first and second order molecular signatures. The important signatures are depicted on Fig. 2-3. Accuracy of prediction on trained sets of molecules compared to descriptors is about the same, though the importance factors are not as distinctive as in case of the descriptors.

significantly better when molecular descriptors are applied. The

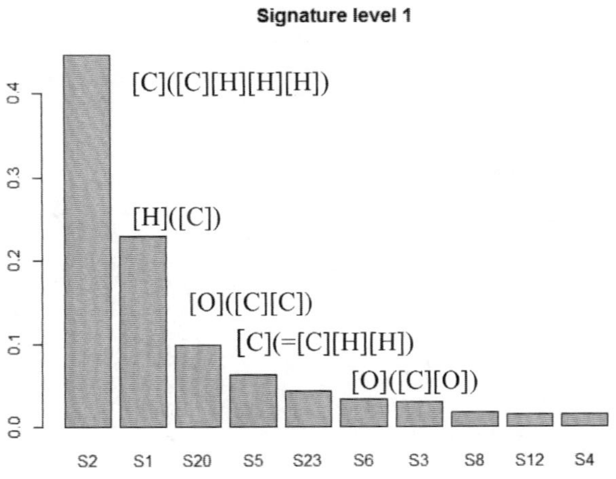

Fig 2. Relative importance of the individual first level signatures for inverse prediction of the selected molecular classes. The first five most important signatures are coded according to SMILES (Simplified Molecular Input Line Entry System).

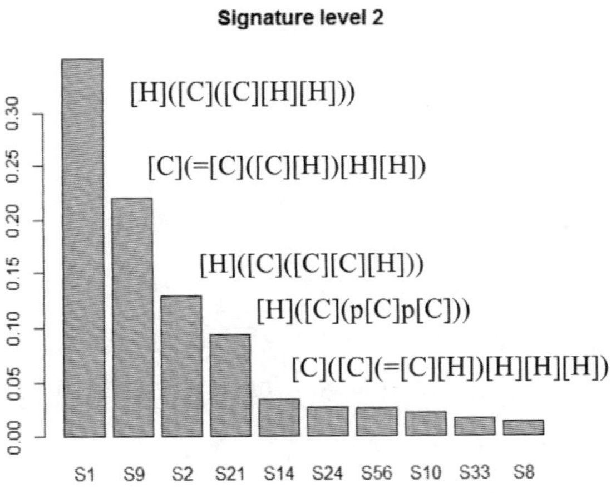

Fig 3. Relative importance of the individual second level signatures for inverse prediction for the selected molecular classes. The first five most important signatures are coded according to SMILES (Simplified Molecular Input Line Entry System).

However, model accuracy on untrained set of molecules, depicted in Fig. 4, is significantly lower for the model with molecular descriptors compared to signatures. The average error with descriptors is about 5% while for signatures about 11%. The "box and whiskers" plots are obtained by error evaluation of ten randomly generated untrained sets.

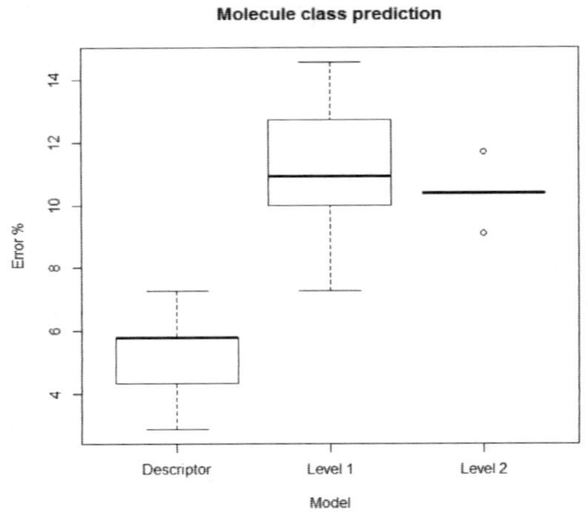

Fig 4. "Box and whiskers" plot for the model validation (relative errors) for inverse prediction of the molecular classes based on the set of untrained molecular descriptors and the first and second level of molecular signatures.

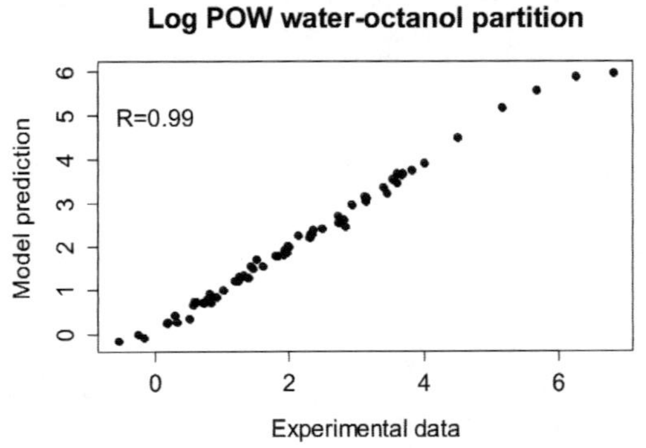

Fig 5. Predictions of the partition octanol-water coefficients POW, for the specified molecular classes of organic molecules, are based on the random forest model with the molecular descriptors.

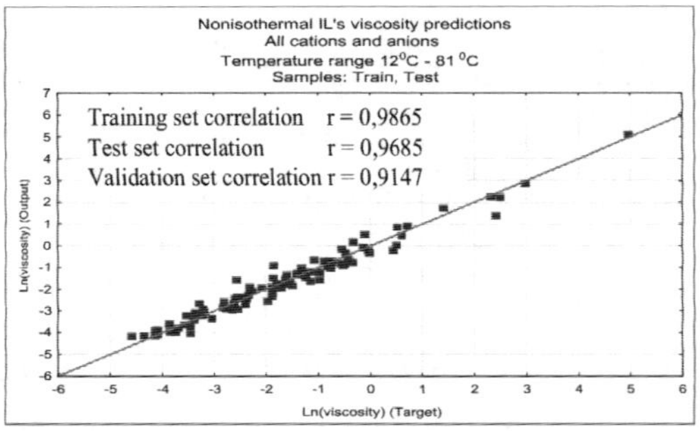

Fig 6. Predictions of dynamic viscosities for the specified groups of ionic liquids (ILs) based on the random forest model with the molecular descriptors. The viscosities cover the temperature range from 12 – 81 ^{0}C and five orders of magnitude.

In Fig. 4 can be observed that dispersion of errors (quantiles) for the second level molecular signatures is negligible, though two outliers are present.

For prediction of molecular properties applied are random forest models based on molecular descriptors. For all of the organic molecules, included in the seven classes, evaluated are octanol-water partition coefficients POW, shown in Fig. 5. The range of POW values is very high, on log scale fro 0 to 6, but the obtained accuracy of the model predictions is high (Pearson correlation R=0.99). The same approach is applied for prediction of ionic liquid viscosity data in temperature range from 12 to 81 ^{0}C. Since temperature effects on viscosity are intrinsically nonlinear, they represent a modelling challenge. However, the decision trees are very efficient for accounting nonlinearities, though nonlinearity is reflected in non-orthogonal decision planes, rendering trees with significantly increased number of branches. For the range of 6 order of magnitude the accuracy of prediction on trained set is R = 0.98 and on untrained set R = 0.91.

Conclusions

Compared are ensemble decision trees models for molecular feature class and property prediction based on extensive list of molecular descriptors and first and second level of extended valances or signatures.

The aim was to test effects of regularization on ensemble of CART decision tree models. Proven is effectiveness of the extreme gradient boosting algorithm in cases with extensive list of (molecular) features.

Obtained is increased accuracy on untrained sets of data, and more importantly, obtained are parsimonious lists for ranking of variable importance lists. This enables further significant reduction in development of robust and simple decision tree models.

The reduction in number of molecular features applied for decision tree models may greatly simplify the challenging problem of inverse modeling based on molecular descriptors and/or signatures.

References

1. J. Sangster, J. Phys. Chem. Ref. Data, **18**(3)(1989) 1111-1227
2. The UFT/ Merck Ionic Liquids Biological Effects Database, www.il-eco.uft.uni-bremen.de
3. NIST, Ionic Liquids Database Standard Reference Database #147, <ilthermo.bouder.-nist.gov /ILThermo/mainmenu.uix>
4. R. Todeschini and V.Consonni: "Molecular Descriptors for Chemoinformatics", (2 volumes), WILEY-VCH,Weinheim (Germany) 2009.
5. Chun Wei Yap, PaDEL-Descriptor: An Open Source Software to Calculate Molecular Descriptors and Fingerprints, J. Comp. Chem., (2010):**32**: 1466-1474.

6. J.-L. Faulon, D.P. Visco Jr., R.S. Pophale, The signature molecular descriptor. 1. Extended valence sequences vs. topological indices in QSAR and QSPR studies, J. Chem. Inf. Comput. Sci. **43** (2003) 707–720.
7. J.-L. Faulon, C.J. Churchwell, J.D.P. Visco, The signature molecular descriptor. 2. Enumerating molecules from their extended valence sequence, J. Chem. Inf. Comput. Sci. **43** (2003) 721–734.
8. L. Breiman, J. H. Friedman, R. A. Olshen, C.J. Stone, „Classification and Regression Trees", CRC, Boca Raton, 1998.
9. T. Hastie, R. Tibshirani, J. Friedman, The Elements of Statistical Learning, Data Mining, Inference and Prediction, Springer Verlag, Second Edition, New York, 2009.
10. T. Chen, T. He and M. Benesty (2016). xgboost: Extreme Gradient Boosting. R package version 0.4-3. https://CRAN.R-project.org/package=xgboost
11. R Development Core Team, , R: A language and environment for statistical computing. R, Vienna, Austria,www.R-project.org 2015.

444478 Biogas Reforming with Addition of Carbon Dioxide to Syngas over Ni-Based Bimodal Pore Catalyst

Abstract:

Methane dry reforming of (MDR) has been widely studied due to its important applications in producing of syngas ($CO + H_2$), removal of two greenhouse gases, and upgrading of biogas (mainly composed of CH_2 and CO_2) into value added chemicals. Compared to the noble metal catalysts, nickel-based catalysts are suitable for industrial scaled-up MDR due to their high activity and low cost. However, nickel-based catalysts have been encountering a significant challenge that is the active metal sintering and coke deposition, leading to the catalyst deactivation and then the poor stability. Nevertheless, it has been reported that the deactivation of nickel catalysts can be suppressed by adding promoters like strong Lewis bases with the enhancement of chemisorb CO_2, such as MgO and CaO, and like lanthanide elements with the capacity of oxygen storage and release, such as CeO_2 and La_2O_3. Al_2O_3 is a commonly applied catalyst carrier own to its good pore size dispersion, high specific surface area and high mechanic strength.

Herein, we report that the bimodal NiCeMgAl catalysts with different NiO-loading were synthesized by the refluxed co-precipitation method and were evaluated for MDR performance by using real biogas from anaerobic digestion of food wastes. The sample containing 15wt% NiO-loading (Ni_{15}CeMgAl) was found to be active enough at 750 °C with a high CH_4 conversion of around 90%. Ni_{15}CeMgAl catalysts kept their bimodal porosity after reduction and MDR process. The evolution of the Ni_{15}CeMgAl catalyst before and after MDR process was investigated by SEM, TEM, XRD, and TGA techniques. The correlation between the structure evolution and catalytic performance change was also discussed.

444994 Multiscale Simulation of Proteins Motion in the Microchannel

Abstract:

Protein adsorption is an important phenomenon, which plays a prominent role in chemical, industrial and medical areas. In this work, the dynamic behavior of the egg-white lysozyme has been investigated in a micro-channel with a multiscale modeling approach. The Molecular Dynamic simulation, free energy computation and Langevion Dynamics (LD) simulations have been combined to study lysozyme adsorption process on Au (111) surface. The gold surface has strong attraction to proteins. The protein-surface adsorption or desorption free energies is calculated. From the case studies of LD simulations, the surface has effects on the protein dynamic adsorption process. By the comparison of LD, MD and the experiments, the time step is concluded as an important factor to study protein adsorption behavior at the water-surface interface. Fine time step at the scale of 1 picosecond is necessary to provide enough accuracy to resolve proteins' dynamics trajectories.

444827 Syngas to Ethylene Glycol: A Technology Overview

Abstract:

SABIC is the world largest MEG producer petrochemical company and most of its MEG is being produced through the hydrolysis of EO. Ethylene oxide (EO) is a simplest form of epoxides and due to its molecular structure, acts as a versatile intermediate for a variety of chemicals such as; monoethylene glycol (MEG), di- and triethylene glycols, ethoxylates, ethanol amines, and glycol ethers. The growing demand of MEG owing to its downstream utilization in polyester industry has prompted the research of an efficient, cost effective, and environmentally benign EO production process.

The major feedstock for MEG are ethylene and syngas. The heavy dependence of MEG producers on ethylene and its scarcity in some parts of the world promoted the need of alternate feedstock for highly demanded MEG. The technological developments utilizing syngas as a feedstock will be discussed along with the commercial analysis and future growth in EO/EG business.

445029 Electrospun Nanocomposite Fibers for the Removal of Heavy Metals in Water

Abstract:

Uniform electrospun nanofibers with small diameters were successfully prepared by systematically optimizing the electrospinning parameters including polymer concentration, salt loading, applied voltage, tip-to-collector distance, and flow rate. These parameters are shown to affect the fiber morphology and the addition of iron (III) nitrate salt was found to be able to reduce the fiber size. Due to the conductive nature of iron salt in polymer solution, even a relative lower voltage of 6 kV could generate a stable fluid jet and yield the nanofibers. The thermal properties of the composite fibers were investigated by thermal gravimetric analysis (TGA) and differential scanning calorimetry (DSC). Rheological studies were also conducted. The composite nanofibers were further carbonized to carbon nanofibers. Applications of the nanofibers in environment remediation have shown promise.

445004 Grand Challenges in Future Energy and Chemicals: Climate Change and Sustainability As Drivers for New Technology and Innovation

Abstract:

Global demand for energy will double by 2050, driven by increasing population and affluence especially in emerging economies. While fossil fuels will continue to underpin the energy supply for the near future, concern over impact on climate change continues to grow, driving interest in expanded use of renewables and development of carbon capture and sequestration or re-use solutions. In addition to CO_2, water use and air emissions are critical issues to be addressed in future and current energy supply, as is energy storage for effective use of renewables. This presentation will provide an overview of what industry is doing today to address these issues, including the factors and metrics used in decision making.

444825 Significance of Energy Intensity and GHG Emission Reduction Towards Sustainability of a Petrochemical Plant

Abstract:

Sustainability is an important pillar in today's petrochemical industry for improving process efficacy by reducing specific energy consumption, greenhouse gas (GHG) emissions, and wastewater. SABIC (Saudi Basic Industries Corporation) like other giant petrochemical companies has set stringent sustainability targets and investing huge resources to transform its plants from the present position to the first quartile in each category.

SABIC is the world largest producer of ethylene oxide and ethylene glycol. For each ton of glycol produced, 0.2 to 0.5 tons of CO_2 is emitted as GHG. The significance of sustainability "in general" while the methodology adopted to reduce energy intensity, carbon footprint, and wastewater "in particular", will be discussed thoroughly in our talk.

444591 Lehigh University: Crude Lights End

Abstract:

A Light Ends Separation unit was required to process 607.08 kmol/h of light hydrocarbon material from the Pipestill column in a large refinery. Ethane, propane, isobutane, n-butane, mixed pentanes, and heavy naphtha components were separated by a series of distillation columns to meet given product purities. Conventional distillation columns, absorption columns, and divided wall columns were considered as potential separation methods of these components. An optimum scheme was chosen that included one absorption column, two conventional distillation columns, and a divided wall column, meeting all product purity specifications. Dynamic simulations of step changes in crude stock flow rate and composition, in Aspen Dynamics, were performed on separation schemes optimized in Aspen Plus on a Total Annualized Cost (TAC) basis. The plant was able to appropriately handle the given step changes, converging to a new steady state in a maximum of 11 hours and 12 hours for changes in feed flow rate and feed composition steps, respectively.

445006 University of Houston: Design of a Propylene Glycol Plant

Abstract:

Propylene is the second most common petrochemical feedstock in the chemical processing industry. It is the building block for a variety of products, such as polypropylene, propylene oxide, and propylene glycol. The propylene glycol market is currently growing at eight percent per annum and exceeds two million tons per year. The production capacity of propylene glycol in the world reached 1.6 million metric tons (3.5 billion pounds) in 2010. Among the many manufacturers are Dow, Lyondell, Sinopec, Shell, BASF, Repsol, Huntsman, and others. Dow Chemical has approximately half of the world production. The goal of this project is to design a plant that will produce 90,000 metric tons/yr of propylene glycol: 79,200 metric tons/yr will be propylene glycol (MPG) and 10,800 metric tons/yr will be dipropylene glycol (DPG). This presentation includes the design of the plant and the process economics of the plant.

444362 Depolymerization of Lignin in the Production of High-Valued Chemicals By Ethanolysis

Abstract:

To effectively utilize lignin for valuable chemical feedstocks, it is imperative to develop an efficient and environmentally pathways to depolymerize lignin. One-step depolymeriaztion of lignin in supercritical condition with the help of catalysts consider as one promising method for lignin valorization, and many catalysts had been tested such as NaOH, nickel-based catalysts, MoO_3/SiO_2, and CuMgAlOx. Here, we report the use of zeolites to depolymerize lignin depolymerization in supercritical ethanol. At the presence of zeolite, the lignin conversion gets improved with increased aromatic products, particularly those high-value phenols such as guaiacol and alkanesubstituted guaiacols.

443945 Biomass Conversion Technologies for Liquid and Gaseous Fuels, Chemicals and Materials

Abstract:

The Thermochemical Process Development Unit (TCPDU) at the National Renewable Energy Laboratory is used to demonstrate biomass to fuels technologies at an industrially relevant scale. The pilot plant can be configured to operate in multiple pathways including gasification and pyrolysis. Pyrolysis operations are conducted under oxygen-deficient environments and generate biochar as a byproduct. The biochar created under these conditions will self-heat and smolder when exposed to oxygen. This attribute can cause post handling of the char difficult and poses a potential safety hazard. In order to render the char stable for transport and use in secondary industries such as boiler fuel or soil enhancements, the char must be passivated. A unique passivation system was designed, implemented, and tested in the TCPDU using low percentage of oxygen to successfully yield a stable biochar.

16AIChE
Spring Meeting, Houston, TX

Using modelling to compare and optimize continuous chromatographic processes

Jay Yun
Ypso-Facto Inc.
One Broadway 14th Floor, Cambridge, MA 02142, USA
jay.yun@ypso-facto.com

Laurent David, Jean-Pierre Corriou, Roger-Marc Nicoud
Ypso-Facto
10, Viaduc Kennedy, 54000 Nancy, France
laurent.david@ypso-facto.com

Prepared for Presentation at
American Institute of Chemical Engineers
2016 Spring Meeting
11th Global Congress on Process Safety
Houston, Texas
April 10-14, 2016

UNPUBLISHED

Using modeling to compare and optimize continuous chromatographic processes

Presenter/Principal Author
Jay Yun
Ypso-Facto Inc.
One Broadway 14th Floor, Cambridge, MA 02142, USA
jay.yun@ypso-facto.com

Laurent David, Jean-Pierre Corriou, Roger-Marc Nicoud
Ypso-Facto
10, Viaduc Kennedy, 54000 Nancy, France
roger-marc.nicoud@ypso-facto.com

Keywords: sugar, separation, chromatography, HFCS, fructose, glucose, true moving bed, simulated moving bed

Abstract

Sugar purification is one of the largest applications of continuous chromatography. As a typical example, fructose is separated by chromatography from glucose to prepare High Fructose Corn Syrup (HFCS). This takes place in a process where HFCS 55 (55% fructose purity) is produced from HFCS 42 (42% fructose). The separation is ensured by a Simulated Moving Bed (SMB) filled with cationic resins. The sugar market is extremely competitive, thus the precise and robust design of the system to maximize recovery, productivity and minimize water consumption is of utmost importance. The "conventional process" is compared with a bypass process in which fructose is first obtained at 90 % purity at the extract stream of the chromatography prior to be diluted with the feed mixture.

1. Introduction

High Fructose Corn Syrup (HFCS) is increasingly used in the food and beverage industry. The starting product is corn which is progressively transformed by several physical and chemical steps into a glucose syrup, and finally isomerized to lead to HFCS 42, a fructose/glucose mixture with 42% fructose at around 800 g DS (dry substance)/l concentration. To increase the sweetening effect, higher fructose concentrations have been required by the food and beverage industry, originally as HFCS 55 (55% fructose) used in beverages or more recently HFCS 90 that can be used as such in pastries. The purification is typically performed by chromatography [1,2,3]. Instead of obtaining HFCS 55 directly from HFCS 42, HFCS 90 (90% fructose) can be used which is then diluted by a blending with HFCS 42 to finally obtain HFCS 55. "Bypass chromatography" is opposed to "conventional chromatography" where the required 55% purity fructose is obtained directly at the outlet of the chromatographic process. The design of the batch bypass chromatography has been studied by [4].

In the present study, the uses of the continuous Simulated Moving Bed (SMB) in both conventional and by pass modes are compared. Our goal is to show how a rationale methodology based on process modeling allows deriving important conclusions.

2. Process generalities

2.1 Bypass continuous chromatography process

For the sake of simplicity, it is assumed that the feed mixture contains only glucose and fructose. After the isomerization system assumed to produce a 800 g DS /l mixture, HFCS 42 is fed to the SMB which produces HFCS 90 at the extract side (Figure 1). To obtain the required purity (55% in Figure 1), a bypass is used to dilute HFCS 90 with HFCS 42. The second stream of the continuous chromatographic separation, the raffinate, is recycled to the isomerization reactor.

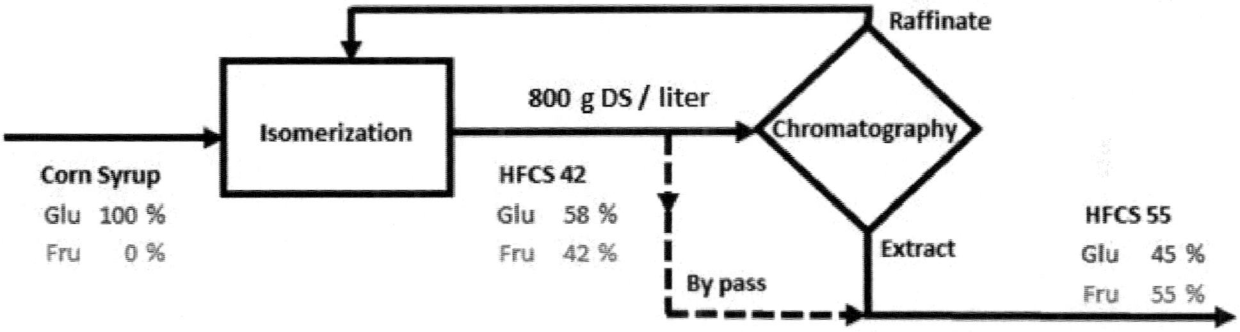

Figure 1: Manufacturing principle of HFCS55.

In the conventional mode, the SMB delivers directly the fructose at 55 % and no bypass is used.

2.2 Presentation of the TMB

Before designing the SMB, it is useful to design the True Moving Bed (TMB) which is an idealized chromatographic process where the liquid and the solid (chromatographic medium) circulate counter-currently. The reader is referred to [1] for general considerations on TMB and SMB and to [5] for detailed modeling considerations.

The TMB is composed of four zones: Zone I is the desorption zone located between the elution (water) and extract line. Zone II and III are separation zones, respectively located between Extract/Feed and Feed/Raffinate lines. Zone IV is the adsorption zone located between Raffinate and Eluent lines. The different zones can have different lengths.

For a given chromatographic medium, feed composition and eluent, the design parameters of the TMB system are the lumped solid flow rate $\bar{Q}$, two inlet flow rates (feed and eluent), two outlet flow rates (extract and raffinate), and the fluid flow rate in zone I (fluid flow rates in subsequent zones are then calculated knowing the inlet/outlet flow rates). The column diameter, zone lengths are mainly consequences of the above mentioned flow rates.

In a TMB, the four internal flow rates (*i*=I,.., IV) secure the separation. The four external flow rates associated with the Feed/Eluent inlet streams and of the Extract/Raffinate outlet streams result from the following fluid balances:

$$Q_F^{tmb} = Q_{III}^{tmb} - Q_{II}^{tmb} \quad , \quad Q_E^{tmb} = Q_I^{tmb} - Q_{IV}^{tmb} \qquad \text{[Eq. 1]}$$

$$Q_{Ext}^{tmb} = Q_I^{tmb} - Q_{II}^{tmb} \quad , \quad Q_{Raf}^{tmb} = Q_{III}^{tmb} - Q_{IV}^{tmb} \qquad \text{[Eq. 2]}$$

In a TMB, the fluid flow rates are typically normalized by the resin flow rate, resulting in the so called m ratio, so that the ratios m_i for the different zones are defined as:

$$m_i = \frac{Q_i^{tmb}}{\bar{Q}} \qquad \text{[Eq. 3]}$$

where $\bar{Q}$ is the lumped solid flow rate. In a similar manner,the *m* ratios are defined for the feed and eluent inlet stream as:

$$m_F = \frac{Q_F^{tmb}}{\bar{Q}} \quad , \quad m_E = \frac{Q_E^{tmb}}{\bar{Q}} \qquad \text{[Eq. 4]}$$

2.3 Presentation of the SMB

The Simulated Moving Bed is the practical implementation of the counter-current TMB process [1,5]: the circulation of the solid is simulated by a cyclic commutation of the inlet (feed and eluent) and outlet (extract and raffinate) ports. The commutations of the inlet and outlet lines are performed simultaneously: at each time period ΔT^{smb}, the two inlet lines and the two outlet lines are moved by one position.

The configuration adopted for the SMB in this study is 1-2-2-1, *i.e.* one column in zone I, two columns in zones II and III, one column in zone IV.

For a given chromatographic medium, feed composition and eluent, the design parameters of the SMB system are the shift time period ΔT^{smb}, two inlet flow rates (feed and eluent), two outlet flow rates (extract and raffinate) and the fluid flow rate in zone I (fluid flow rates in subsequent zones are then calculated knowing the inlet/outlet flow rates). The columns diameter and length (and thus the column volume V_{col}), the number of columns in each zone are largely consequences of the above determined flow rates.

It can be shown [5] that the SMB shift period and the flow rates in zone *i*=I,..,IV are related to the TMB flow rates by:

$$\Delta T^{smb} = \frac{(1-\varepsilon_e).V_{col}}{\bar{Q}} \qquad \text{[Eq. 5]}$$

$$Q_i^{smb} = Q_i^{tmb} + \varepsilon_e \frac{V_{col}}{\Delta T^{smb}} = Q_i^{tmb} + \frac{\varepsilon_e}{1-\varepsilon_e}\bar{Q} \qquad \text{[Eq. 6]}$$

As soon as internal flow rates are determined, external flow rates are easily deduced and the following relations are obtained:

$$Q_F^{smb} = Q_F^{tmb} \quad , \quad Q_E^{smb} = Q_E^{tmb} \qquad \text{[Eq. 7]}$$

$$Q_{Ext}^{smb} = Q_{Ext}^{tmb} \quad , \quad Q_{Raf}^{smb} = Q_{Raf}^{tmb} \qquad \text{[Eq. 8]}$$

If the TMB parameters are known, the SMB parameters are thus immediately obtained.

3. *Physico-chemical properties of the fructose-glucose mixture*

Fructose-glucose mixtures can be separated by means of ionic resins [2,3]. In the present case, a laboratory chromatographic column of diameter 1.6 cm, length 100 cm (V_{col} = 200 ml) and extragranular porosity ε_e=0.4, filled with Dowex Monosphere 99Ca resin with a particle size of 320 µm, is used with water at 65°C as eluent. The feed is a 42% fructose, 58% glucose mixture at a total concentration of 800 g/l. Even if the adsorption isotherms of fructose and glucose are slightly anti-Langmuirian [6], the behavior of fructose and glucose is close to linear and they are assumed to follow the linear adsorption equation:

$$\bar{C_j} = \bar{K_j}\, C_j \qquad \text{[Eq. 9]}$$

where C_j is the concentration of component j (fructose or glucose) in the solution, $\bar{C_j}$ is the lumped concentration of j in the resin and $\bar{K_j}$ is the lumped Henry's coefficient [5].

Assuming the fluid flow rate Q to be constant, the mean retention time of component j is given by:

$$t_{Rj} = \frac{V_{Rj}}{Q} = t_0\left(1 + \frac{1-\varepsilon_e}{\varepsilon_e}.\bar{K_j}\right) \qquad \text{[Eq. 10]}$$

where $t_0 = \frac{V}{Q} = \frac{\varepsilon_e . V_{col}}{Q}$ is the "zero retention time".

Determining the t_{Rj} allows determining the lumped Henry's coefficients of fructose and glucose equal to $\bar{K}_{glu} = 0.33$ and $\bar{K}_{fru} = 0.55$.

The chromatographic columns are typically characterized by a number of plates N. The number of plates is normally solute dependent and can be calculated from the variance and mean retention time of the peaks. For symmetrical peaks, the mid height width $w_{1/2}$ can be used instead of peak variance so that:

$$N_j = 5.54\left(\frac{t_{Rj}}{w_{1/2}}\right)^2 \qquad \text{[Eq. 11]}$$

Experience shows that the $HETP_j$ is approximately linearly related to the interstitial velocity u_e according to the simplified Van Deemter equation:

$$HETP_j = \frac{L_{col}}{N_j} \approx b_j + c_j.u_e \qquad \text{[Eq. 12]}$$

b_j is typically three to four particle diameters. c_j depends on particle size, pore size, intragranular porosity, diffusion coefficients.

The number of plates has been determined on the fructose peak (the more broadened), leading to 30 plates on the 100 cm length column at a flow rate of 10 ml/min. This corresponds to a $HEPT$ of 33 mm on fructose thus to about 100 particle diameters. This shows a strong mass transfer limitation and it is compatible with $c_j = 0.0044$ h.

Finally, if the chromatographic media is made of hard particles, Darcy's law allows to write:

$$\Delta P = \delta_e u_e L_{col} \qquad \text{[Eq. 13]}$$

where δ_e is a proportionality constant depending on fluid viscosity (thus concentration) and bed permeability. Performing a few pressure drop measurements with a feed concentration of 200 g/l (intermediate between feed and eluent concentration), it leads to $\delta_e \approx 0.04$ bar.h.m^{-2}. A more precise approach would consist in considering the influence of concentration on viscosity and thus on δ_e.

4. *System design*

We assume that the outlet stream of the isomerization reactor delivers a constant fructose purity of 42% and a constant dry substance content of 800 DS kg/l. The objective is to produce HFCS 55 at a reference flow rate of 100 kg DS/h. The conventional mode where SMB delivers 55 % fructose will be compared to the bypass mode where SMB delivers a 90 % pure fructose fraction that is blended with feed coming from isomerization.
For simplification purposes, it will be considered that the solutions contain only fructose and glucose.

4.1 Design of the TMB: looking for a reference point

Idealized TMB systems able to continuously produce fructose at a purity of either 90 % or 55% and at a fructose recovery of 90 % are considered. The process flow diagrams allowing to produce 100 kg DS/h HFCS 55, either without bypass or with bypass, are given in Figure 2.

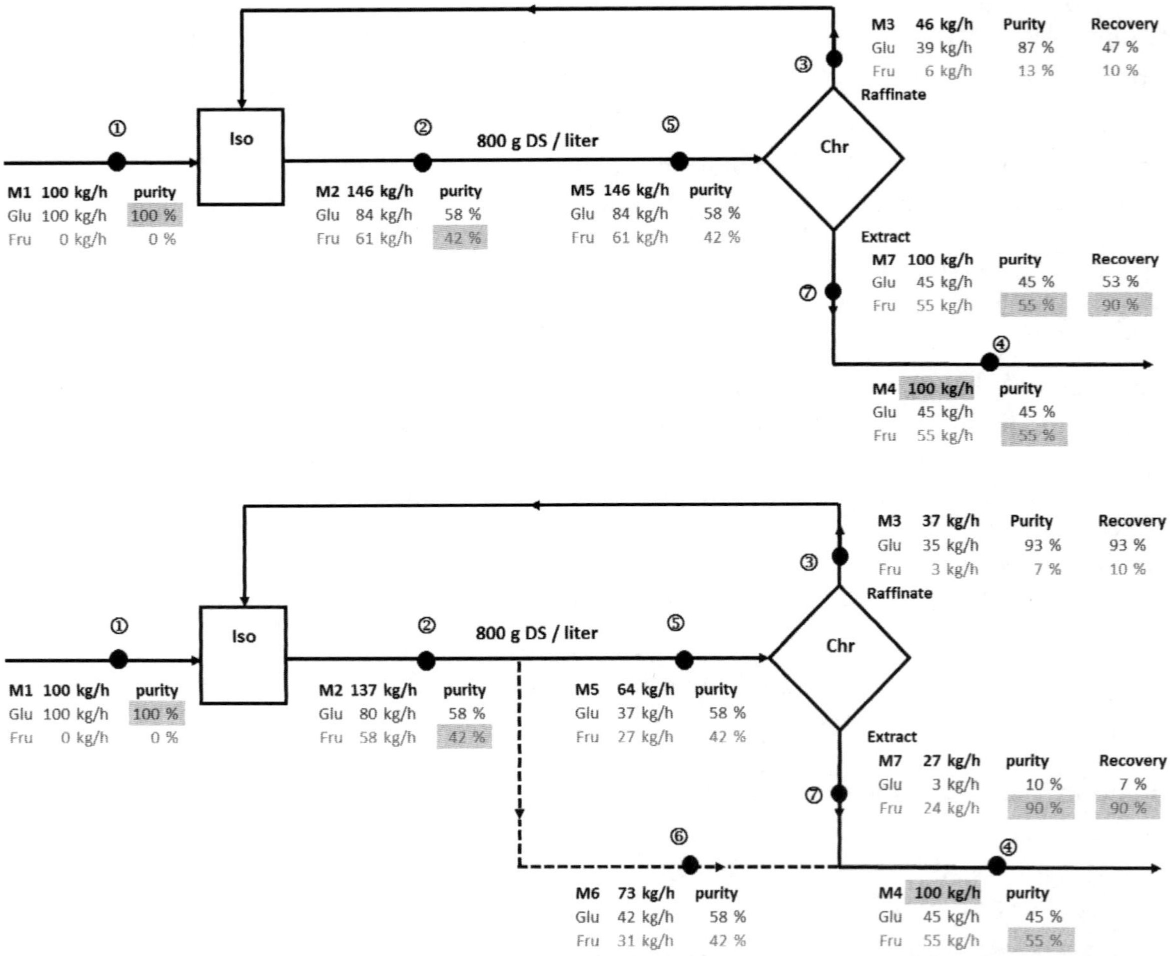

Figure 2: Process flow diagrams. Top: conventional process. Bottom; bypass process.

It appears that TMB in the conventional mode has to perform an easy job (raising fructose purity from 42 % to 55 %) on a 146 kg DS/h mass flow rate and that the TMB in the bypass mode has to

perform a more complex job (raising fructose purity from 42 % to 90 %) but on a smaller mass flow rate of 64 kg DS/h.
In terms of liquid flow rates, the TMB in the conventional mode has to process a feed flow rate of $Q_F = 182.5$ l/h, whereas the TMB in the bypass mode has to process a feed flow rate of $Q_F =$ 80 l/h (91 l/h being by passed).
The problem is now to design the TMB systems. As a starting point, it will be assumed that the TMBs are equivalent to an infinite number of plates. This is clearly a very idealized picture, but this will prove to be a very useful intermediate step for complete process design. Based on the equilibrium model theory [5], ChromWorks [7] is able to give the (m_2,m_3) zone that allows to get 90 % (55 %) fructose purity at 90 % recovery. These domains are given in Figure 3.

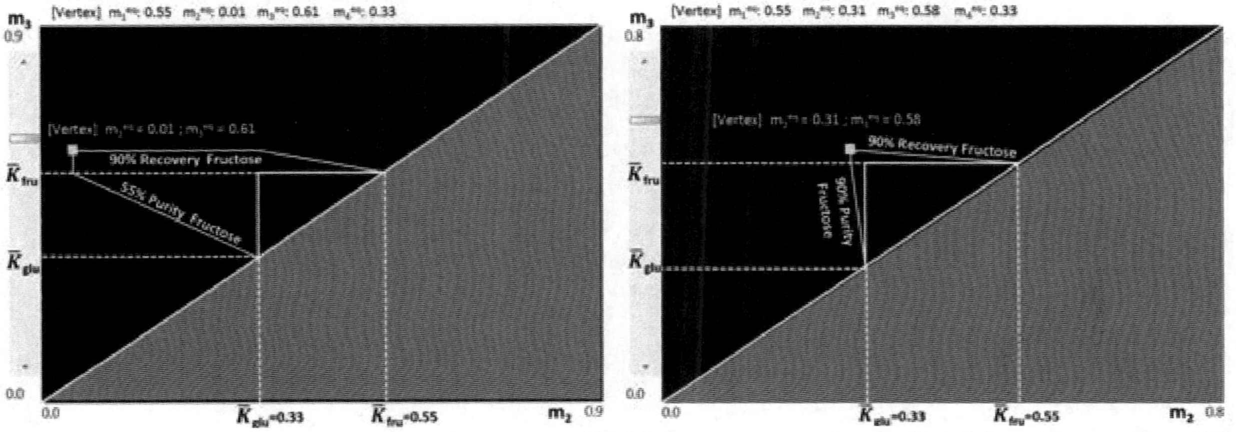

Figure 3: (m_2, m_3) domain allowing to get 55 % fructose purity (left, without bypass) or 90 % fructose purity (right, with bypass) at 90 % recovery. Design provided by ChromWorks.

At first, we select the (m_2,m_3) pair allowing to reach the target purity and recovery and at the same time maximizing the normalized feed injection ratio m_F. This pair is named (m_2^{eq},m_3^{eq}). The superscript "*eq*" is used as the reference to the equilibrium model assuming a column of infinite efficiency. In addition, the (m_1^{eq},m_4^{eq}) ratios allowing a stabilization of the fructose front in zone 1 and a stabilization of the glucose front in zone 4 for a system of an infinite efficiency are determined. It yields:

- For a 55% fructose purity and 90 % recovery

$$m_1^{eq} = 0.55, \quad m_2^{eq} = 0.01, \quad m_3^{eq} = 0.61, \quad m_4^{eq} = 0.33 \qquad \text{[Eq. 14]}$$

hence $m_F^{eq} = 0.60$ and $m_E^{eq} = 0.22$

- For a 90% fructose purity and 90 % recovery

$$m_1^{eq} = 0.55, \quad m_2^{eq} = 0.31, \quad m_3^{eq} = 0.58, \quad m_4^{eq} = 0.33 \qquad \text{[Eq. 15]}$$

hence $m_F^{eq} = 0.27$ and $m_E^{eq} = 0.22$

Let us remind that this set of m^{eq} values would allow to get exactly the target purity and recovery on systems of infinite efficiency. Knowing the m^{eq} values, based on Eqs [1-4], all the TMB flow rates required to process the imposed feed flow rates can be calculated. The flow rates in zone I calculated for the TMB without and with bypass are almost identical around 165 l/h. Thus, similar system sizes can be expected. Moreover, the eluent consumption is also almost identical at around 66 l/h for both strategies.

4.2 Calculation of the SMB: using the reference point for more precise process design

A real SMB being associated with finite efficiency (in terms of plate number), some safety margins with respect to the above determined m_F^{eq} and m_E^{eq} must be taken in order to reach the target purity and recovery. The safety margins with respect to feed and eluent flow rates are defined:

$$\eta_F = \frac{m_F}{m_F^{eq}}, \quad \eta_E = \frac{m_E}{m_E^{eq}} \qquad \text{[Eq. 16]}$$

where m_F and m_E are the real m ratios used in the SMB system. $\eta_F>1$ is the safety margin at the feed injection level and $\eta_E<1$ is the safety margin at the eluent injection level. For the sake of simplicity, $\eta_F = 1/\eta_E$ is imposed and the associated margins will be evenly distributed between m_2 and m_3 for η_F and between m_1 and m_4 for η_E.

For a given $\eta_F = 1/\eta_E$, the ratios m_i are determined from m_i^{eq} given above. From ratios m_i, the TMB flow rates are known, and the SMB flow rates are calculated with Eqs [5] and [6]. The SMB systems are then simulated using the Mixing Cell (MC) model of ChromWorks [7] and the required numbers of plates per column allowing to reach the target purity and recovery are determined. The influence of the number of plates per column on the purity without bypass and with bypass is shown in Figure 4.

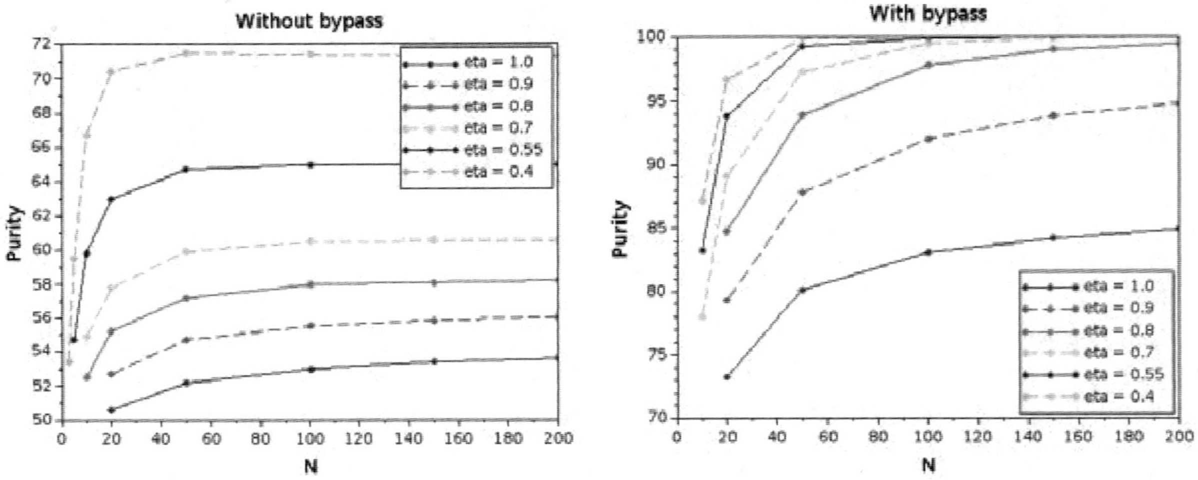

Figure 4: Influence of the number of plates per column on the purity without bypass (left), with bypass (right). Simulations performed by ChromWorks with MC-Equil model.

In agreement with our assumptions, the target purity (55 % without bypass or 90 % with bypass) and target recovery (90 %) are obtained for systems of infinite efficiencies when $\eta_F = 1$. The lower η_F, the higher the safety margin (and eluent consumption !), the lower the required number of plates to reach the required purity and recovery (Table 1).

Table 1: Number of plates per column required for the 1-2-2-1 SMB to reach target fructose purity (55 % for conventional or 90 % with bypass) and target fructose recovery (90 %).

	Conventional						Bypass					
η_F	1	0.9	0.8	0.7	0.55	0.4	1	0.9	0.8	0.7	0.55	0.4
N^{req}	Infinity	72	26	17	11	8	Infinity	76	37	21	16	12

It is now assumed that the systems are operated at 6 bar, corresponding to a pressure drop of 1 bar per column. If the pressure drop and the required number of plates per column N^{req} are imposed, Eqs [12] and [13] give two constraints that can be solved with respect to the column length L_{col} and the mean interstitial velocity u_e (average of the four zone velocities). Note that the total system length is $L_{syst} = 6L_{col}$.

The systems are then sized by taking into account the feed flow rates Q_F to be processed. For each system, the mean internal flow rate Q_{mean} in the SMB is defined as the mean of the four internal flow rates weighted by the zones lengths. The column cross section is then calculated by $\Omega_{col} = Q_{mean}^{SMB} / \varepsilon_e u_e$.

The key SMB design parameters, as determined with the above defined methodology, are given in Tables 2 and 3 for different values of the feed safety margin η_F. Two performance parameters are considered: φ_{55} is the productivity expressed in kg of total HFCS 55 produced per day and per system volume (both systems are designed to produce 100 kg/h of HFCS 55 DS). Υ_{55} is the volume of eluent used per kg of HFCS 55 DS produced.

Table 2: Influence of η_F on the design parameters of the conventional 1-2-2-1 SMB operated at 6 bar total pressure drop. Production of 100 kg HFCS 55 DS /h (without bypass).

η_F	1.0	0.9	0.8	0.7	0.55	0.4
Q_F (l.h^{-1})	182.5	182.5	182.5	182.5	182.5	182.5
Q_E (l.h^{-1})	66.9	89.7	119.4	159.3	251.0	421.8
$\bar{Q}$ (l.h^{-1})	304.2	338.0	380.2	434.5	553.0	760.4
Q_{mean}^{SMB} (l.h^{-1})	310	348	396	458	594	832
L_{col} (m)	NA	2.8	1.7	1.4	1.1	0.9
u_e (m.h^{-1})	NA	8.9	14.8	18.3	22.7	26.7
V_{col} (l)	NA	275.9	113.3	85.6	71.8	73.3
φ_{55} (kg/l/day)	NA	8.7	21.2	28.0	33.4	32.8
Υ_{55} (l/kg)	0.7	0.9	1.2	1.6	2.5	4.2

Table 3: Influence of η_F on the characteristic design parameters of the bypass 1-2-2-1 SMB operated at 6 bar total pressure drop. Production of 100 kg HFCS 55 DS /h (after bypass).

η_F	1.0	0.9	0.8	0.7	0.55	0.4
Q_F (l.h^{-1})	80	80	80	80	80	80
Q_E (l.h^{-1})	65.2	87.4	116.3	155.2	244.5	410.9
$\bar{Q}$ (l.h^{-1})	296.3	329.2	370.4	423.3	538.7	740.7
Q_{mean}^{SMB} (l.h^{-1})	329	366	413	473	605	838
L_{col} (m)	NA	2.9	2.0	1.5	1.3	1.1
u_e (m.h^{-1})	NA	8.6	12.4	16.4	18.8	21.8
V_{col} (l)	NA	306.0	167.9	109.2	106.5	110.6
φ_{55} (kg/l/day)	NA	7.8	14.3	22.0	22.5	21.7
Υ_{55} (l/kg)	0.7	0.9	1.2	1.6	2.4	4.1

The information given in Tables 2 and 3 contains very interesting results:

- Whatever the process, the productivity reaches a maximum with respect to η_F. This is a consequence of a compromise to be found between injecting as close as possible to the maximum level permitted by thermodynamics (represented by m_i^{eq}) and thus to design a system associated with a very significant number of plates, or to take safety margins with respect to the level permitted by thermodynamics and to enjoy freedom to operate with less efficient systems.

- Irrespective of the process (conventional or bypass), the maximum of productivity is obtained for η_F values of about 0.5.
- The eluent consumption increases very significantly with η_F. That could be a killer for sugar separations.
- The optimum value of η_F can only be found by optimizing an economical function: productivity is maximized for $\eta_F \approx 0.5$ but a minimum eluent consumption requires $\eta_F \approx 1.0$.
- For a given η_F, the performances of the two processes are remarkably close: both systems have almost identical sizes for producing 100 kg HFCS 55 DS per hour, and this is obtained with similar eluent consumption.

Finally, based on this analysis, choosing between the "conventional" or the "bypass" option is not evident. Both systems deliver close productivity and eluent consumption. A significant advantage of the bypass process is related to its greater flexibility: when, for instance, feed composition varies, the quality of the end product (HFCS 55) is simply adjusted by tuning a bypass ratio.

5 Conclusion

The production of HFCS 55 can be performed by two different SMB based chromatographic operations, one producing directly this quality, the other one producing an overpurified intermediate HFCS 90 which is later blended by use of a bypass. The performances of the two chromatographic systems have been compared using the simulation software Chromworks. By means of a relatively simple approach, a rationale methodology for comparing the different processes has been proposed. Semi-quantitatively, the two different processes, "conventional" or "bypass" lead to pretty similar productivities and eluent consumption. The final design of the SMB process must depend on the selected compromise between productivity and eluent consumption (including isomerization) as well as on other factors like the ability to cope with feed variability.

6 References

[1] Nicoud, R.M., "The Amazing Ability of Continuous Chromatography to Adapt to a Moving Environment", Ind. Eng. Chem. Res., 53, 3755−3765, 2014

[2] "Deashing and Mixed Bed Polishing of High Fructose Corn Syrup with DOWEX Ion Exchange Resins", Dow Chemical

[3] "Chromatographic separation of fructose and glucose with DOWEX MONOSPHERE ion exchange resins", Dow Chemical

[4] Siitonen J., Sainio T., Rajendran A., "Bypass chromatography – design and analysis of an improved strategy for operating batch chromatography processes," *Journal of Chromatography A*, 1230, 77-92, 2012

[5] Nicoud R.M., "*Chromatographic Processes, Modeling, Simulation and Design*", Cambridge University Press, Cambridge, 2015

[6] Seidel-Morgenstern A., Schmidt-Traub H., Michel M., Epping A., Jupke A., "Preparative chromatography", Schmidt-Traub H., Schulte M., Seidel-Morgenstern A. (Eds), 2nd Edition, Wiley-VCH, Weinheim, Germany, 2012

[7] Chromworks™, A software for chromatographic separations, Ypso-Facto, Nancy, France

445429 Air-Cooled Heat Exchanger Reliability

Abstract:

Air-Cooled Heat Exchangers (ACHE's), sometimes called Air Coolers or Fin-Fans®, are used in a variety of applications in refineries, petrochemical plants, gas treating plants, compressor stations, power plants, and other industrial applications. Many times these processes are critical to the operation of the plant and an outage to the ACHE's or operation at less than design duty will cause the plant to shut down or operate at reduced capacity. The impact on the plant can be severe and may result in significant loss of revenue to the operator. This paper will describe the operation of ACHE's and identify critical components that could fail and result in reduced cooling capacity. We will also identify maintenance practices and procedures that would either reduce the probability of component failure or reduced heat exchange capability.

16AIChE
Spring Meeting, Houston, TX

Think beyond efficiency gain when adding an air-preheater to a process fired heater

Sultan Ahamad
Bechtel Corporation
Houston, TX, USA
sahamad@bechtel.com

Rimon Vallavanatt
Bechtel Corporation
Houston, TX, USA
rvallava@bechtel.com

Prepared for Presentation at
American Institute of Chemical Engineers
2016 Spring Meeting
11th Global Congress on Process Safety
Houston, Texas
April 10-14, 2016

UNPUBLISHED

Think beyond efficiency gain when adding an air-preheater to a process fired heater

Sultan Ahamad
Bechtel Corporation
Houston, TX, USA
sahamad@bechtel.com

Rimon Vallavanatt
Bechtel Corporation
Houston, TX, USA
rvallava@bechtel.com

Keywords: Energy Efficiency, Process Fired heater, Air-Preheater, Run length

Abstract

An air-preheater (APH) augmented to a fired heater increases the efficiency, thereby reducing the fuel consumption. The fuel saving is primarily due to recycling of low quality heat (from lower temperature flue gas) to the radiant section by means of an APH. The recycled heat to the radiant section results in an increase of radiant heat flux and higher NOx emission. The increased radiant heat flux reduces fired heater run length, which must be evaluated for new fired heaters as well as fired heater revamps.

There are several critical parameters (e.g. radiant heat flux, heater run length, NOx emission, APH size, convection section size, preheated air temperature and adiabatic flame temperature) which should be considered and evaluated for a well-designed fired heater. Evaluation of every fired heater design parameter is critical. Any oversight may adversely affect fired heater performance resulting in decreased fired heater capacity, run length, efficiency etc.

An integrated analysis of an air-preheater and fired heater system is essential for a successful operation of fired heater with an air-preheat system. An effort has been made in this paper to summarize and highlight the major parameters related to APH size and duty which will affect the overall fired heater performance. The recommendation and discussion in the paper shall also be emphasized using a case study.

1. Introduction

Installing an air-preheater (APH) for a fired heater is a common practice to increase the efficiency of fired heater, thereby reducing the fuel consumption. The fuel saving is primarily due to an increase in sensible heat of combustion air which recycles the heat from low temperature flue gas to the radiant section of fired heater. Fuel saving also comes from improvement in overall combustion efficiency. The extent of fuel saving depends on many factors including flue gas temperature, excess air and fuel composition.

In many cases of new fired heater installation as well as revamp, the energy efficiency improvement scheme has been looked at in isolation rather than analyzing the entire fired heater system performance. In a typical fired heater, air-preheaters are generally designed for about 15-20% additional capacity over and above the fired heater design condition. The designer or operators generally tend to use this additional capacity to compensate for any degradation in the convection section or during fired heater revamp for capacity increase. Also, during revamp of a fired heater, additional surface area or a larger APH is installed. These options of recovering additional heat in APH is relatively easy revamp exercise and a lucrative addition to a fired heater. However, the addition of new APH to a fired heater or size increase of an APH may adversely affect fired heater performance especially fired heater run length.

Air-preheater is recovers the residual low quality heat from cold flue gas and transfer it to the combustion air. The heat from the air-preheater is recycled in the radiant section, which results in an increase of radiant heat flux and higher NOx emission. This increase in radiant heat flux translates into reduction in fired heater run length. The reduction in fired heater run length must to be carefully evaluated for new fired heaters as well as during revamp. Run length becomes even more important for critical services like coker furnace etc.

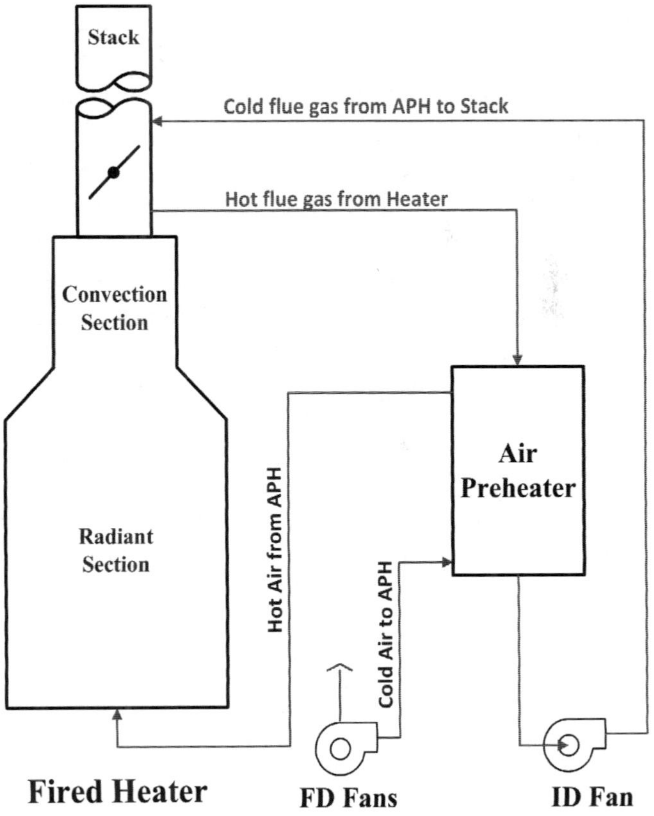

Figure-1 : Fired heater with an APH System

2. Air preheating system

A typical fired heater air-preheating system largely consists of air-preheater, induced draft fan (ID fan), forced draft fans (FD fans), ducting and dampers. The flue gas from the fired heater stack is diverted to an air-preheater where it exchanges heat with combustion air. The cold flue gas from the air-preheater is sent back to the stack through an ID fan. FD fans supply combustion air to the air-preheater. The air is heated in the air-preheater and hot air is supplied to fired heater burners. Figure-1 shows a simple schematic of a fired heater with an air pre-heating system.

The primary purpose for using an air-preheater in a process fired heater is to increase the overall thermal efficiency thereby saving on fuel consumption in a process plant. The potential for energy savings increases with increase in fired heater size. A typical combustion air-preheater will increase the fired heater efficiency by approximately 10%. In most of the cases, the pay-back period on installation of an air pre-heater is less than 3 years.

Fuel generally contains H_2S or Sulfur which converts into SO_2/SO_3. The air-preheater's heat transfer surface is subject to cold end corrosion attack caused by condensation of Sulfur Tri-Oxide (SO_3) which results in air-preheater leakage. The amount of Sulfur in fuel limits the temperature up to which flue gas can be cooled in an air pre-heater.

3. Fired heater efficiency and air preheating system

The heat available from combustion of fuel for heating is equal to flue gas enthalpy change from adiabatic flame temperature and stack temperature. The target flue gas stack temperature shall be determined based on fuel H_2S or Sulfur contents (flue gas sulfur dew point temperature). If the flue gas stack exit temperature is higher than the target flue gas stack temperature, then the heat available in flue gas can be used for efficiency improvements of the fired heater. The integration of fired heater with air-preheater determines the maximum recoverable heat from flue gas.

The net efficiency of a fired heater is equal to total heat absorbed divided by total heat input. The heat absorbed is equal to total heat input minus the total losses. The net thermal heater efficiency can be calculated by following:

$$\eta = \frac{Total\ Heat\ Input - (Stack + Setting)\ Losses}{Total\ Heat\ Input} x100 \quad (1)$$

$$\eta = \frac{LHV + H_a + H_f + H_m - H_s - H_L}{LHV + H_a + H_f + H_m} x100 \quad (2)$$

Where:

η = Net thermal efficiency, %
LHV = Lower heating value of the fuel, Btu/lb of fuel
Ha = Sensible heat of air, Btu/lb of fuel
Hf = Sensible heat of fuel, Btu/lb of fuel
Hm = Sensible heat of atomizing media, Btu/lb of fuel
Hs = Stack heat losses, Btu/lb of fuel
H_L = Setting loss, Btu/lb of fuel

Figure-2 provides an estimated efficiency for a case study. It can be noted from this figure that degree of combustion air preheat does not affect the fuel consumption (i.e. fired heater efficiency). It is the flue gas stack temperature which determines overall fired heater efficiency. Whether flue gas is cooled in the convection section or air-preheater, is irrelevant for fired heater efficiency but the same is very important for overall fired heater health.

For optimum performance of a fired heater, the proper air-preheater size and duty selection along with proper fired heater design is very important. Every fired heater design parameter is critical for optimum performance. Any oversight may adversely affect fired heater performance resulting in diminishing of fired heater capacity, run length or efficiency. There are several critical parameters, some of these are listed below, which should be considered and evaluated for a well-designed fired heater with air-preheater for a successful run length:

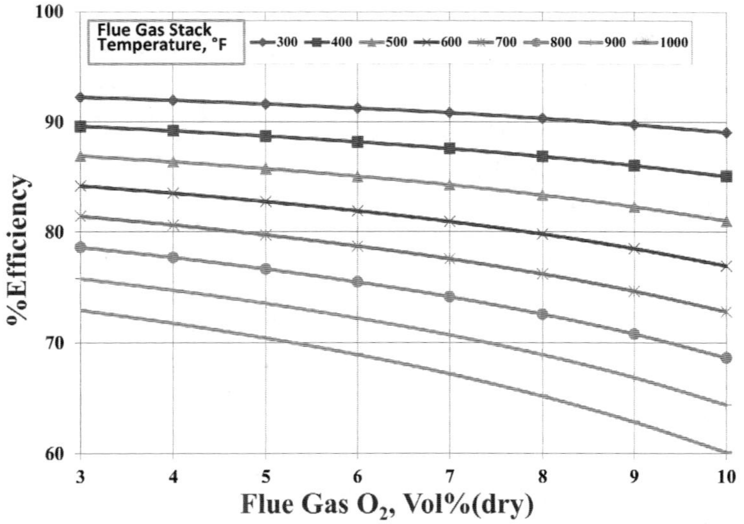

Figure-2 : Fired heater Efficiency

- Evaluate the effect of air-preheater size and duty on radiant heat flux and hence on fired heater run length
- Evaluate the effect of convection section size on heater run length
- Evaluate the effect of preheated air temperature on adiabatic flame temperature and consequently on fired heater run length

4. Effect of air-preheater size / duty on radiant heat flux and run length

Average radiant heat flux is the amount of heat absorbed per unit heat transfer area in the radiant section. There are recommended average radiant heat flux limits for fired heaters based upon process service and fired heater configuration. Average heat flux is a single point indicator of operating risk in a process fired heater. The chances of problems occurring in a fired heater increases with an increase in radiant heat flux. An air-preheater recovering the residual heat from cold flue gas is essentially recycling this heat into the radiant section. This recycled heat results into an increase of radiant heat flux.

Figure-3 shows a summary of case study indicating a proportional increase in average radiant flux with an increase in air-preheater absorbed duty. It should be noted that overall efficiency of the fired heater with air-preheater system is kept the same for this case study. As the air-preheater size increases, the convection section size decreases, so that the overall fired heater efficiency is maintained the same for all data point shown in this figure.

Figure-4 shows a summary of same case study indicating an increase in relative coking rate with increase in air-preheater absorbed duty.

Based on the data from the case study and as shown in these figures, it can be concluded that it is important to reduce the air-preheater size by increasing the heat transfer area in the convection section. The goal is to cool down the flue gas to the extent possible in the convection section. A larger convection section reduces the air-preheater duty thereby resulting in lower radiant heat flux and hence longer fired heater run length.

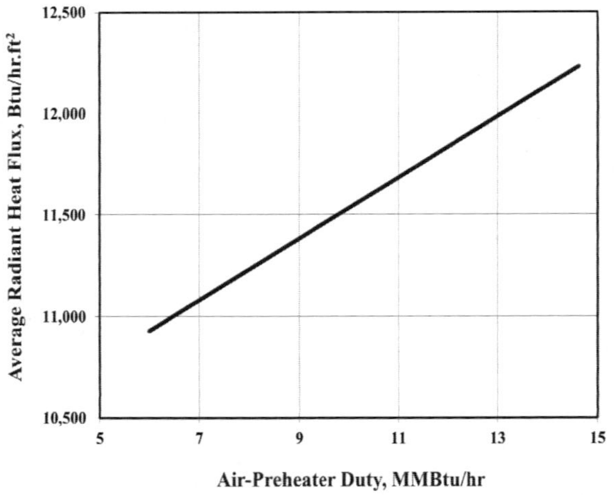

Figure-3 : Effect of APH Size on Radiant Heat Flux

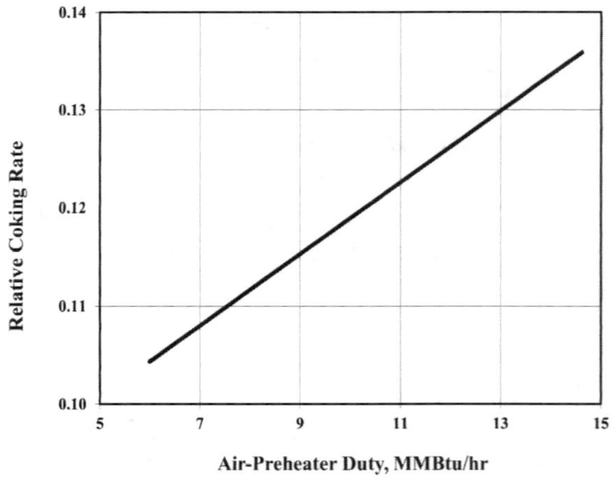

Figure-4 : Effect of APH Size on Coking Rate

5. Degree of air-preheat and adiabatic flame temperature

The adiabatic flame temperature (AFT) is defined as the temperature of combustion products that would result with no heat loss. Any radiative, convective, or conductive heat loss, which will result in lower combustion product temperature, is not included while estimating adiabatic flame temperature.

In a fired heater, heat is provided by combustion energy of fuel and sensible heat of fuel and air. The combustion air is available at ambient temperature and is preheated to air-preheat temperature in an air-preheater. The combustion of fuel in a fired heater results in a flue gas at a flame temperature. Therefore, as combustion air temperature increases, adiabatic flame temperature also increases with it. Figure-5 shows a summary of the effect of combustion air temperature on adiabatic flame temperature for a case study with a typical natural gas with 15% excess air.

The heat transfer calculation in the radiant section is calculated with the widely used Lobo-Evans method. The equation for radiant heat transfer is as follows:

$$\frac{q_r}{\propto A_{cp} F} = 0.173\left[\left(\frac{T_g}{100}\right)^4 - \left(\frac{T_s}{100}\right)^4\right] + 7(T_g - T_s) \qquad (3)$$

Where:

q_r = *Total heat absorbed in radiant section, Btu/hr*

T_g = *Average radiating flue gas temperature in radiant section, °R*

T_s = *Cold surface (tube metal) temperature, °R*

αA_{cp} = *Equivalent cold plane area, ft²*
F = *Overall exchange factor*

The amount of heat transfer in the radiant section shall be equal to the heat provided by flue gases in dropping the temperature from the adiabatic flame temperature to the flue gas temperature leaving radiant section (also known as bridgewall temperature). For the simplification of this analysis, the difference between average radiating flue gas temperature in the radiant section and bridgewall temperature is ignored.

$$q_r = m_{fg} * c_p * (T_f - T_g) \qquad (4)$$

Where:
m_{fg} = *Total flue gas flow rate in radiant section, lb/hr*
T_f = *Adiabatic flame temperature, °R*
c_p = *Specific heat of combustion gases at average temperature, Btu/lb/°F.*

From the above two equations (3 & 4), the following equations can be derived.

$$\frac{m_{fg}c_p(T_f - T_g)}{\alpha A_{cp}F} = 0.173\left[\left(\frac{T_g}{100}\right)^4 - \left(\frac{T_s}{100}\right)^4\right] + 7(T_g - T_s) \qquad (5)$$

$$Radiant\ heat\ flux = \frac{q_r}{A} = \frac{m_{fg}c_p(T_f - T_g)}{A} \qquad (6)$$

It is very clear from the above equations that the increase in adiabatic flame temperature will increase the radiant heat flux as well as bridgewall temperature, both adversely affecting the performance and run length of the heater. Figure-6 shows a summary of case study showing effect of increase in adiabatic flame temperature on radiant heat flux and bridgewall temperature.

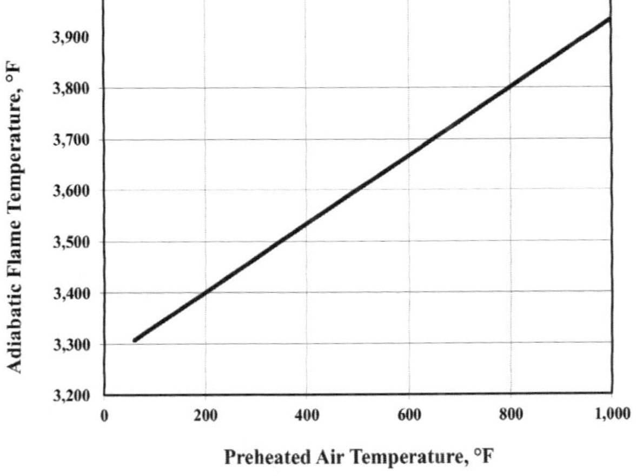

Figure-5 : Effect of air temperature on Adiabatic Flame Temperature for a typical Natural Gas with 15% Excess Air

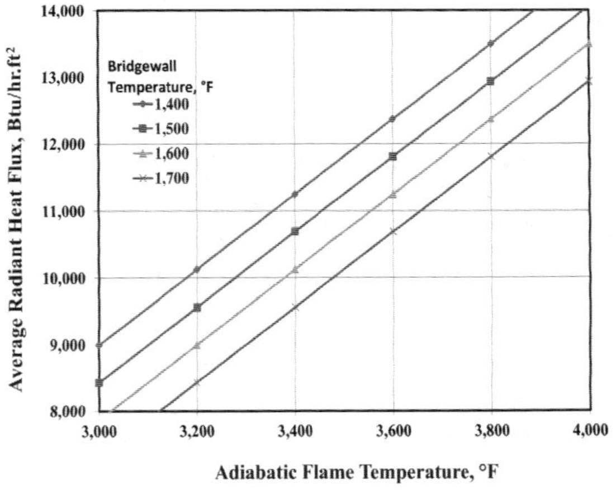

Figure-6 : Case Study - Effect of Adiabatic Flame Temperature on Radiant Heat flux

6. Convection section size

The size of convection section plays a major role in fired heater performance. Generally it is expensive to provide additional heat transfer area in convection section compared to recovering the same amount of heat in an air-preheater. It is even comparatively more expensive to recover heat in the convection section, if fired heater is using higher grade metallurgy (e.g. stainless steel tubes). It is primarily due to cost concerns that a fired heater is designed with higher flue gas temperature approach (defined as flue gas temperature leaving convection minus process inlet temperature). A higher temperature approach will result in a larger air-preheater absorbed duty for the same fired heater efficiency. It is recommended to keep the flue gas temperature approach between 50-100°F.

The convection section absorbs approximately 30-40% of the total fired heater absorbed heat duty in a typical fired heater. In many cases during revamp, designers try to use additional available capacity in the air-preheater or install additional surface area in an air-preheater. Also, as the convection section surface degrades with time, the heat absorbed in the convection section reduces, resulting in higher air-preheater duty. However, the increase in air-preheater duty should be avoided as it will increase the radiant heat flux. To avoid increase in air-preheater duty during fired heater revamp, the convection section heat transfer surface can be augmented with additional surface. Addition of convection section heat transfer surface will increase the convection section absorbed duty resulting in lower duty of the air-preheater. A smaller air-preheater duty will recycle less heat into radiant firebox which in turn will keep the radiant heat flux lower.

Convection section heat transfer area can be increased by several ways during revamp. Some of the most frequently used revamp options are as follows:

- Installation of additional tube rows (e.g. use of future row and /or installation of additional rows by modifying the convection breaching section)
- Replacing bare tubes with extended-surface tubes, if available.
- Increase extension ratio of extended surface i.e. increase extended-surface area per unit length of tube by changing the configuration of extended surface.
- Increasing number of tubes per rows by increasing convection section width. This is an expensive but very effective revamp option.

There are several long term advantages of using a bigger convection and smaller air-preheater. Some of these advantages are:

- Lower radiant heat flux resulting into less coking and longer run length
- Smaller hot flue gas and hot air ducting.
- Smaller fan sizes – lower fan operating cost.

7. Case Study

In this case study, two designs of a fired heater are compared side by side. Design-1 is with a smaller convection section and design-2 is with a larger convection section. Both these designs have the same absorbed heat duty and overall efficiency. To keep the same overall efficiency, design-1 (with smaller convection section) requires a larger air pre-heater. Similarly, design-2 requires a smaller air pre-heater. The key performance parameters are listed in Table-1. Following are some of the key advantages of design-2 over design-1 presented in this case study:

- Lower radiant heat flux: As design-2 has a relatively lower radiant heat flux, this will result in lower coking rate of process fluid inside the tube therefore resulting in longer fired heater run length
- Smaller hot flue gas and hot air ducting: Since the flue gas temperature leaving the convection section is lower in this design, therefore, it will have a small flue gas duct cross-sectional area. Similarly, this design will have a lower hot air temperature leaving air-preheater resulting in small cross-sectional area for hot air duct.
- Smaller fan sizes: With lower hot flue gas and hot air temperature along with smaller air-preheater, this design will have small fan. This will save not only on installation cost, but also on fan operating cost.
- Lower NOx: Fired heater NOx emission has a direct relationship with hot air temperature. As combustion air temperature is lower for design-2, it will result in lower NOx emission.

As it can be seen from table-1, there are many advantages of using a larger convection design. In summary, it will result in lower radiant heat flux, longer run length, lower NOx emission, and lower duct sizes etc.

Table-1 : Case Study Summary

Parameter	**Design-1: Smaller convection**	**Design-2: Larger convection**
Absorbed duty, MMBtu/hr	160	160
Efficiency, %	90.7	90.7
Air-preheater duty	26.7	11.6
Flue gas temperature leaving radiant section, °F	1,570	1,518
Flue gas temperature leaving convection section, °F	768	519
Average radiant heat flux, Btu/hr.ft^2	10,965	10,090
Hot air temperature to heater, °F	607	306
Estimated NOx reduction	-	83%
Estimated increase in run length	-	13%
Hot flue gas duct cross-section area, ft^2	37.1	29.6
Hot air duct cross-section area, ft^2	29.3	21.1

8. Conclusion

An addition of air-preheater to a fired heater always helps in increasing the fired heater efficiency thereby reducing plant fuel consumption. However, preheated air temperature does not affect the fuel consumption (i.e. fired heater efficiency). It is the flue gas stack temperature which determines overall fired heater efficiency. Whether flue gas is cooled in the convection section or an air-preheater, is irrelevant for fired heater efficiency but the same is very important for overall fired heater health. An air-preheater is essentially recovering low quality heat from cold flue gas and recycling it to the radiant section, which is the most severe operating zone of the heater. If the entire system is not carefully designed, it can adversely affect the overall performance of the heater. The fired heater critical parameters such as radiant heat flux, NOx emission and run length reduction must be carefully evaluated for new heater designs as well as revamps. An integrated analysis of air-preheater and fired heater system is essential for a successful operation of fired heater with an air-preheat system.

9. Biography

Sultan Ahamad is a Sr. Fired Heater Specialist at Bechtel Corp. in Houston, Texas. He has more than 18 years of experience in the design, engineering and troubleshooting of fired heaters and combustion systems for the refining, petrochemical and LNG industries. He graduated from the Indian Institute of Technology in Roorkee, India, with a degree in chemical engineering. He is a member of API Sub-Committee on Heat Transfer Equipment. He has published and presented several papers on fired heaters and related subjects.

Rimon Vallavanatt is Sr. Principal Engineer at Bechtel Corp. in Houston, Texas. He has more than 40 years of experience in the design, engineering and troubleshooting of fired heaters, thermal oxidizers, boilers and flares. He graduated from the University of Kerala in India with a degree in mechanical engineering. He also received a degree in industrial engineering from St. Mary's University in San Antonio, Texas. Mr. Vallavanatt is a registered professional engineer in the state of Texas, and he has served on the American Petroleum Institute's subcommittee on heat transfer equipment for the past 30 years.

KOBELCO

2016 AIChE Spring Meeting & 12th Global Congress on Process Safety
Innovation in-and Next Gen.- Heat Transfer Technologies I

(77c)Design Method and Application for Diffusion Bonded Compact Heat Exchanger (DCHE)

Koji Noishiki, Dr
8:50-9:25AM
April 12th, 2016
KOBE STEEL, LTD

1

KOBELCO

Contents

1) Background Technology of DCHE
2) DCHE construction
3) Mechanical issue
4) Performance issue
5) Application --- Hydrogen Re-fueling Station, Offshore use
6) Conclusion

DCHE is our trade name and is same as PCHE (Printed Circuit Heat Exchanger)

2

KOBELCO

Main Product in Takasago Equipment Plant

3

KOBELCO

Market trend of heat exchanger

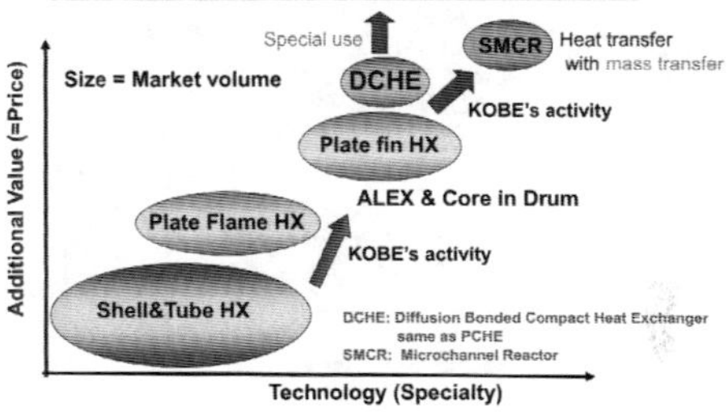

4

KOBELCO

Diffusion bonded Compact Heat Exchanger (DCHE)

5

KOBELCO

Kobe Steel, Ltd Background to technology

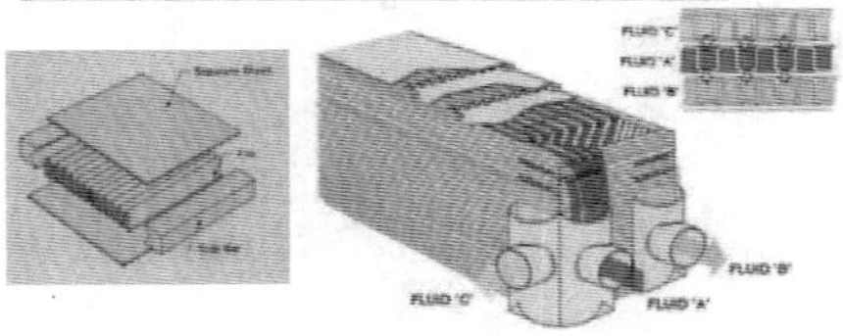

Over 50 years experience as major manufacturer of compact heat exchanger (ALEX=Brazed Aluminum Plate-fin type).

Extensive range of technologies to design and fabricate heat exchangers of this type.

Has now developed the micro or milli channel heat exchanger (DCHE) and reactor (SMCR) based on this technology.

6

KOBELCO

APPLICABLE RANGE of ALEX

Brazed Aluminum Plate-fin Heat Exchanger is applicable to the following condition ....
- Design Temperature : **Lower than 200°C**
- Design Pressure : **Lower than 130BarG**
- Fluid Composition : **Non-Corrosive against AL No plugging material**

The applicable range (for ASME code) ....

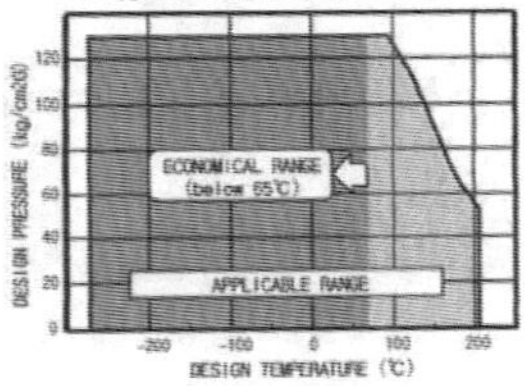

7

KOBELCO

Recommendation at Operation for BAHX -ALPEMA-

Startup/shutdown rate
<2 deg C / minutes

Temperature difference between streams
Single Phase: <50 deg C
Two Phase: <30 deg C

Many user would like to use Compact Heat exchanger for more critical condition

8

KOBELCO

Features & Application of DCHE

<Features>

a) Diffusion bonding
DCHE can withstand high pressure (over 95MPaG)

b) Micro-channel equipment
High thermal performance (approximately 1/10th of Shell &Tube type)

c) Stacked type
DCHE can handle multi-stream in one block

<Application>

a) Gas cooler for offshore use

b) Special use for high pressure, high temperature or large temperature difference

9

KOBELCO

Fabrication flow of DCHE

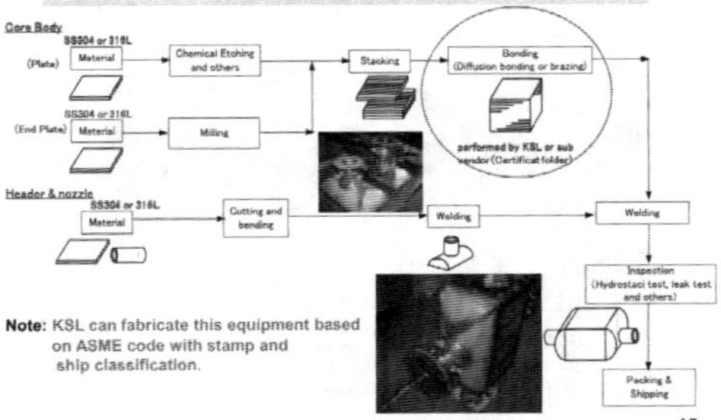

Note: KSL can fabricate this equipment based on ASME code with stamp and ship classification.

10

KOBELCO

Bonding method

Brazing
(Aluminum and others)

Diffusion bonding
(Stainless Steel)

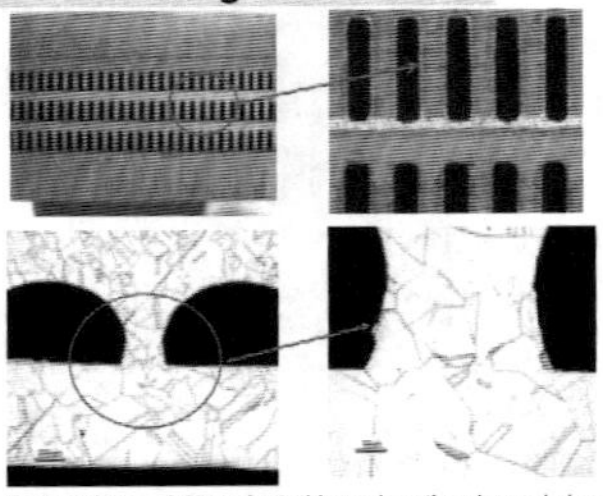

Bonding method and material is selected based on the channel size, corrosion resistance, required strength and other relevant factors.

11

KOBELCO

How to handle multi-stream at one block?

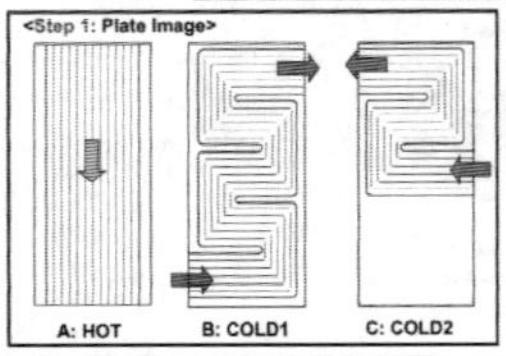

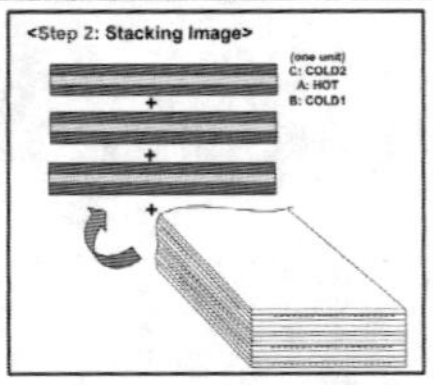

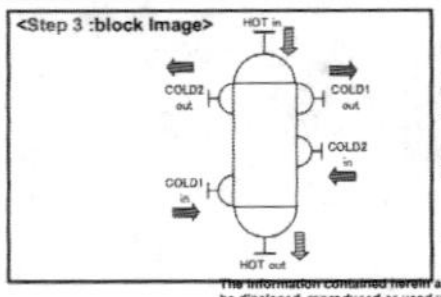

Step1:
Channel arrange shall be decided based on inlet and outlet temperature.
Step2:
Required plates number are stacked based on heat duty and pressure drop.
Step3:
Header & nozzle are attached to core body

⇒DCHE can handle multi-stream at one block.

12

KOBELCO

How to fabricate large size DCHE?

<Fabrication step>

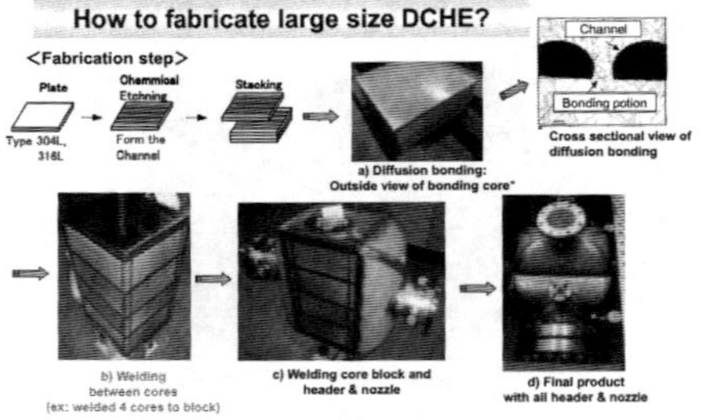

Cross sectional view of diffusion bonding

a) Diffusion bonding: Outside view of bonding core"

b) Welding between cores (ex: welded 4 cores to block)

c) Welding core block and header & nozzle

d) Final product with all header & nozzle

Note: One bonding core size is limited from the etching plate size or bonding height. However, the more larger size core is fabricated by using welded cores (Step b)).

13

KOBELCO

Mechanical issue

KOBELCO

New Requirements in ASME for diffusion bonding

There were the special requirement for Diffusion bonding in Code Case(2437-1, 2621-1). But these requirements are limited to the special application.

Based on the customers requests,
the general requirements are specified in ASME 2011a **as follows,**
ASME Sec. VIII Div.1 "Diffusion bonding" at Mandatory Appendix-42
Sec. IX "Diffusion Welding" in QW-185

Requirement for production tests (only key issues):
aa) Test block shall be a minimum of a 200mm x200mmx 50 interface planes
bb) free from cracks and shall show no incomplete bond or porosity on or adjacent to the bond lines
cc) After the bonding, mechanical properties (TS,YS Hardness) shall match the ASME requirements (same as base metal).

Further more, M.A.W.P. (Maximum Allowable Working Pressure) for the device will be decided by the results of the bursting test in UG-101.

15

KOBELCO

Bursting test Results (sample)

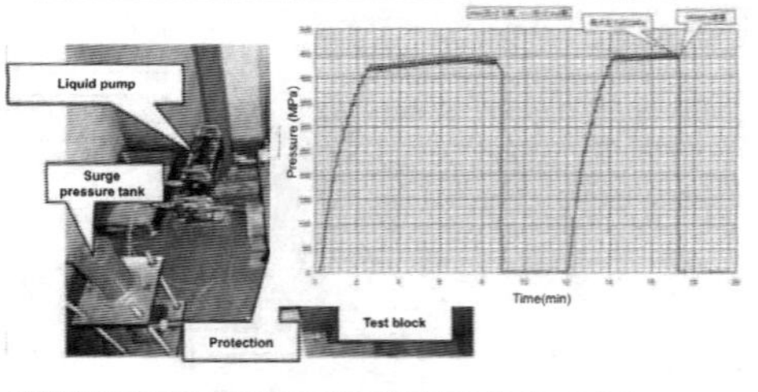

MAWP (Maximum Working Pressure) will be decided based on bursting test (ASME UG-101).

16

KOBELCO

Demo heat exchanger based on ASME code

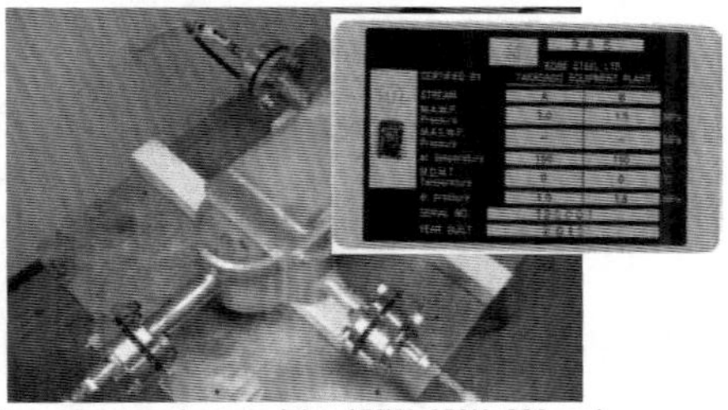

Demo heat exchanger (size 150Wx150Hx300mm)

The information contained herein are the property of Kobe Steel, Ltd. They shall not be disclosed ,reproduced or used without the prior written consent of Kobe Steel.

17

KOBELCO

Performance issue

The information contained herein are the property of Kobe Steel, Ltd. They shall not be disclosed ,reproduced or used without the prior written consent of Kobe Steel.

KOBELCO

Comparison between ALEX and DCHE

All design parameters for ALEX and DCHE are basically the same.

Hence heat transfer and pressure drop performance will be calculated using the same design tool as for ALEX.

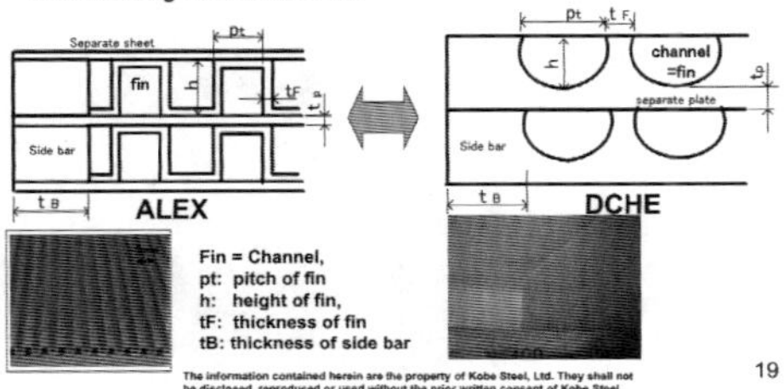

The information contained herein are the property of Kobe Steel, Ltd. They shall not be disclosed ,reproduced or used without the prior written consent of Kobe Steel.

19

KOBELCO

DCHE performance check
(After cooler for Compressor)

Shell&Tube
Φ650 × 4200TL-TL (5500kg)

DCHE
330X350X650 (700kg)

Design Pressure: 8MPaG (Hydrostatic test: 12MPaG)
Code: Japanese High Pressure Gas Safety Law (with stamp)

Kobe Steel tests confirm the mechanical and thermal performance of this DCHE to be superior to S&T.
*Performance Design tool: Kobe Steel software, Rating calculation: MUSE
(DCHE:S&T) Volume: 1:10, Weight: 1:8

The information contained herein are the property of Kobe Steel, Ltd. They shall not be disclosed ,reproduced or used without the prior written consent of Kobe Steel.

20

KOBELCO

For hydrogen station

The information contained herein are the property of Kobe Steel, Ltd. They shall not be disclosed ,reproduced or used without the prior written consent of Kobe Steel.

KOBELCO

Application for hydrogen station

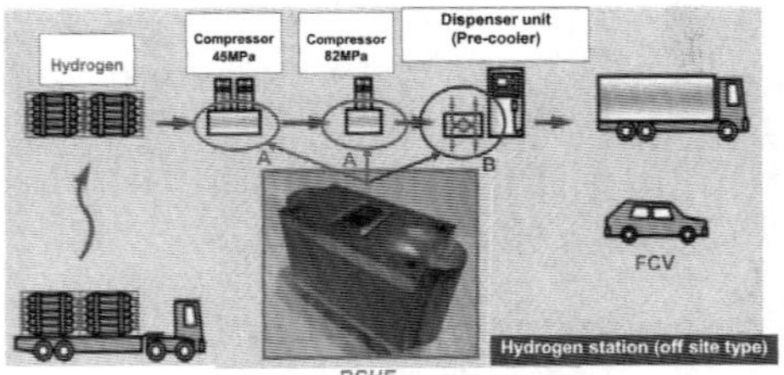

DCHE

A: cooler for compressor
B: pre-cooler for dispenser

The information contained herein are the property of Kobe Steel, Ltd. They shall not be disclosed ,reproduced or used without the prior written consent of Kobe Steel.

22

KOBELCO

Fatigue test

The following fatigue test is applied to DCHE.
There is no indication of leakage or crack after this test.

<Test1: for aftercooler>
Medium: Water
Temp: 15 degC
Pressure rage: 87.5MPa⇔1MPa
Cyclic times: 1,000,000

<Test2: for pre-cooler>
Medium: Hydrogen
Temp: - 40 degC
Pressure rage: 92MPa⇔0.6MPa
Cyclic times: 70,000

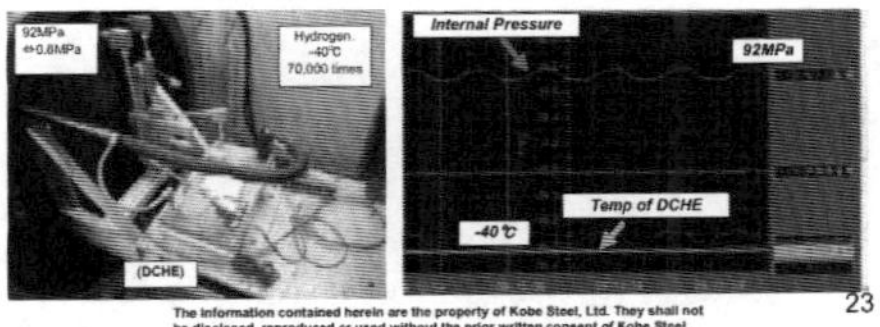

The information contained herein are the property of Kobe Steel, Ltd. They shall not be disclosed ,reproduced or used without the prior written consent of Kobe Steel.

23

KOBELCO

Example for aftercooler

CONFIDENTIAL

• Double tube type
• DCHE

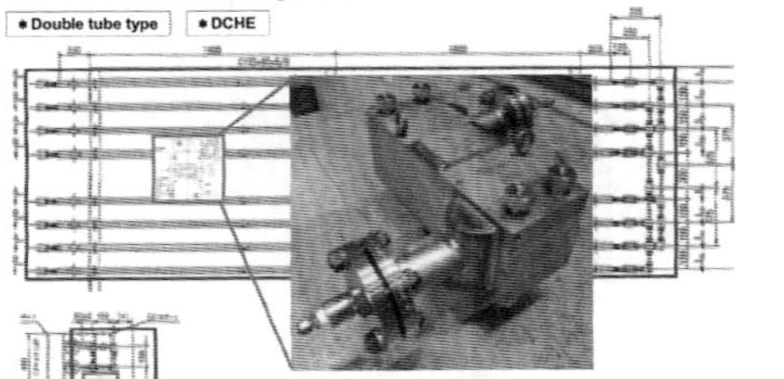

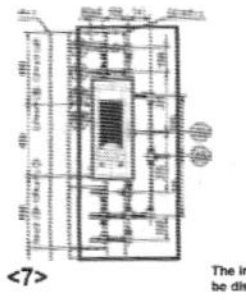

Feature of DCHE
* Compact(approximately 1/30～1/100)
* Reduce the number of connection
(high reliability and low cost)

<7>

The information contained herein are the property of Kobe Steel, Ltd. They shall not be disclosed ,reproduced or used without the prior written consent of Kobe Steel.

24

KOBELCO

Example for pre-cooler

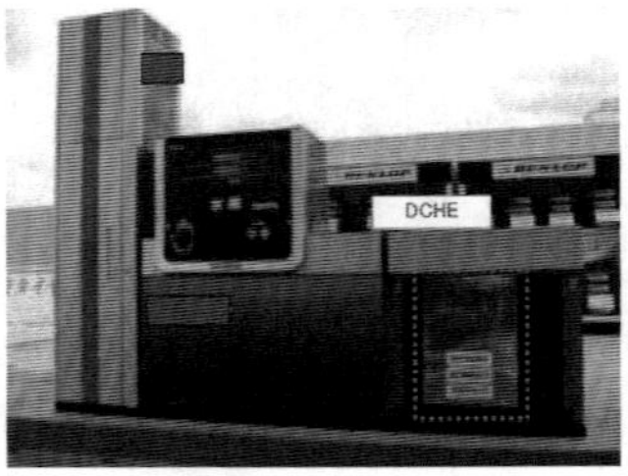

Feature of DCHE

* **Compact** (approximately 1/10)
* **Quick start** (minimize heat capacity of Heat Exchanger)

KOBELCO

For offshore use

KOBELCO

Offshore use (BOG Re-liquefaction)

Size: 580W × 1360H × 1200L(mm), 6ton
(Design pressure:32MPa, Design temperature:-196℃ ～+60℃)

※DNV ship classification＋ASME code with stamp
(Basically, design and inspection was based on ASME)

KOBELCO

Applications for DCHE

<Offshore>

* **Cooler for compressor**
* **Natural gas service (FLNG, FPSO, BOG use)**

<Onshore>

* **High temperature use (Heat recovery unit)**
* **Cyclic use (daily change or many shut down cases)**
* **High temperature difference between hot and cold stream**

KSL would like to find new markets for DCHE with customer.

Note: In general, aluminum brazed plate-fin heat exchanger is unsuitable for the above applications

442709 Propylene Glycol Plant Design

Abstract:

After ethylene, propylene (propene) is the most important raw material used in the production of organic chemicals. It is usually known as "the crown prince of petrochemicals". As an olefin, propylene is a reactive compound that can react with many common substances making it a very attractive feedstock. Its major use is in the production of polypropylene, a polymer widely used in the world. However, propylene is also used in many other applications such as acrylonitrile, propylene oxide, propylene glycol, and many others.

Propylene oxide is even more reactive that propylene. Propylene oxide is the feedstock for propylene glycols. Propylene glycol is used as an ingredient in products such as coolants, antifreeze, aircraft deicing fluids, hydraulic and brake fluids, heat transfer fluids, food, fragrances, and pharmaceuticals. It can also be used as a raw material for other chemicals such as unsaturated polyesters resins used in paints, varnishes, and solvents that are used in printing inks and urethanes.

The production capacity of propylene glycol in the world reached 1.6 million metric tons (3.5 billion pounds) in 2010. However, in 2013, the global production of propylene glycol was 8% per year and exceeded 2 million tons. Among the many manufacturers are Dow, Lyondell, Sinopec, Shell, BASF, Repsol, Huntsman, and others. Our goal is to design a plant that will produce 90,000 metric tons/yr. 78,300 metric tons/yr will be Monopropylene Glycol (MPG) and 11,700 metric tons/yr will be Dipropylene Glycol (DPG).

443830 Senior Design Project: Propylene Glycol Plant

Abstract:

The 2016 senior class at the University of Houston is designing a propylene glycol plant by non-catalytic hydration of propylene oxide at high temperature and pressure. This design has a capacity of 90,000 metric tons/year and achieves a specified production quantity, product selectivity, and product quality for mono-, di-, tripropylene glycol.

Alternative process flow diagram, heat and material balances, equipment sizing, and economic evaluation are presented. Computational tools including VBA and MATLAB routines are used for the flash calculations, reactor and column design equations. Moreover, commercial software packages including Aspen Plus and CHEMCAD are used for parametric and topological optimization.

Both isothermal and adiabatic operation is considered for a various reactor configurations. Multi-effect evaporators are employed to concentrate the glycol mixture and recycle the excess water. The glycols products are then separated to meet specifications via vacuum distillation. Ultimately, equipment performance, raw material and utility usage, optimum heat integration and HSE considerations are integrated to provide a technically feasible, profitable, and socially responsible design.

444063 Economic Optimization and Hazard Analysis of a Hydrofluoric Acid Catalyzed Alkylation Process

Abstract:

A hydrofluoric acid catalyzed alkylation process was modeled using four series/parallel reactions starting with propylene and isobutane. A target of 5,000 barrels per day of 95% pure 2,2-dimethylpentane was specified. The product was also defined to contain less than 1% of heavies (1-nonene and 2,2-dimethyl-octane). The system was modeled using the process simulator ChemCAD and priced using cost equations from Turton, Bailie, Whiting and Shaeiwitz (2012). Both topological and parametric optimizations were considered. The recycle streams increased the complexity of the simulator significantly, but were required to accurately model the concentration build-up in the system. The impact of reactor residence time, separator recycle purity, and feed ratio on the economics will be presented. Process hazards will be identified and suggestions to mitigate hazards will be presented.

443536 Process Development for High Octane Number Gasoline Production from Lignin

Abstract:

Biofuels production from lignocellulosic biomass will reduce the nation's reliance on imported fossil fuels while yielding economic benefits. This project developed chemical methods for processing lignin through the steps of dissolution, depolymerization, deoxygenation, and separation.

Ionic liquids were used as solvent and lignin was catalytically depolymerized in a hydrogen environment generated in-situ from formic acid decomposition to prevent repolymerization of the fragmented molecules. In the hydrodeoxygenation step, both model compounds and lignin depolymerized oligomers were studied. A series of catalysts, including Pd/alumina, Ni/alumina, Ni-Mo/alumina, and some custom-prepared catalysts were tested for the catalytic hydrodeoxygenation activities. Two catalysts were found to be very reactive and contaminants tolerant for hydrogenation of the depolymerized lignin oligomers to high octane number aromatics.

444386 Comparison of Entrained-Flow and Fluidized-Bed Reactors for Production of Fast Pyrolysis Oils

Abstract:

The National Renewable Energy Laboratory (NREL) uses both entrained-flow and fluidized-bed reactors for the development and demonstration of biomass-derived fuel technologies. Fluid dynamic characteristics between the two types of reactors may result in differing heat transfer rates to the biomass particles, thereby changing the final pyrolysis oil quality when operated under similar conditions. The objective of this work was to understand the difference between the two types of reactors, so that product quality from the bench-scale systems can be reproduced at the pilot scale.

NREL has two bench-scale fluidized bed reactor systems and one entrained-flow pilot-scale reactor system from which pyrolysis data may be compared. A 2-inch diameter bench-scale fluidized bed reactor (0.5 kg/h feed rate) is used for feedstock selection and in- and ex-situ vapor-phase upgrading research. A larger bench-scale unit, attached to a Davison Circulating Riser used for catalyst development, employs a 2-inch diameter, fluidized bed reactor as the pyrolyzer (5 kg/h). Finally, a new, entrained-flow reactor for pyrolysis was installed in the existing Thermochemical Process Development Unit (TCPDU), a 0.5 ton/day (30 kg/h) biomass-to-fuels pilot plant used to demonstrate biomass conversion technologies at an industrially-relevant scale. The systems are similar in overall process design, control capability, pyrolysis oil collection, and analytical capabilities.

A simplified, 2D computational fluid dynamics (CFD) model was created to calculate heat transfer rates in the pilot-scale entrained-flow reactor and was validated using experimental data. The model shows that the entrained-flow reactor is capable of attaining the heating rates required for fast pyrolysis, despite the lack of heat transfer medium present in fluidized bed reactors.

Finally, data from the three reactor systems was used to compare the two reactor types at similar run conditions and similar feedstocks. Real-time analytical data included hot vapor analysis from Molecular Beam Mass Spectrometers (MBMS), and permanent gases were analyzed using online gas chromatography (micro-GCs) and non-dispersive infrared (NDIR) detectors. Raw pyrolysis oil was collected, and ultimate, proximate, ^{13}C NMR, water content, viscosity, carbonyl, and acid analyses were conducted on several samples. Oil and light gas yields and mass balances from the systems were also evaluated.

443062 Towards a RE(DE)Finery Model for Green Chemicals, Fuels and Energy

Abstract:

Towards a RE(DE)FINERY model for Green Chemicals, Fuels and Energy

Jan van Breugel (Corbion/BBI, NL), Erick Fernandes (Worldbank, USA), Gustavo Valença Paim (UNICAMP, BR), Adrie Straathof (TU Delft, NL) and Luuk van der Wielen[1] (BE-Basic and TU Delft, NL)

Biobased production of chemicals, fuels and energy –when done right- can make significant positive contributions to abating emissions and other climate impacts often associated to fossil alternatives, to food security and poverty, can provide (impact) investment alternatives, and overall facilitate sustainable development (SCOPE, 2015[2]). Key requirements are robust financial feasibility, also in case of unfavourable economic environments such as in case of (todays) low prices of fossil alternatives, and (near) complete utilisation of the feedstock to avoid wastes and emissions. With the first lignocellulosic biorefineries (for ethanol) coming on line, and technological maturity is emerging, it is high time to investigate how they blend in (or disrupt) a landscape of chemicals and fuels industry that is predominantly organised in port-industry clusters (Rotterdam, Singapore, Houston, Sjanghai, others) worldwide, that are connected to networks of downstream industries (packaging, coating, car, others).

The biobased opportunity requires a redefinition of the refinery concept – the REDEFINERY-, to the characteristics of biobased and other renewable feedstocks to allow as much as possible full utilisation of all bio-mass of the feedstock, and not only its energy content as is the case in current biofuels plants. This is particularly important for products that have no alternatives such as chemicals, materials and liquid fuels for aviation, marine and long-distance/heavy duty applications.

During the presentation, we will discuss

- overall yield and efficiency requirements, how specific technology/feedstock /product combinations contribute to these objectives and how new applications and technologies open new opportunities such as bioconcrete and other materials. Attractive and current examples will be presented, in the framework of a rough innovation roadmap .
- We also discuss how and which REDEFINERY structures also financially enable the robust large scale development of a biobased economy. This also deals with establishing specific (intermediate) platforms that can be converted efficiently into a wide range of chemical and fuel products.
- A last item is the role of logistics. Since biobased feedstocks need to be collected from large areas, the infrastructural requirements are fairly different from those of the fossil industry where extraction is in general from a single or a few wells, or from the agro-food/feed industry that has already some large scale food-oriented processing complexes operational.

All of this should be seen in the framework of the local opportunities that regional implementation provides. These are different in the Brazilian, and e.g. the European context, allowing specific and differential developments. Although this is work-in-progress, we plan to discuss some thoughts for further discussion.

---

[1] President of BE-Basic Foundation and Distinguished Professor for Biobased Economy, Delft University of Technology. Julianalaan 67, 2628 BC Delft, The Netherlands. E: L.A.M.vanderWielen@tudelft.nl

[2] Downloadable report (open access): http://bioenfapesp.org/scopebioenergy/index.php/chapters

442525 Deja Vu All over Again: Another Look at Evaporative Cooling

Abstract:

All chemical processing plants have one common feature: they possess cooling water towers. In fact, cooling water towers are ubiquitous in our 21st Century. Unfortunately, most chemical engineers have a weak understanding of cooling water tower operation --- through no fault of their own, but rather because cooling water technology is generally not taught in universities these days. There are a variety of reasons for this outcome, but the major reason is that labortory-sized and pilot plant-sized cooling water towers are not readily available.

However, we do not need a scale replica of a cooling water tower to understand its operation and to identify those features of its operation most relevant to technology transfer. To gain a thorough understanding of cooling water tower operation, we simply need to employ technology developed centuries ago; i.e., we can simulate cooling water tower operation using evaporative cooling from porous, clay pots.

This presentation discusses evaporative cooling in light of cooling water tower operation. We develop a mathematical model of evaporative cooling and discuss the experimental results that support our model. We relate these results to water tower operation.

444041 Design & Development of a Novel PV-Fluid Hybrid Energy System

Abstract:

Embedded Generation has been described as a "paradigm shift" in the way in which electricity is produced with the focus of power production shifting away from large centralized generation plants to production of heat and power close to the point of use. An emerging technology that may play a part in the evolution of this new paradigm is the novel PV-Fluid hybrid system which can simultaneous produce electric and thermal power. The PV-Fluid System discussed here are PV-water and PV-air. To increase the electrical output, the system is further integrated with concentrators (V-Trough, Compound Parabolic Collector, Double booster Side mirror etc.). This paper will develop and test a simple mathematical model of above hybrid system. The different types of concentrated PV/Fluid which are fabricated are compared for their performance and cost. The result is that out of different PV-Fluid compared CPC-PV-Water is found to be most efficient although V-Trough-PV-Water hybrid energy system is cost effective. Further improvement in output is also discussed with the use of automatic and manual tracking. The advantages of PV-Fluid hybrid Energy systems are we can get simultaneous two powers; efficiency is increased, lesser cost, less space requirement, portable, reduced transmission/distribution losses and continuous power generation. The development of these PV-Fluid hybrid energy systems provides the scope for the local power production on the buildings and making the part of the home grid-independent.

444272 Thermal Energy Storage System Based on a Pphase Change Material for Automotive Applications

Abstract:

A latent heat based thermal energy storage (TES) device was developed and tested for use in cold engine start up under cold winter conditions. 3.6 kg of a paraffin wax was used as a phase change material to store heat and a 15 kg aluminum block was used as a simulated engine block. A radiator coolant was then circulated between the thermal energy storage device and the engine block during the heat discharging tests which were conducted in a freezer kept at -15 to -20 degrees C. An analytical model was also developed to simulate the engine block heat up tests. The test data indicated that the TES device initially at 60 degrees C could heat up the aluminum block by 20 to 30 degrees C in 5 minutes. A sensitivity study conducted with the analytical model indicated the coolant flow rate to have the largest effect on the heat up speed.

Rating Model for Plate-fin Exchangers for NGL Recovery Units

Danielle Goldbeck, Kathy Bigger, and Thomas Walter

Linde Engineering North America Inc., Tulsa, OK, USA

The production of cryogenic natural gas liquid recovery units is a continuously growing industry. As natural gas liquids become more valuable, the technology to process these liquids to produce higher value products will continue to advance. In this highly competitive industry, it becomes important to design such facilities to be thermally efficient and cost effective. Additionally, as standardization of cryogenic recovery units and revamping of existing facilities becomes more common place, plants are more likely to be exposed to feed conditions for which they were not originally designed. Therefore, it has become crucial to be able to predict plant performance in the most accurate way possible.

As the natural gas liquefaction industry continues to become more competitive, companies are not only starting to standardize their designs in order to cut down on both capital and operating expenses, but also, and perhaps more urgently so, to reduce the process plant delivery schedule. One of the more difficult challenges that is a result of standardizing is rating this "standard" product for conditions for which it was not necessarily optimized. Plant operators and owners continue to try to push the operating envelope as much as possible to maximize profits. As a result, these standard plants may see a variety of compositions of feed gasses at higher capacities than that for which they were originally designed. Process engineers are then faced with the task of regularly rating the standard product for these wide ranges of operating conditions while still being able to predict the overall performance accurately.

Major influencers of plant performance are the efficiency and prediction accuracy of the heat transfer equipment performance. Cryogenic plate-fin heat exchangers are commonly used as the core heat exchanger units in natural gas liquefaction processes. They are typically preferred over other types due to their high thermal efficiency, compactness, low capital, installation, and operating costs, and ability to handle a wide range of services [1]. These complex exchangers are the "heart" of the cryogenic unit. Understanding the performance capabilities of these plate-fin heat exchangers can lead to a more accurate prediction of the performance of the entire cryogenic unit. A process engineer who can predict the performance of the plate-fin exchangers will have better confidence in the simulated results of the entire plant. This becomes even more important if performance guarantees are involved.

However, the ability to accurately predict the performance of these exchangers has proven to be a challenging task as the design information of the plate-fin exchanger must be available from the heat exchanger supplier and the process engineer must have a solid understanding of the mechanical and thermal-hydraulic design of heat exchanger. Many process simulation tools commonly used throughout the industry have moved towards developing rigorous rating tools for these often complex exchangers. Effective use of these tools allows more accurate predictions about the plant performance.

There are a variety of ways in which one can "rate" plate-fin heat exchangers. One method is to compare the design UA value, or the product of the overall heat transfer coefficient, U, and the heat transfer area, A, to the UA value under the "off-design" conditions using the ratio of the base and desired mass or volumetric flow raised to an exponent value as a correction factor. This method is simple and does not require vendor information or a rigorous rating tool, but it does not prove to be the most accurate method. The UA method proves to be less accurate for condensing and boiling streams, especially as the conditions move further away from the design point. Without taking into account the details about the mechanical design and geometry, the UA method assumes that the heat transfer coefficient is constant throughout its flow path through the exchanger, which is not correct [2]. Another method, and perhaps the most accurate, is to input the physical geometry details into a stand-alone exchanger rating tool. This method requires much more information about the exchanger design such as fin geometry, feed distribution, nozzle design, and physical dimensions. Once these details are known or reasonably estimated, accurate predictions can be generated. Incorporating the geometry of the exchanger allows the prediction of the exchanger performance to reflect how the exchanger will actually perform given the specified inputs.

Comparing the two methods shows the importance and benefits of understanding the physical geometry of the exchanger.

Table 1 illustrates the two methods side by side in a rating case for a natural liquids recovery unit operating to recover ethane. The feed used for the rating case only differs from design conditions in a slight change of feed composition.

Table 1. Method Comparison with Feed Composition Change

Method	Calculated Recovery of Ethane, %
UA Comparison Method	96.89%
Exchanger Geometry Method	96.40%

Using the UA Comparison Method, the recovery of Ethane is over-predicted, though only slightly, by less than one percent (1.0%). However, looking at another example of comparing the same two methods, but with feed conditions further away from design, it is apparent that the UA Comparison Method has a small range of applicability.

Table 2 shows another rating case for a NGL plant, this time with not only a differing feed composition, but also an increase in plant through-put.

Table 2. Method Comparison with Feed and Capacity Change

Method	Calculated Recovery of Ethane, %
UA Comparison Method	95.51%
Exchanger Geometry Method	92.64%

The simpler, far more convenient method proves to be much less accurate the further the conditions move away from the design point. The UA Comparison Method has over-predicted the plant performance by almost three percent (3.0 %). This is a much larger difference that could lead to failure to fulfill contractual guarantees, or to the production of fewer product than financially expected (in this example case, a decrease of approximately 280 barrels per day of NGL product).

Using a stand-alone rigorous exchanger rating tool proves to be the most accurate method available. However, by taking this a step further, one can improve results and boost confidence in overall plant performance. Incorporating the geometry of the exchanger into the rigorous rating tool and then integrating the results with the entire simulation allows the outputs of the exchanger to now feed the other units downstream and reflect how the whole plant will actually perform under the feed conditions. This rigorous model of integration with the entire process simulation can be used to predict plant performance under off-design conditions.

Most of the processes to recover natural gas liquids are quite complex and typically involve more than one heat transfer unit. Therefore, the output of one exchanger can influence the input of another exchanger and vice versa. Using a stand-alone rigorous exchanger rating tool, one cannot link more than one exchanger together to see how the inputs and outputs influence each other. Typically, even the exchanger design vendor cannot predict the performance of multiple exchangers interacting. Only by integration of the rigorous exchanger model into the complete process simulation, can one truly understand the limitations and capabilities of the units and the plant as a whole.

Typically a standard NGL plant consists of at least two multi-pass plate-fin exchangers. One exchanger, known as the Gas/Gas Exchanger, is the heat integration of the feed gas, the hot stream, with three (3) internally generated process cold streams. The second exchanger, known as the Subcooler, consists of the further processed feed stream cross exchanging heat with three (3) streams that come from either the overheads or bottoms of distillation columns. These exchangers can be seen in Figure 1.

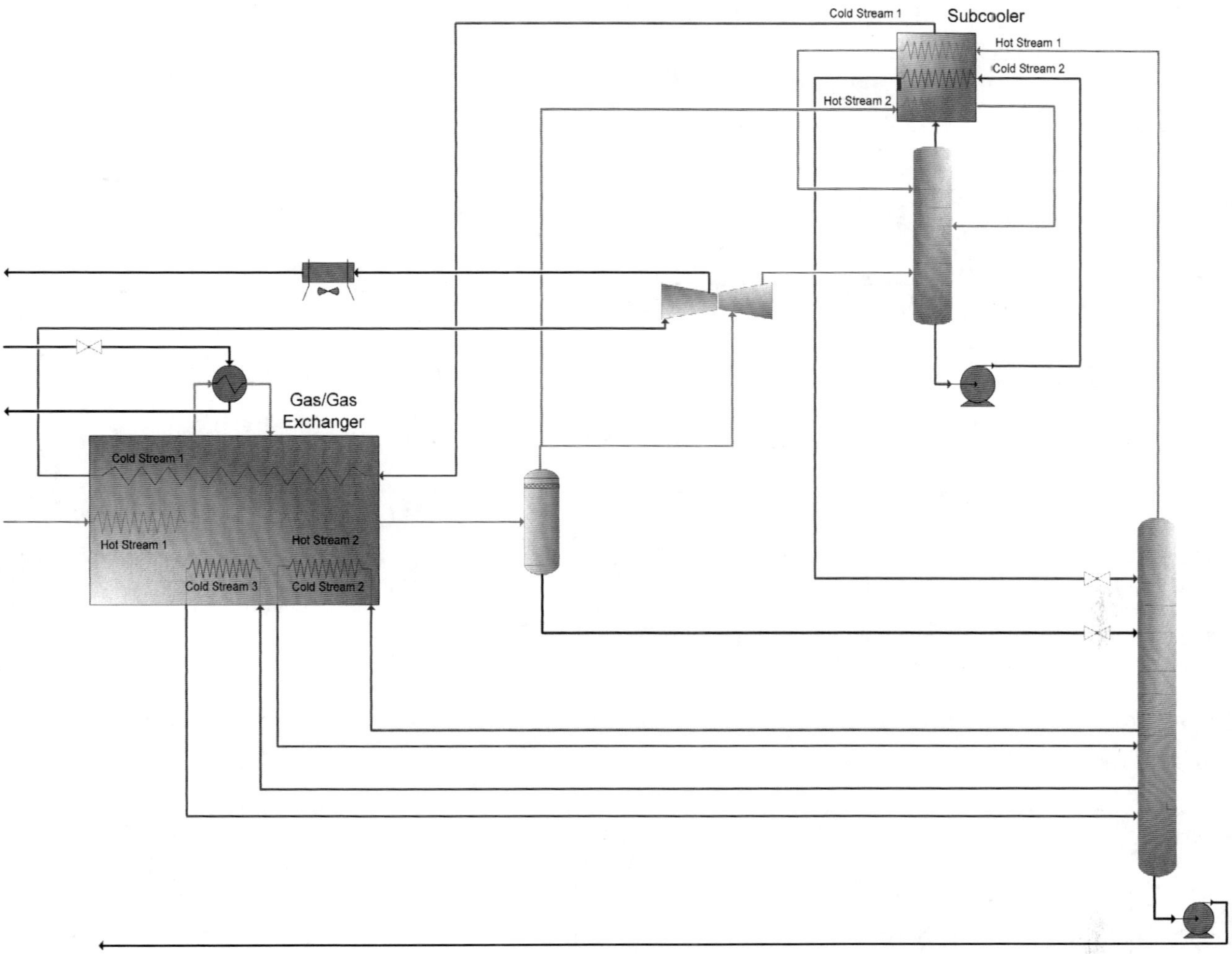

Figure 1 . Enhanced CRYO-PLUS™ PFD for NGL Recovery Process (Patent Pending)

By observation, there are streams that pass through both units, and there are streams that are outputs of one exchanger that feed downstream equipment, such as columns, that then influence the feed streams to the other unit. Thus, these two exchangers directly impact one another.

Table 3 depicts the differences in the inlet and outlet temperatures of the exchangers based on what the vendor predicts using the physical geometry of a stand-alone tool versus a rating tool integrated in the process simulation.

Table 3. Tool Comparison Illustrated with Temperatures

Gas/Gas Exchanger	**Stand-alone Rigorous Exchanger Tool**		**Integrated Tool in Process Simulation**	
Pass	Inlet Temp., °F/(°C)	Outlet Temp., °F/(°C)	Inlet Temp., °F/(°C)	Outlet Temp., °F/(°C)
Hot Stream 1	119.4 (48.6)	47.7 (8.7)	119.4 (48.6)	47.1 (8.4)
Hot Stream 2	11.7 (-11.3)	-29.5 (-34.1)	11.7 (-11.3)	-28.1 (-33.3)
Cold Stream 1	-71.2 (-57.3)	78.4 (25.8)	-71.4 (-57.5)	77.6 (25.3)
Cold Stream 2	-36.6 (-38.1)	4.2 (-15.4)	-35.6 (-37.6)	3.8 (-15.8)
Cold Stream 3	64.6 (18.1)	81.3 (27.4)	63.9 (17.7)	81.3 (27.4)

Subcooler Exchanger	**Stand-alone Rigorous Exchanger Tool**		**Integrated Tool in Process Simulation**	
Pass	Inlet Temp., °F/(°C)	Outlet Temp., °F/(°C)	Inlet Temp., °F/(°C)	Outlet Temp., °F/(°C)
Cold Stream 1	-167.1 (-110.6)	-69.0 (-56.1)	-162.4 (-108)	-71.4 (-57.5)
Hot Stream 1	-82.7 (-63.7)	-135.8 (-93.2)	-81.8 (-63.2)	-135.5 (-93.0)
Hot Stream 2	-30.0 (-34.4)	-152.1 (-102.3)	-28.1 (-33.3)	-150.4 (-101.4)
Cold Stream 2	-127.3 (-88.5)	-93.9 (-69.9)	-124.8 (-87.1)	-94.7 (-70.4)

At first glance, there seem to be little differences in the temperatures reported between the two tools. However, these small differences can cause a larger impact in overall plant performance. For example, the Hot Stream 2 of the Gas/Gas Exchanger is the same stream as the Hot Stream 2 of the Subcooler. Therefore, the outlet temperature of the Gas/Gas Exchanger unit should be the same as the inlet to the Subcooler, as it is under the conditions using the integrated tool in the process simulation. Since the vendor is not able to link these two exchangers, they manually enter the inlet streams as noted on the process datasheet they receive. In this case, the process datasheet reflected a -30.0°F temperature on both the outlet temperature of the Gas/Gas Exchanger and the inlet stream to the Subcooler for Hot Stream 2. The exchanger was not able to achieve the -30.0°F due to thermal and physical constraints, and so, the outlet temperature achieved is -29.5°F per the vendor. However, the -30.0°F is still assumed as the inlet temperature to the second downstream exchanger because the two exchangers are not able to impact one another in the stand-alone tool.

Additionally, the Cold Stream 1 of the Subcooler is the input of the Cold Stream 1 to the Gas/Gas Exchanger, meaning the outlet temperature of the stream from the Subcooler should match the inlet stream temperature to the Gas/Gas Exchanger. If the rigorous exchanger tool is integrated in the process simulation as a whole, this is correctly represented. Again, looking at the stand-alone rating tool, these temperatures do not match, and therefore, slightly better heat transfer is assumed to be achieved.

Table 4 shows how all these seemingly small, temperature differences impact the calculated recoveries of the NGL product.

Table 4. Ethane Recovery at Design

Method	Calculated Recovery of Ethane, %
Stand-alone Rigorous Exchanger Tool	94.61%
Integrated Tool in Process Simulation	93.10%

The stand-alone rigorous exchanger tool over-predicted the plant performance by 1.50% Ethane recovery. In this ever growing, competitive industry, even that small of a percentage can make or break a contract or project award. The above example was performed at design conditions. As conditions move further away from design, these differences results in even greater discrepancies.

Table 5 shows the NGL recovery calculated, again using the same two methods and same plant configuration, but with feed conditions differing only in the feed gas composition.

Table 5. Ethane Recovery at Off-Design

Method	Calculated Recovery of Ethane, %
Stand-Alone Rigorous Exchanger Tool	96.21%
Integrated Tool in Process Simulation	93.75%

A change in composition alone reflected a 2.5% decrease in Ethane recovery. More significant changes in feed such as inlet pressure, temperature, or plant capacity could have an even greater impact. The market is turning to standardizing these processes, making it more common for plants to run in "off-design" conditions from the start of the plant life. Thus, it is crucial to implement the rigorous rating tool model with the process simulation. This tool will provide the process engineer with more confidence in the simulation results under both design and off-design conditions.

Conclusion

A rigorous rating model for plate-fin exchangers for NGL recovery units is highly useful in this competitive industry. The rating model consists of entering the exchanger geometry details into the chosen rigorous rating tool and then integrating the tool with the entire process simulation. By using the rigorous rating model, one can predict the performance of plate-fin exchangers and ultimately reliably predict and optimize the performance of the plant, allowing for a more predictable, robust and cost effective design.

References

[1] ALPEMA, *The Standards of the Brazed Aluminium Plate-Fin Heat Exchanger Manufacturers' Association*, Third Edition, 2010. www.alpema.org

[2] Thulukkanam, Kuppan, *Heat Exchanger Design Handbook*. Second Edition. CRC Press Taylor & Francis Group. Boca Raton, FL. 2013.

Additional References not cited:

1. Sekulic, Dusan P. and Shah, Ramesh K., *Fundamentals of Heat Exchanger Design*. John Wiley and Sons, Inc. Hoboken, New Jersey. 2003.

444271 Modeling of Thermal Gradient Sublimation

Abstract:

Many electronic applications, such as organic light-emitting devices (OLEDs), require low molecular-weight organic materials of high purity. Industrially, this purification is typically accomplished via thermal gradient sublimation. In this process, the crude material is heated and sublimed, and the resulting vapor is transported through a temperature controlled tube until it deposits downstream on a cooler section of the tube, where it is later collected manually. Scale-up and optimization of thermal gradient sublimation have been hindered by difficulties in modeling the transport and deposition of the rarified vapors involved.

In this presentation, we consider the challenges in predicting the deposition location and rate for a given tube setup and temperature profile. We present an overview of the various regimes experienced during transport of the vapor from sublimation to deposition, including viscous flow, volume diffusion, and Knudsen diffusion, and we develop an overall transport model that incorporates each of these mechanisms based on the kinetic theory of gases. Finally, we explore the predicted deposition behavior of the developed model and demonstrate how it can be used to tune and optimize sublimation processes to maximize both purity and yield of high-value materials.

444000 Flare Minimization for an Ethylene Plant Shutdown Via Plant-Wide Dynamic Simulation

Abstract:

There is a great amount of raw materials sent to be flared during the ethylene plant start-up and shutdown, it generates much burned-off gases which including NOx, VOCs, CO, $CO_2$and so on. This not only increases operating cost, but also causes environment pollution which is highly concerned by local public and regulatory agency. Flare Minimization can reduce pollutant generation, save raw materials, and also lower the energy costs. The development of flare minimization strategy for an ethylene plant is quite crucial and beneficial for both plant and local environment.

In this paper, a dynamic simulation model has been built to study flare minimization opportunities of a front-end de-ethanizer ethylene plant shutdown operations. To minimize the shutdown flaring and smooth the plant shutdown operation, various operational improvements have been initiated and examined in this paper via the developed dynamic simulation model. For example, such an idea has been tested by dynamic simulation that shutting down all furnaces first and then recycling vapors from downstream columns to the CGC (Crack Gas Compressor) inlet to necessitate CGC running meanwhile enable flare source recovery. After that, there are much less hydrocarbons left in downstream vessels after CGC is shut down and before the nitrogen purge operation. Compared with the normal procedure, which is CGC shutdown first and then all the furnaces shutdown, significant flare emissions have been saved.

443612 Statistical and Neural Network Modeling of FLARE Performance and Estimation of Controllable Parameters

Abstract:

Recent EPA proposed rules (parts of 40CFR63.11 & 60.18) apply maximum allowable flare tip velocity (400 ft/s), smokeless flaring, and combustion zone heating value at a minimum value of 270 BTU/scf for all steam assisted, air assisted and non-assisted flares including emergency release to achieve high combustion and destruction efficiency (CE/DE). The proposal was based on analysis conducted on a limited set of data from steam assisted and air assisted flares. In this study, the flare soot emission, visible emission scale, DE, and CE test data from 1983 to 2014 (including 1983/1984 EPA, 2010 TCEQ, 2009/2010 Marathon Detroit/Texas City, and 2014 Carleton University) were analyzed. This study aims at developing nonlinear statistical models for flare efficiencies and opacity. All CE data were corrected for soot emissions. Curve fitting and Neural Network tool boxes in MATLAB were used in modeling. Multivariate statistical distribution functions and neural network models were developed for steam and air assisted flares burning propylene, propane, natural gas, methane and ethylene. Since the flare test data includes larger range of exit velocity which affects plume volume, opacity is normalized with plume diameter. The input variables include vent gas combustibility (combustion zone net heating value (NHVcz)/ lower flammability limit (LFLcz)/ combustible concentration (Ccz)), vent gas species (olefins, alkynes, aromatics, hydrogen, nitrogen, carbon to hydrogen ratio), and operating parameters (steam/air assists, and exit velocity). 3-D contour plots were developed to estimate the desirable range of controllable operating variables. The goal is to estimate the desirable operating set point with a high CE >96.5% (or DE >98%) and low opacity (10% for steam assist & 40% for air assist). Neural network models developed for CE and normalized opacity showed high coefficient of determination (R^2) values 0.91 and 0.95 respectively for steam assisted flare data.

Keywords: Propylene, Propane, Methane, Ethylene, Natural gas, visible emissions, Flare performance, Lower flammability limit, Net heating value at combustion zone, Neural network models.

LOW-NOX BURNER RETROFIT CHALLENGE

By

Craig Lieb
Applications Engineer
Zeeco, Inc.
Rutland, UK

Ryan Roberts
Project Engineering Manager
Zeeco, Inc.
Broken Arrow, OK, U.S.A.

Ryan Roberts has more than 20 years of combined combustion experience for the process burner industry. His experience includes 5 years of working as a Project Engineer at Zeeco on thermal oxidizer, flare, and burner projects before moving into the process burner group. Ryan served as a Project Engineer as well as a Project Manager for the first 19 years of his career at Zeeco, and has experience in a wide range of burner projects for both the process and power industries. He has had the good fortune to travel extensively with Zeeco for startup and commissioning of different burner projects around the world. Ryan has been a member of the Zeeco team since 1995 and currently serves as the Project Engineering Manager for the process burner group. He graduated from the University of Oklahoma's Mechanical Engineering Program in 1995.

Presented at the

2016 Spring Meeting and 12th Global Congress on Process Safety
Houston, Texas -- April 10-14, 2016 (April 12, 2016)

Introduction

As increasingly stringent NOx requirements are implemented around the world, end users with fired heaters and industrial boilers are forced to update existing fired equipment to meet regulations. Both new applications and retrofits projects may need Ultra-Low NOx or Next Generation Low NOx Burners to meet the ever-tightening emissions requirements. Recently, Zeeco accepted a retrofit challenge to replace 34 conventional burners at a European refinery with a capacity of 11.5 million tonnes per year. The project objectives were to significantly reduce NOx production, increase combustion efficiency, and eliminate flame impingement on the heater process tubing. Zeeco's GLSF Min-Emissions Ultra-Low NOx Burner was the selected equipment to achieve these goals.

Background

The existing raw gas burners installed on the client's forced draft, pre-heated air, crude heater were proving troublesome. The existing burners were producing NOx values well above 200 mg/Nm2 [100 ppmv]; the combustion in the firebox was poorly distributed across the burners; and in several places, the burner flames were long enough to impinge on the short heater roof tubes, causing process tube coking. To eliminate these issues, the client decided to retrofit the crude heater with Ultra-Low NOx burners. Zeeco was selected to provide 34 GLSF-8 Round Flame, Min-Emissions Ultra-Low NOx Burners. The heater is a twin-celled cabin heater with a sloped roof leading to a common, central, convection section. There is one row of 17 burners installed in each heater cell. A rendering of the cross-sectional view of the heater can be seen in Figure 1 below along with the original heater drawing for reference.

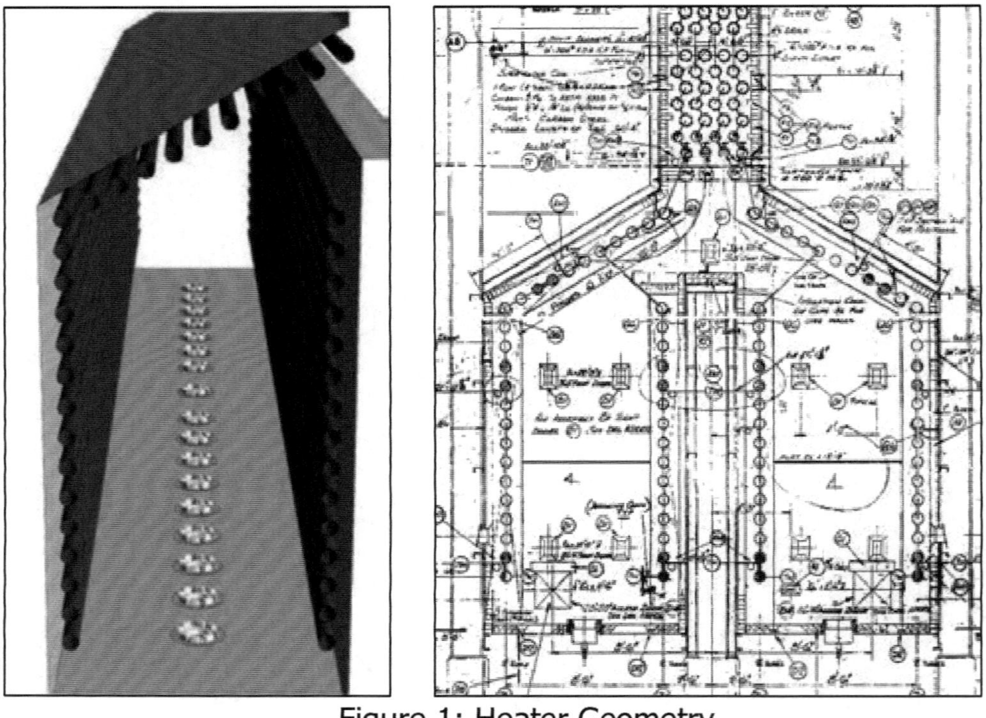

Figure 1: Heater Geometry

Project Phases

This project was completed in three main phases. The first phase included the design and drafting of the burners, and the completion of two geometry models for Computational Fluid Dynamics (CFD) simulation. The two CFD models included the existing forced draft, preheated air ducting system, and half of the heater firebox geometry with the row of 17 burners. Phase two included the completion of a combustion test performed at Zeeco's combustion testing facility in Tulsa, Oklahoma. The test was conducted with two GLSF-8 Min-Emissions burners installed in one of Zeeco's cabin-style heaters. Phase three included the installation and commissioning of the burners on site, as well as site acceptance emissions measurements.

Phase One – Burner Design and CFD Modeling

The geometry of the existing combustion air ducting system required a custom burner design to match the existing combustion air inlet and the heater mounting dimensions. The customized burner design varied in several physical aspects from Zeeco's standard burner design. The final burner design geometry can be seen below in Figure 2.

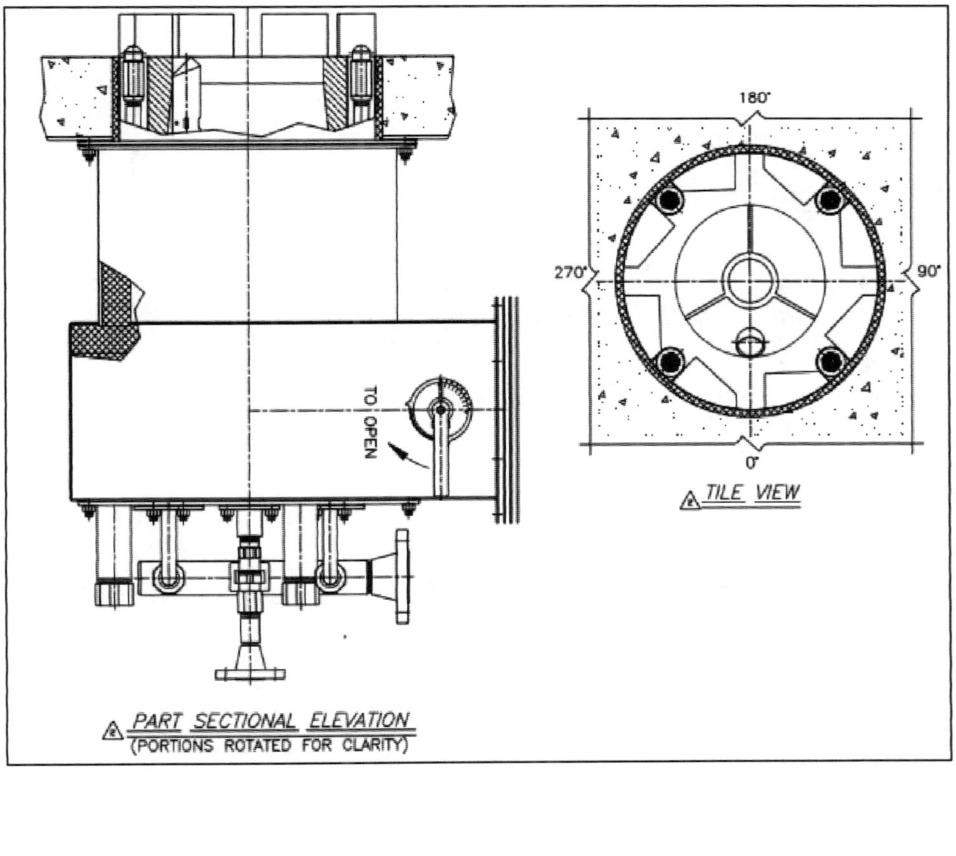

Figure 2: Final Burner Design

In order to examine and then improve the air flow distribution in the existing forced draft ducting system, Zeeco modeled the full furnace ducting geometry. Several iterations were simulated in order to design and locate specific baffle plates and turning vanes. Examining the original geometry showed the air flow distribution to each burner was uneven, varying by as much as +13% and -30% mass flow [kg/hr] across the 34 burners. This was deemed unacceptable for the successful operation of the new Ultra-Low NOx burners. New turning vanes and baffle plates were implemented into the model geometry to improve the airflow distribution. At the conclusion of the iterations, the designed airflow to each burner improved to within the desired +/- 3% [kg/hr]. Zeeco created and supplied the client with a detailed CFD report showing the locations and dimensions of the designed turning vanes and baffles so the ducting modifications could be performed during shutdown. A snapshot comparison of the velocity profile through the ducting before and after the implementation of new turning vanes and baffles is shown in Figures 3 & 4.

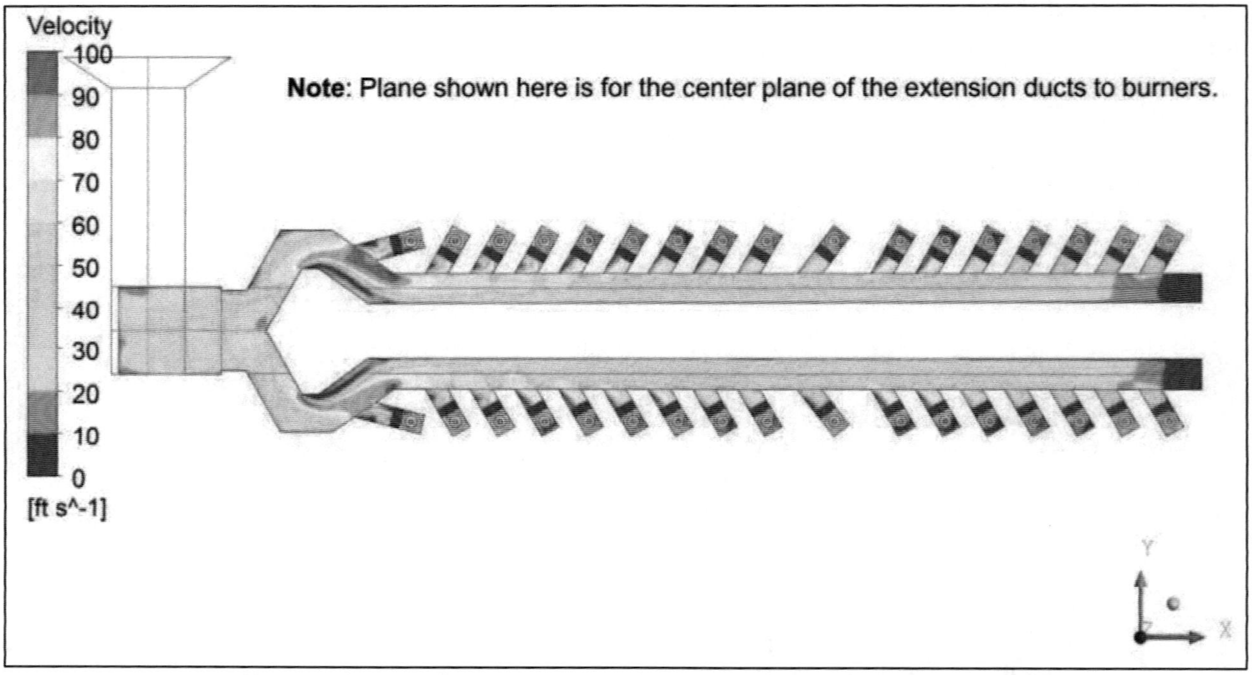

Figure 3: Combustion Air Velocity Profile – Existing Geometry

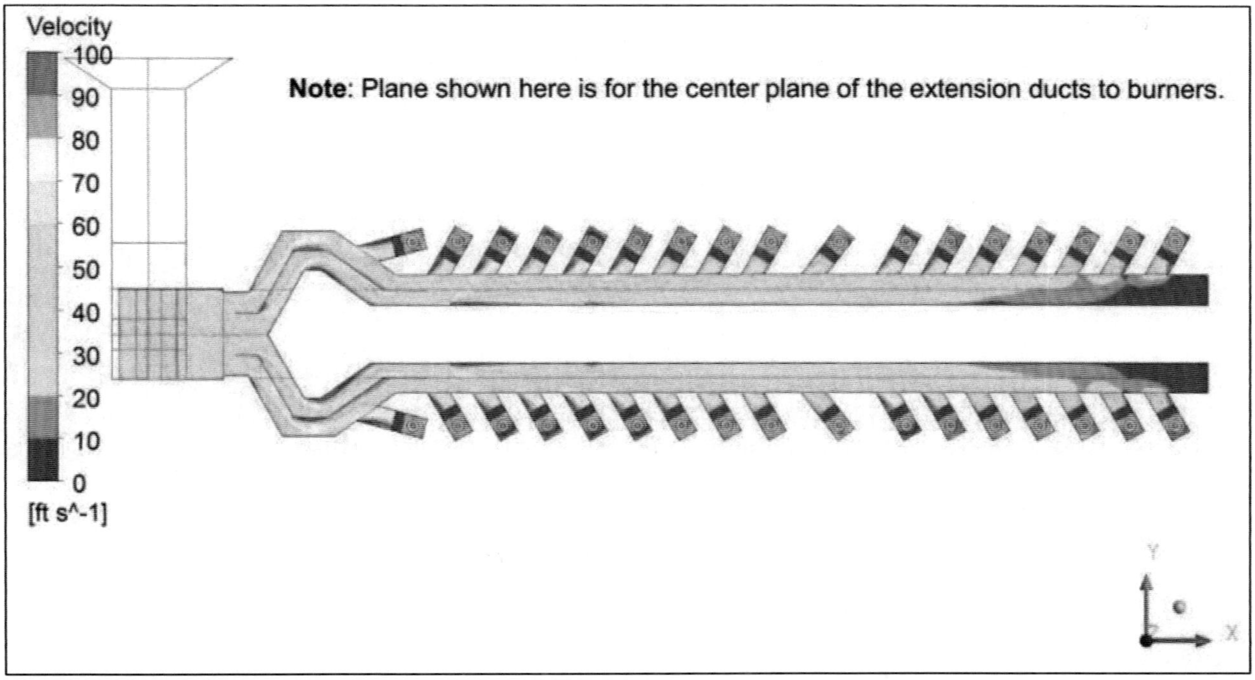

Figure 4: Combustion Air Velocity Profile – Modified Geometry

A CFD of the heater firebox was modeled to ensure proper flame profiles and heat distribution to the process tubes. The model included half of the heater geometry and 17 ZEECO® burners due to heater symmetry. The flue gas patterns and flame dimensions were evaluated and were deemed acceptable by both Zeeco and the client. The predicted flame length as based on 2150 ppm CO (dry) iso-surface was 8.19 feet.

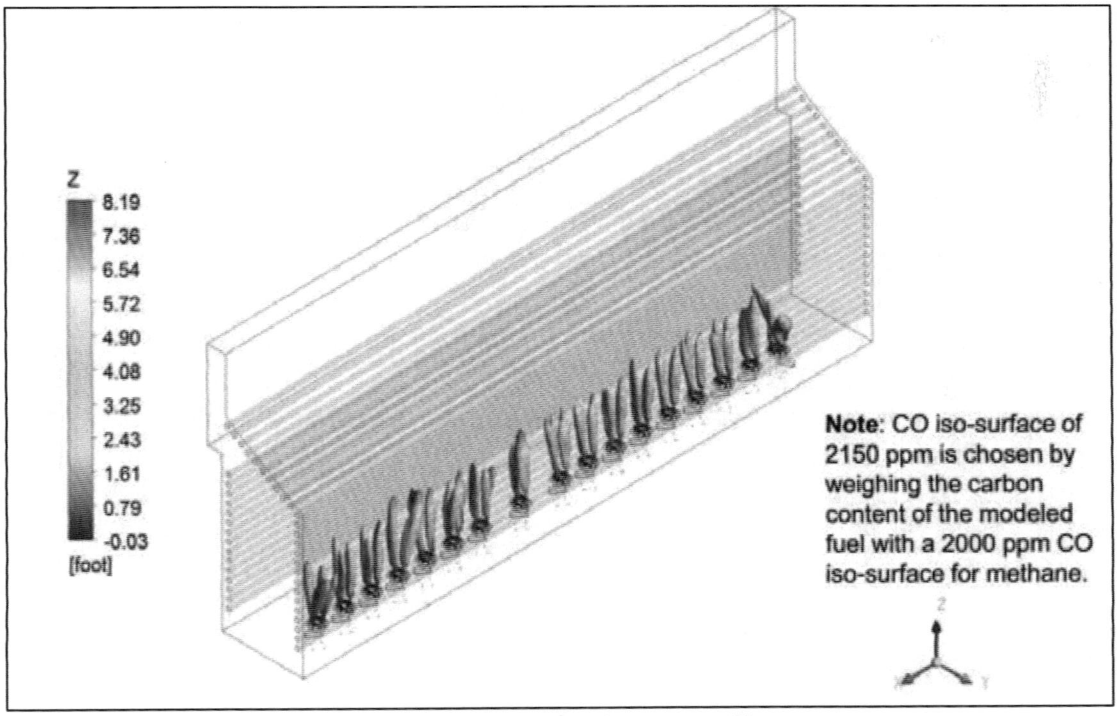

Figure 5: CFD Flame Profiles

All burner design features and CFD geometries and results were approved by the client's engineers.

Phase Two – Combustion Testing

Zeeco's test facility engineers performed a witnessed test for members of the client's engineering team. The goal of the combustion test was to confirm predicted emissions values when firing the fuel cases specified, and to confirm the flame stability and flame dimensions. The challenges included modifying the Zeeco test furnace to match the burner spacing, modifying the heater internal dimensions and insulation to match the furnace temperature of the site operation, and simulating the required fuel compositions. Twenty seven test points were performed to confirm the required burner performance. Test results were as follows:

Table 1: Burner Test Fuels

BURNER TEST FUELS

GAS FUELS	LHV	SP.GR.	FUEL B	FUEL E	FUEL F	FUEL G
NATURAL GAS	937.4	0.5982	60	46	32	100
HYDROGEN	273.8	0.0696	26	24	48	
PROPANE	2314.9	1.5226	14	30	20	
TOTAL			100	100	100	100
LHV (BTU/SCF)			957.7	1191.4	894.4	937.4
SP. GR.			0.590	0.749	0.529	0.598
MW			17.09	21.68	15.33	17.33

Test fuels were blended upstream of the burner to accurately match the fuel gases being fired on site. The chosen test fuels were then fired in a range of conditions to simulate the variety of site operation conditions defined by the client's engineers. The fuel gas pressure, combustion air pressure and temperature, and furnace flue gas oxygen percentage were all varied according to agreed upon testing procedures so the burner performance could be confirmed. The NOx emissions burner test figures are shown below for the maximum firing rate of each dedicated test fuel.

Table 2: Test Results at Maximum Firing Rates

Fuel Case	Heat Release [MMBTU/HR]	Measured Furnace Temperature [F]	Measured O2 %	Measured NOx [ppmv]	Corrected NOx [ppmv] @ 3% O2 and 1790F
Fuel B	6.98	1776	3.0	27.0	27.5
Fuel E	6.98	1759	2.9	27.4	27.9
Fuel F	6.98	1740	3.0	30.4	31.6
Note: A single Fuel G test point was tested to prove flame stability on N.G.					

The testing results, along with the results from other test points, proved the GLSF-8 Round Flame Min Emissions Burner would successfully achieve the required emissions levels while maintaining a desirable flame profile in the furnace.

Phase Three – Installation & Site Acceptance Testing

Upon delivery of the equipment, Zeeco performed several site supervision trips to ensure proper installation of the burners. All mechanical dimensions of the burner design proved to match the existing heater cutout and combustion air ducting mounting. The burners were successfully started up in June, 2015 and the site acceptance test was performed one month later in July, 2015. A review of the heater revealed that the heater was operating at a higher excess air rate than the designed 15%. In fact, the heater was operating around 25% to 30% excess air, which correlates to 5.25% to 6.3% oxygen in the flue gas. Additionally, the gaps around the process tube penetrations were not sealed, allowing for tramp air to enter the heater. Tramp air is defined as any uncontrolled air entering the firebox through any locations outside of the burner combustion air stream. Tramp air has been proven to increase the emissions produced by combustion due to the uncontrolled addition of air into the combustion chamber. Zeeco recommended that the client seal these tramp air infiltration points.

During the site acceptance test, emissions were measured with a Testo 350 analyzer at a sample point located just below the common convection section. Emissions measured were observed to be at an acceptable level and were under the 50 ppmv [100 mg/Nm3] requirement set by the end user's local legislation. The values measured over the span of about half an hour ranged from 42 ppmv to 45 ppmv [86 to 92 mg/Nm3]. Both the client and Zeeco agreed that when the tramp air locations were eliminated and the oxygen level in the heater decreased to the designed level of 15%, the NOx emissions would drop to the levels shown in the combustion testing at Zeeco's facility.

Project Review

The three phases of the project required time and careful technical engineering design to achieve the successful results reported by the end user since start up. The scheduling of the project needed to be maintained at a constant pace of progression in order to engineer the customized burner design, generate the CFD results and reports, set up and execute the combustion test, manufacture and deliver the equipment, and finally, execute a successful install and start up. Overall, the project was completed in 44 weeks from start to finish. The equipment was delivered as scheduled, before a scheduled major site turnaround.

REACTIVE DISTILLATION FOR PROCESS INTENSIFICATION – A STUDY OF DESIGN PARAMETERS

Gaurav Arora, Mohammad Shafiei
Process Technology, T&I
SABIC
Sugar Land, TX

AIChE Spring Meeting
Houston, TX
12th April, 2016

CHEMISTRY THAT MATTERS™

AGENDA

- Introduction
- Applications
- Case Study – Irreversible Reaction in Series
- Design Variables
- Results and Discussion
- Conclusions

No. 1

INTRODUCTION

- Reactive Distillation: Combination of reaction and distillation in a single equipment.
- Earliest application dates back to 1920s for esterification processes in a series of patents by Backhaus.[1]
- There were over 200 licensed commercial scale reactive distillation processes by the year 2007, many of which are in commercial operation.

Example: A + B ↔ C + D

$\alpha_C > \alpha_B > \alpha_A > \alpha_D$

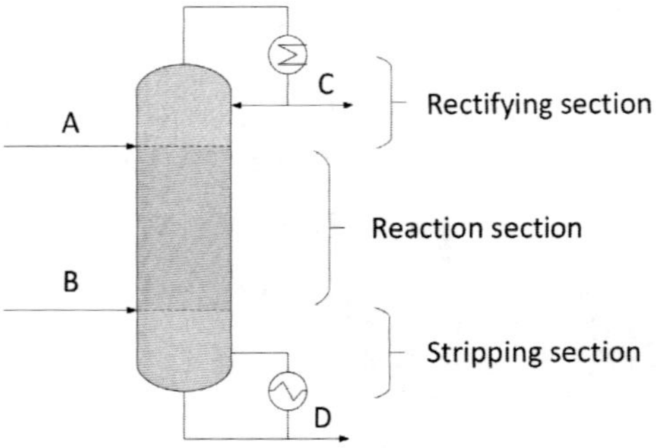

- Advantages
  - Process Simplification
  - Economical Benefits
    - Improved conversion
    - Higher selectivity
    - Improved reliability
    - Capital cost reduction
    - Operating cost savings
  - Environmental, safety benefits
- Constraints
  - Temperature
  - Volatility
  - Reaction Rates
  - Reaction phase
  - Scale-up

[1] US Patents 1400850. 1400851. 1400852

No. 2

APPLICATIONS

- Over 236 types of reaction systems for reactive distillation application.[1]

- Types of reactive distillation systems
  - Reversible or **Irreversible**
  - Catalytic or **Non-catalytic**
  - Catalytic: Homogenous or Heterogeneous
- Types of reaction systems
  - A + B ↔ C +D
  - A + B ↔ C
  - A ↔ B +C
  - A + B ↔ C + D
B+C ↔ D + E
  - **A + B → C**
C + B → D

- Class of problems and Examples
  - Equilibrium limited reactions
    - MTBE, ETBE, TAME, ...
  - **Suppressing side reactions**
    - Diacetone alcohol, EG, ...
  - Azeotropic system
    - Methyl Acetate, ...
  - Temperature control
    - Cumene, EG, ...

[1] W.L. Luyben, C.C. Yu, *Reactive Distillation Design and Control*. John Wiley & Sons, Inc.: Hoboken, New Jersey, 2008.

No. 3

CASE STUDY - IRREVERSIBLE REACTIONS IN SERIES

- Problem definition – Minimize rate of formation of [D] thereby maximizing selectivity for [C] by controlling reactant concentration profile using design variables.
- Class of problem – suppressing side reactions
- Reaction system – Irreversible reactions in series

$A + B \xrightarrow{k_1} C$

$C + B \xrightarrow{k_2} D$

$\alpha_B > \alpha_A > \alpha_C > \alpha_D$

$E_{a1} = E_{a2}$

$k_2 > k1$

$\frac{r_2}{r_1} = \frac{k_2\,[C]}{k_1\,[A]}$

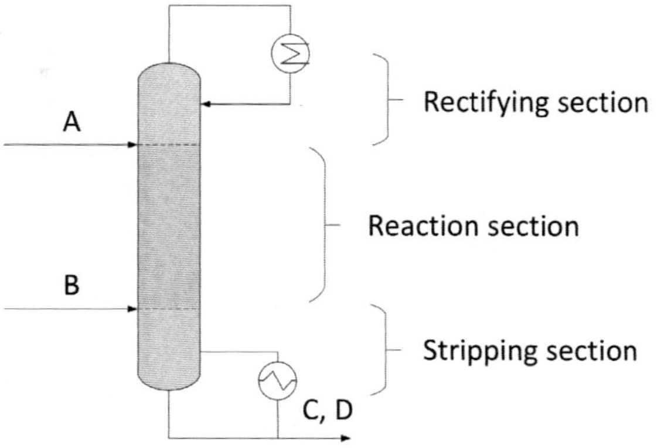

No. 4

DESIGN VARIABLES

- Pressure
  - Conventional distillation: Typically P ↓ ⇨ T↓ ⇨ α ↑
  - Reactive distillation: P ↓ ⇨ T↓ ⇨ α ↑, also *k* ↓ ⇨ hold up volume ↑
- Tray holdup
  - Assume a holdup → design column → calculate liquid height
- Reboiler duty (or distillate rate or distillate/feed ratio etc.)
  - Affects temp profiles, conc. ⇨ phase equilibria along the column ⇨ selectivity
- Number of theoretical stages
  - Higher number of trays does not necessarily reduce the energy requirements.
- Feed location
  - Feed location effects the compositions profiles and therefore the column performance.
- Stoichiometric operation or Excess reactant

No. 5

CONFIGURATIONS

- Feeds fed in exact stoichiometric ratio reduces downstream processing.
- Process configuration with excess feed offers better control of the column.

A + B → C
C + B → D

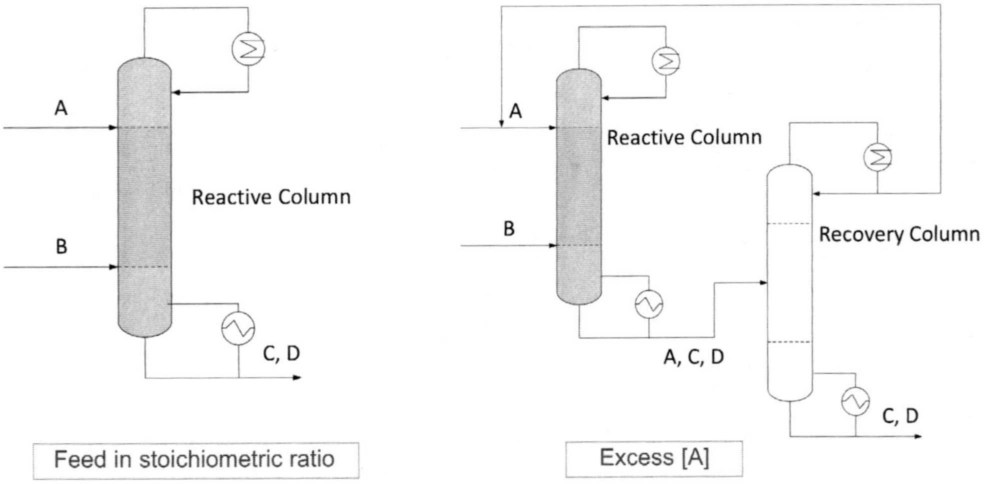

No. 6

NUMBER OF THEORETICAL STAGES

- Reactant [B] gets consumed moving up the column.
- Unreacted [B] in the vapor phase condenses at the top of the column.
- [B] in the condenser results in formation of [C] followed by [B].

A + B → C
C + B → D

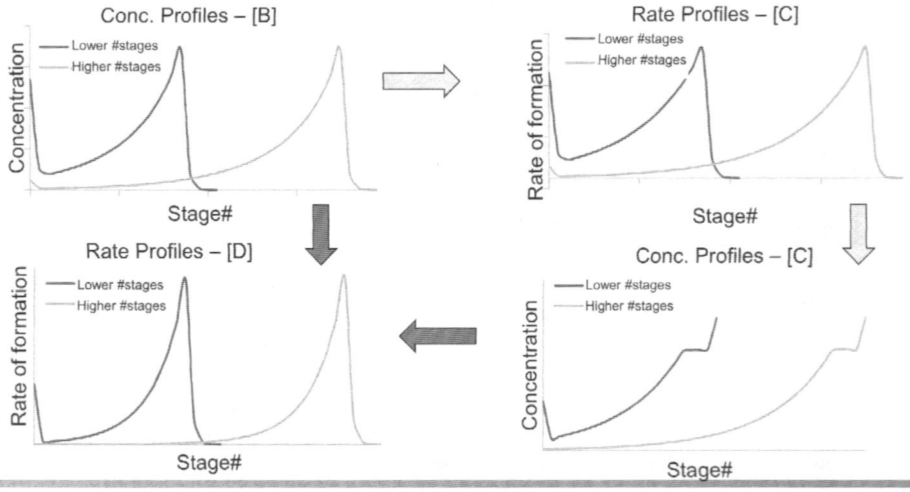

No. 7

NUMBER OF THEORETICAL STAGES

- Higher amount of [B] in the condenser → higher the amount of [D] formed → lower the selectivity for [C]

A + B → C
C + B → D

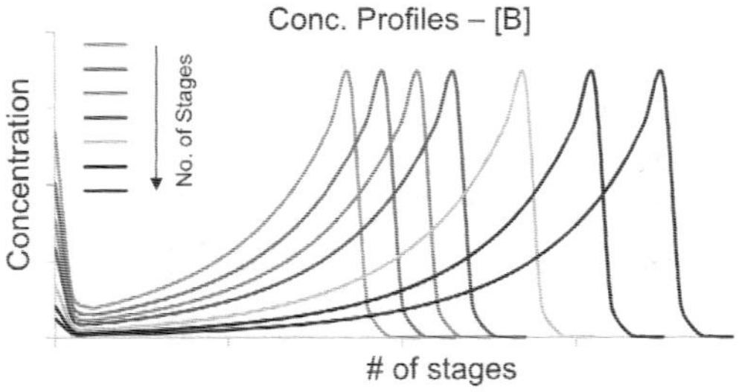

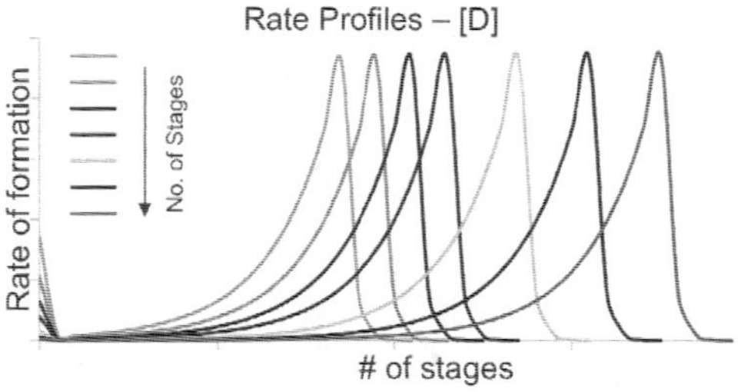

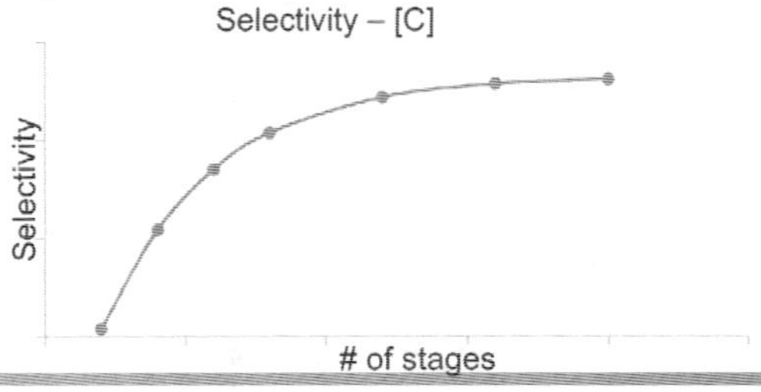

FEED LOCATION – [B]

- Reactant [B] gets consumed moving up the column.
- Unreacted [B] in the vapor phase condenses at the top of the column.
- [B] in the condenser results in formation of [C] followed by [B].

A + B → C
C + B → D

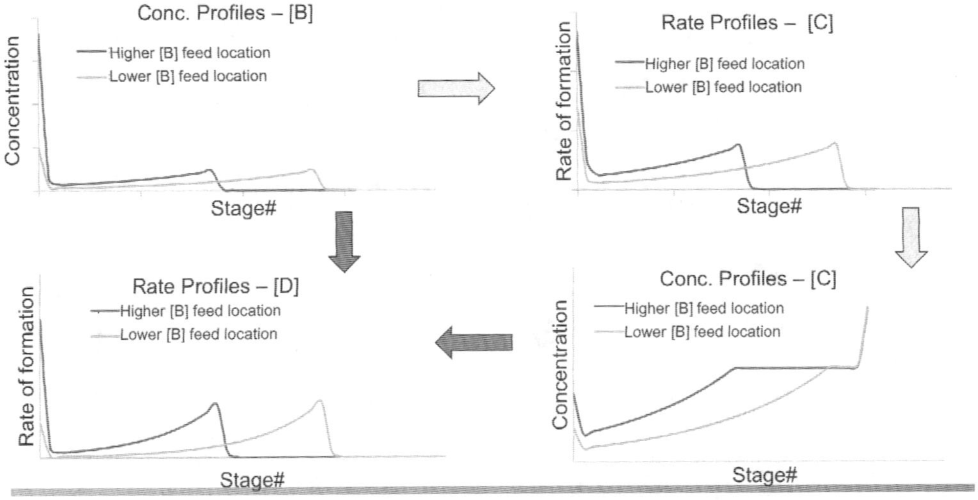

FEED LOCATION – [B]

- Higher the amount of [B] in the condenser → higher the amount of [D] formed → lower the selectivity for [C]

A + B → C
C + B → D

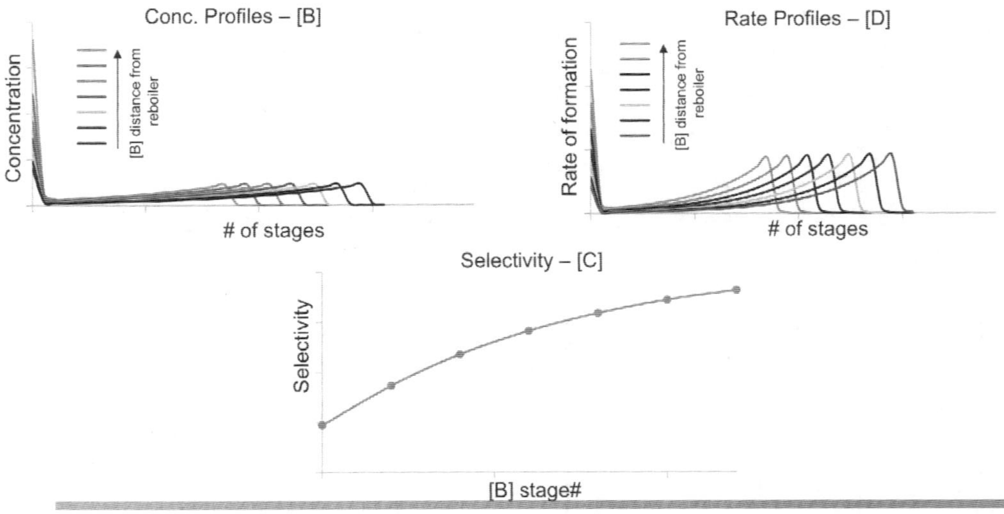

No. 10

FEED LOCATION – [A]

- Higher conc. of [C] on stages above [A] feed location.
- Higher the conc. of [C] → higher the rate of formation of [D] → lower the selectivity for [C]

A + B → C
C + B → D

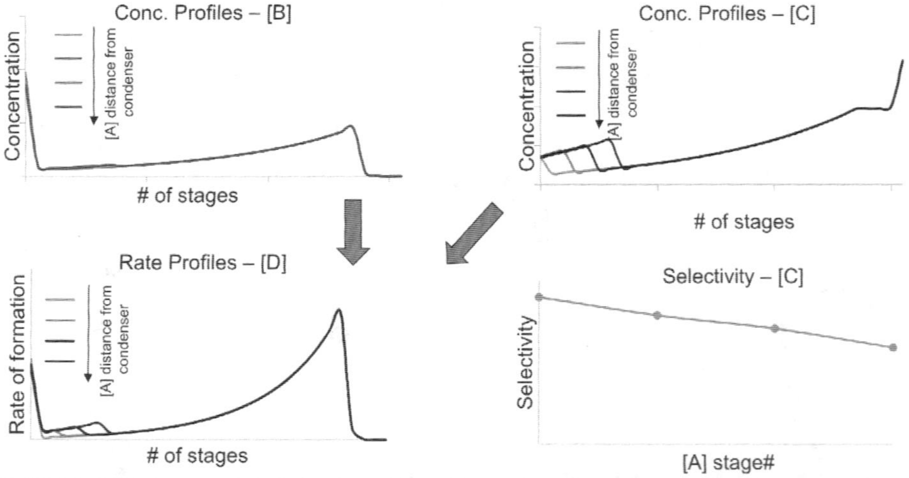

No. 11

REBOILER DUTY

- Lower the duty → lower the vapor traffic in the column
- Larger temperature drop if colder feed is mixed with a lower vapor traffic.
- Colder temp → higher the amount of [B] in the liquid phase → higher the rate of formation of [C] → higher the rate of formation of [D]

A + B → C
C + B → D

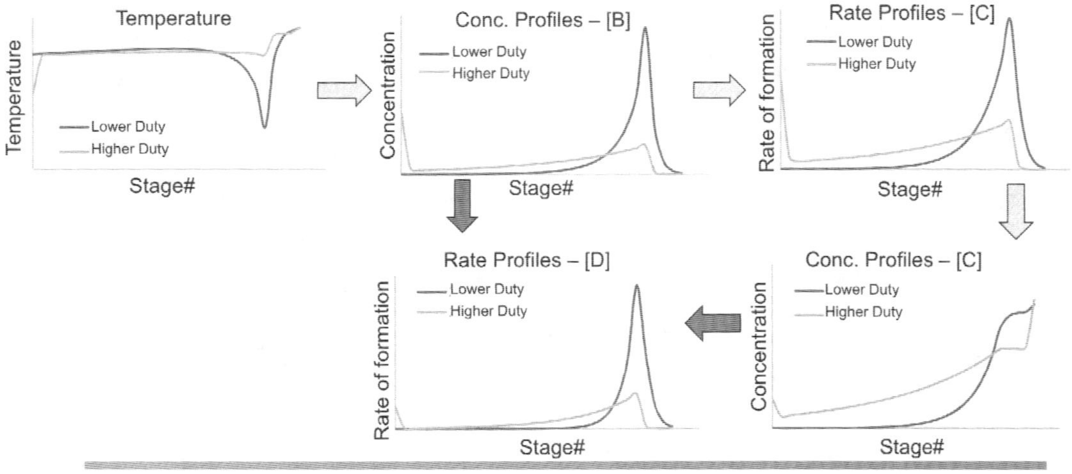

No. 12

REBOILER DUTY

- Higher the amount of [B] at the feed location → higher the rate of formation of [D] → lower the selectivity for [C]

A + B → C
C + B → D

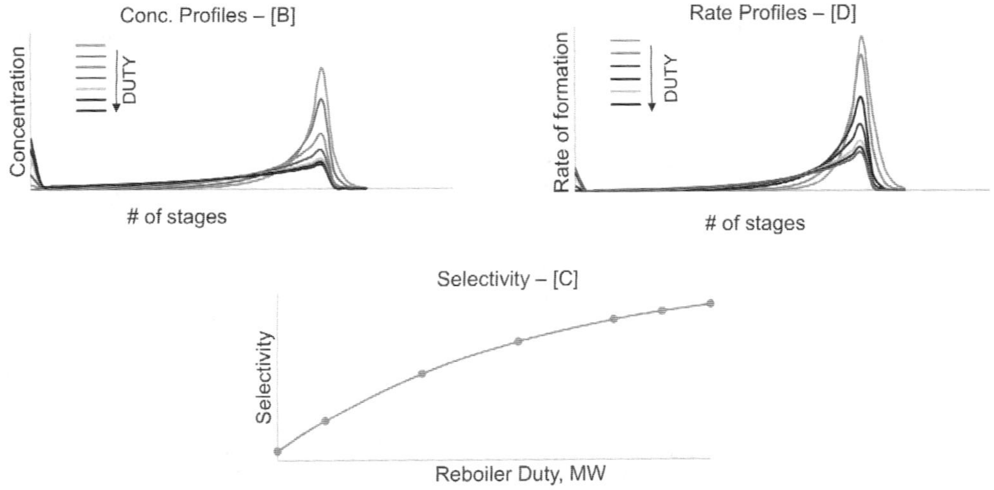

No. 13

FEED FLOW RATIO

A + B → C
C + B → D

- Lower [B] traffic in the column results in higher [C] concentration.
- Higher [C]/[A] → higher r_2/r_1 → lower the selectivity for [C]

$$\frac{r_2}{r_1} = \frac{k_2}{k_1}\frac{[C]}{[A]}$$

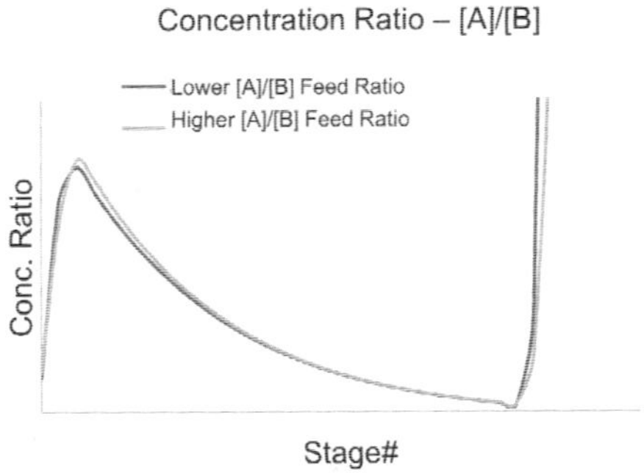

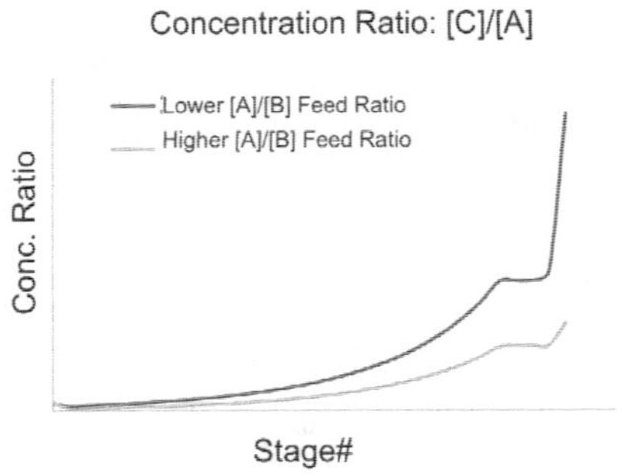

PRESSURE

A + B → C
C + B → D

- Higher pressure → higher temperature → higher rate constants.
- Higher pressure → higher temperature → lower the amount of [B] in liquid.
- Rate of formation of [D] set by competing rate constants and conc. of [B] in liquid.

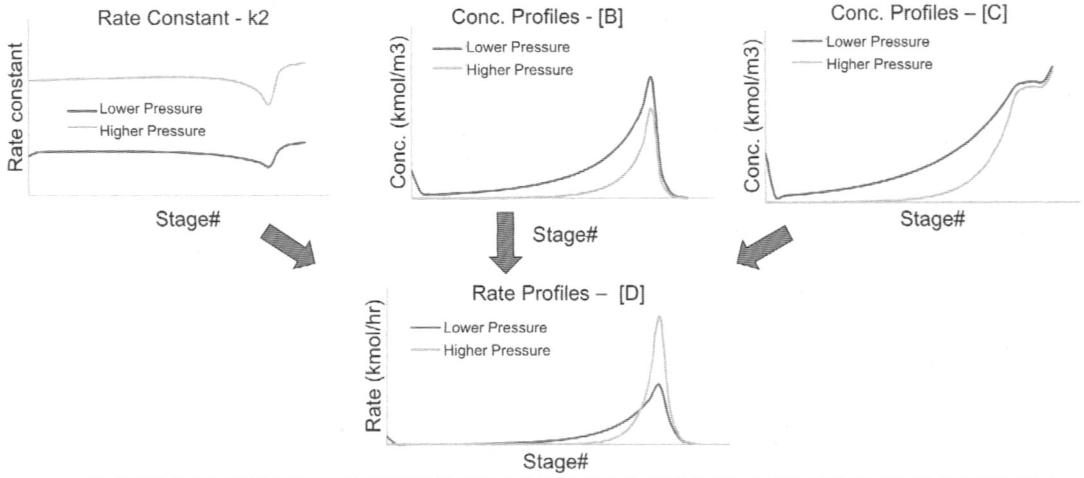

CONCLUSIONS

- Effectivity utilized analysis of concentration and hydraulic profiles in the column to study the effect of design parameters on the reactive distillation performance.
- Analysis allows optimizing the design maximizing economic benefits of the reactive distillation design.
- In general, for a reaction in series system [A + B → C, C + B → D], selectivity for C can be increased by controlling concentrations of the reactants as set by phase equilibria
  - reducing the amount of light component [B] reaching the condenser
  - reducing the ratio of [C]/[A] in the column
  - reducing the concentration of [B] at the feed location
- Similar analysis can be used for obtaining a fundamental understanding of other reactive distillation systems.

No. 16

16AIChE
Spring Meeting, Houston, TX

Design and Operation of B&P Process Podbielniak Liquid-Liquid Separator Centrifuges for AICHE Symposium

B&P Process Equipment and Systems LLC /
B&P Littleford Day LLC
1000 Hess Avenue
Saginaw, MI 48601
sales@bpprocess.com

Presenter: Arthur Bach / B&P Process Engineer

Prepared for Presentation at
American Institute of Chemical Engineers
2016 Spring Meeting
11th Global Congress on Process Safety
Houston, Texas
April 10-14, 2016

UNPUBLISHED

Design and Operation of B&P Process Podbielniak Liquid-Liquid Extractor Separators for AICHE Symposium

B&P Process Equipment and Systems LLC /
B&P Littleford Day LLC
1000 Hess Avenue
Saginaw, MI 48601
sales@bpprocess.com

Presenter: Arthur Bach / B&P Process Engineer

Other Author(s):
Cited B&P Process Equipment Process Engineers involved in research and development of machines.

Keywords: Pod Contactor ®, Podbielniak, Liquid- Liquid contacting centrifuge, continuous process centrifuge, Oil acidulation, pharmaceutical production, Fermentation, separation, extraction, stripping, B&P Process Equipment, Vitamin Extraction, Flavor oils, Essential Oils, Fat, Water-Oil separation

Abstract

The Podbielniak Liquid-Liquid contacting centrifuges are a type of phase separator that uses the enhanced gravitational field for phase separation or contacting for mass transfer. These type of units may be used for a two phase separation of immiscible liquids, distillation applications that could be accomplished by a liquid- liquid solvent extraction (possibly due to temperature concerns for product), tower stripping/ rectification or any gravity settling applications where process purity may benefit due in a higher G field. These may also be considered where tower or tank liquid- liquid phase separation takes up to much plant space or to reduce quantities of a hazardous material held up in process area equipment.

This article will cover some of the concepts for design and examples of operational performance of acetic acid extraction with water from methyl isobutyl ketone and kerosene/ water separation. These are two well known process examples for a Podbielniak Liquid-Liquid contacting centrifuges and other liquid-liquid contacting technologies. Additionally examples of Process implications of centrifuge operation parameters are discussed as are some existing user operation troubleshooting ideas.

Applications include major chemical process steps such as separations, extractions, rectification, ion exchanges, neutralization & washings. Examples are found industry wide in chemical production, Pharmaceutical applications, Petroleum Refining Liquid Ion Exchange, Food industries (Flavor oil extractions) and many other specific examples.

1. Introduction & Background

The Podbielniak Liquid-Liquid contacting centrifuge is a well known example of Liquid-Liquid extraction process equipment which has been in service for many years. It is often referred to as a "Pod Contactor®". It may also be referred to as a Centrifugal Extractor, a Continuous Centrifuge, a Continuous Centrifugal Extractor, or several variations there-in. It is a rotating chamber with 2 phase feeds that takes advantage of the higher gravitational field within the rotating chamber to achieve or improve phase separation, extract a target species or any other chemical process application involving the contact of two immiscible liquid phases.

Initial development originated from, among others, the research and efforts of Dr. Walter J. Podbielniak (1899-1978). Associated patents from him include US Patents 2,004,011 (6/4/1935) & 2,044,996 (6/23/1936) on counter current centrifugal devices consisting of rotating tubes to perform two phase contacting in the 1930s. Additionally, in early 1935 his wife purchased US Patents 1,936,523 & 1,936,524 (11/21/1933) by Adolf Placek for a spiral banded, horizontally rotating chambered device for distillation & absorption. Dr. Todd's patent expanded the application to include contacting of immiscible or semi-immiscible phases in addition to the Gas-Liquid & Vapor-Liquid contacting. Development work was ongoing through the 1930s until 1960 under Dr. Podbielniak's company Podbielniak Centrifugal Super-Contactor Laboratories (and several other iterations) until being sold to Dresser Industries as the Podbielniak Division. Some of the chief processes driving the development came from the use of the Podbielniak separators in penicillin production and Phenol contacting of Lubrication oils, in addition to many others.

Dr. David B. Todd joined the Podbielniak division of Dresser industries in 1963. Under his direction, research and redesign activities were completed to standardize the product lines. The standardized Pod Contactor sizes ultimately became the A-1, B-10, D-18, D-36 & E-48 designations when the division was sold in 1968 to Baker & Perkins, relocating the division and personnel to Saginaw, Michigan; the current B&P Process Equipment site.

Since the initial development and design refinements of these machines, wider chemical industry awareness has decreased. Awareness is kept up by existing users, but it may not be as familiar to outside or newer processes. This paper is to refresh industrial awareness that the Liquid-Liquid contacting centrifuge is a viable and effective technology for any application where two immiscible / partially immiscible liquid phases are to be separated or contacted together for mass transfer and/ or reaction.

2. Concepts for design & operation of Liquid-liquid contacting Centrifuges

2.1 Design

2.1.1 Basics

Figure 1 shows an example of B&P Process Equipment Liquid-Liquid contacting centrifuges:

Figure 1 Example of a B&P D-900 (previous D-36) Liquid- Liquid contactor Centrifuge. Rated Capacity ~68.1 m3/hr (~300 GPM)

Figure 2 illustrates an example of the fluid passage through the unit.

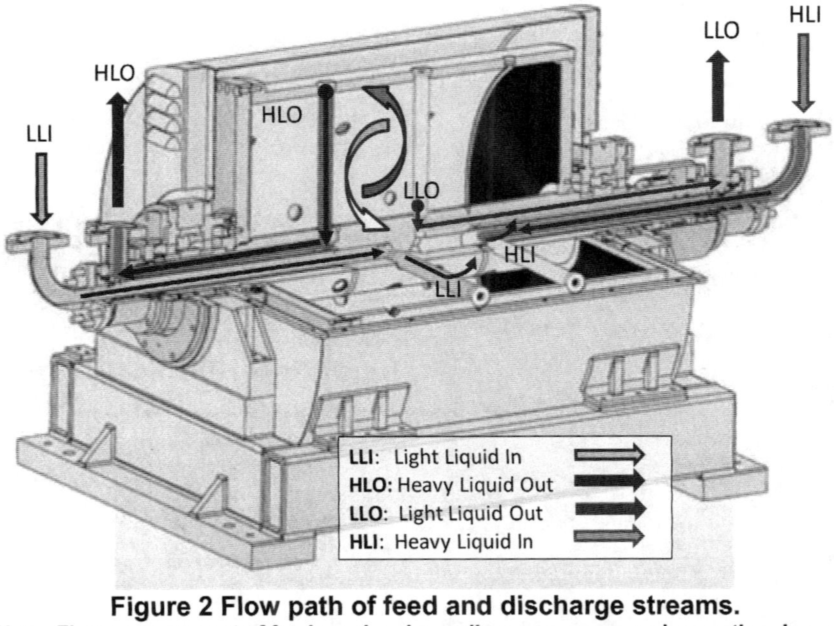

Figure 2 Flow path of feed and discharge streams.
(Note- Flow arrangement of feeds varies depending on process and operational needs)

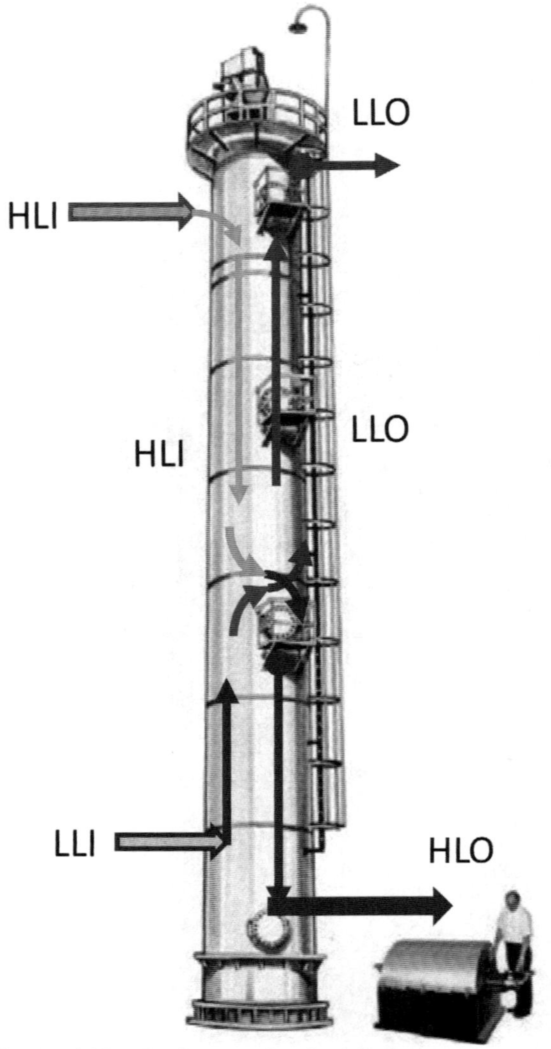

Figure 3 Equivalent stripping Column flow paths

Figure 4 Picture of concentric elements in Pod Contactor Rotor

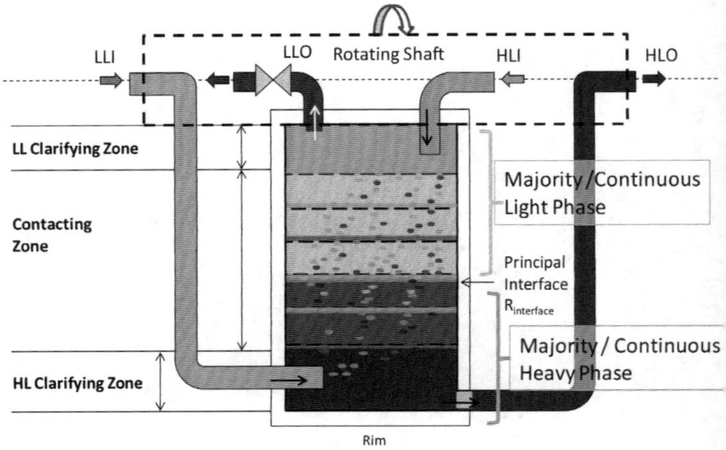

Figure 5 Tower Analogy of Centrifugal Extractor [1, 2]

The radial flow path in a Liquid-Liquid contacting centrifuge is equivalent to a separation tower (or equivalent device) where the Heavy Liquid (HLI) enters near the top of the vessel and works its way past the light phase, through the column trays and to the Heavy Liquid discharge (HLO) on the bottom of the unit. Conversely, the Light Liquid phase will enter (LLI) near the bottom and float is way up to the Light Liquid Out (LLO) discharge.

Physically, the top of the tower corresponds to the rotor shaft discharge path of the LLO; the bottom tower HLO discharge is equivalent to the centrifuge outer rim position; the Pod Contactor HLO discharge path is via discharge tubes which isolates it from the contacting phases and carries it to the shaft to exit the machine.

The Feed streams (LLI & HLI) will enter at feed tubes with discharge holes positioned at radial positions approximately equivalent to the LLI & HLI tower positions

Some contacting centrifuges (such as B&P Centrifugal Extractors) use shaft concentric metal band "elements" (Note Figure 4) to provide multiple individual equilibrium separation stages- parallel to tower trays .

Liquid-Liquid contactors are used in many applications. Table I is a brief example of B&P Process Pod's applications. The full range of processes is in the hundreds or thousands.

Service	Generic case	Some Specific Examples
Petroleum Refinement	Lubrication oil extractions	Color/ impurity removal
	Acid / base/ neutralization treatment of feed stocks for impurity removal	Paraffin Purification, :"White" oil (Hi Purity paraffin), Transformer Oil, Sulfonic Acids
	Acidulation/ caustic treatment for product recovery	Napthenic Acid recovery
Liquid Ion Exchange	Metal recovery	Uranium, Vanadium, Rare Earth Elements
	Acid recovery	Citric Acid recovery
Extraction	Acrylic Ester	
	Aromatic / Flavor oil recovery	Citral (Lemonal), Terpenoids, others
	Pharmaceutical Extractions	Penicillin, Erythromycin, Tylosin, other antibiotics
	Plasticizers	Fumaric Acid /Maleic Acid extraction from DiButyl Phthalate
	Aniline (phenylamine / amino benzene)	
	Butadiene Extraction	Butadiene extraction
	Vitamin recovery	Tocopherol (Vitamin E), Ascorbic Acid (Vitamin C)
Wash	Water Wash	
	Polyols (multi Hydroxyl alcohols, Poly esters)	Precursor for many polymer processes
	Nylon intermediates	
	Polycarbonate wash	

Table I Short Example list of applications [4]

2.1.2 Scaling

Continuous Liquid-Liquid Extraction centrifuge equipment and its understanding remains an under researched area. Historical manufacturers of Liquid-Liquid extractors include Podbielniak Extractors (B&P Process Equipment, formerly: APV Chemical Division, Baker Perkins Inc, Chemical Machinery, Dresser Industries Podbielniak [3]), Alpha-Laval (Sweden), GEA Westfalia Separator GmbH. How scale up is conducted by current manufactures may vary. This presentation covers the current practices of B&P Process Equipment in line with its internal design policies. These practices are from 1st principles, pilot testing and manufacturing experience. Scale up is facilitated by machine designs which accommodate scaled and similar physical concepts such as similar internal liquid velocities, residence times, design available G-fields, pressure operations, etc.

2.1.2.1 Scaling by B&P Process Equipment

Pod Contactor scale up is generally based on testing results from a smaller pilot scale test machine. For best pilot scaling B&P uses the B-250 (~40 GPM) unit. The smaller A-25 model (0.4 GPM) is also used as a proof of concept machine, but scalability is not as well reflected. There is a recent reintroduction of the A-125 model that is expected to scale better to the production size units. Machines may be scaled from one production model to the others by the concept of G-minutes. This is the mathematical product of the multiples of gravity (G's) and the "Total Traffic" residence time (τ_{TT}).

$$\text{G-minutes} = (G's) * \tau_{TT}$$

Equation 1 G-minutes

G's = Rim Centripetal acceleration/ 1 g

τ_{TT} = Residence time for total feed rate

$$\tau_{TT} = \left(\frac{V_i}{Q_{TT}} \right)$$

Equation 2 Total Traffic Residence Time

V_i = Hold up volume of considered Centrifugal Extractor

Q_{TT} = Total Volumetric Feed rate

$$Q_{TT} = Q_{HL} + Q_{LL}$$

Equation 3 Total Volume Traffic

Q_{LL} = Light Liquid Volumetric Feed rate total

Q_{HL} = Heavy Liquid Volumetric Feed rate total

(Equation 3 may be $Q_{LLI} + Q_{HLI}$ or $Q_{LLO} + Q_{HLO}$)

When the G-minutes from one model to the other are matched, the Extraction stages or separation quality is conservatively expected to be the same. This is supported by many years of observed operating experience. Centrifuge capacity varies from process to process and scale up continues to rely on initial pilot scale trials.

Historical process capacity information may also be used for scaling of similar chemical processes provided it is both available and a close match for process operation conditions. Historical Data does have its limits though. Chemically similar processes can behave differently due to process subtleties, affecting system capacity. Additionally what is defined as acceptable product for one user to may not be acceptable for another user. And certainly, new processes may not have data or operating conditions available.

Machine design also accommodates a "Traffic Velocity" which is set to establish heavy & light liquid fluid velocities at the Pod Contactor elements. Traffic Velocity is set to values known to accomplish effective process performance- IE emulsion prevention, effective mass transfer, and phase separation.

$$\upsilon_{\text{Total Traffic}} = \frac{Q_{TT}}{\Sigma A_{\text{Flow Area}}}$$

Equation 4 Traffic Velocity calculation

υ_{TT} = Total Traffic Velocity (for individual element)
Q_{TT} = Total Volumetric Feed rate
$\Sigma A_{\text{Flow area}}$= Σ element flow area

Effective process performance judgment is based on observed process characteristics for a process and/ or chemical families. Some of these families include pharmaceutical/ fermentation separations, vitamin extractions from fats or oils, petroleum refinement oil acidulations, clean fluid extractions or other generic families of processes. Other process performance considerations include viscosity or solids content that may or may not be an additional specific process concern.

When B&P refers to Centrifugal Extractor process capacity, the reference is to a specific well known Liquid- Liquid Extraction process: Methyl-Isobutyl Ketone, Acetic Acid and Water. It is a clean service (IE no solids) with a nominal SG difference (ΔSG) of 0.2- which is typical for many water- oil type systems. For a fermentation process separation, capacity is expected to be less than ½ of clean service capacity. Penicillin separation process is the "benchmark" for fermentation services. Different chemical processes can have different capacity ratings. Pod Contactors are able to accommodate down to 0.01 ΔSG, although a capacity reduction occurs as ΔSG gets very small.

2.1.3 Extraction efficiency and stages

The extraction efficiency of centrifugal extractor operation is quantified using the same methods used by other extraction or stripping processes equipment. Theoretical Transfer Stages can be used to quantify process efficiency and be determined by thermodynamic process practices such as these:

- Counter current theoretical stages from the right triangle method from system equilibrium.
- The shortcut McCabe-Thiele stepwise calculations can be used by stepping off the number of stages between the operating line and the equilibrium line.
- The NTU (numbers of transfer units) may also be used to express efficiency.

Figure 6 is an example of the A-25 model centrifugal extractor demonstrated capacity for Water solvent Extraction of Acetic Acid from Methyl Isobutyl Ketone Feed- often used as a bench mark for quantifying centrifugal extractor capacity and comparing with other Liquid-Liquid Extraction behaviors:

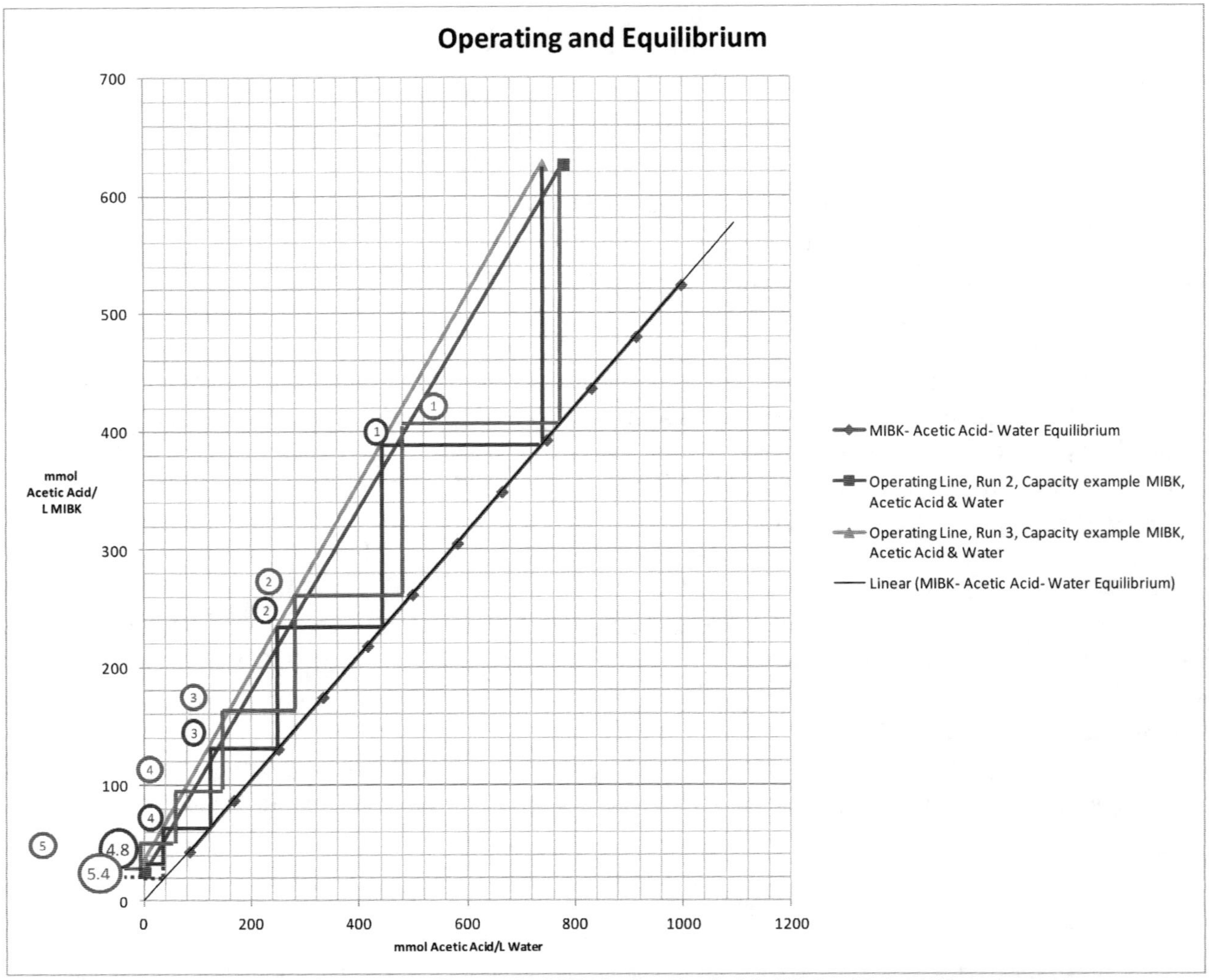

Figure 6 A-1 Pod Contactor Example MIBK- Acetic Acid- Water Extraction stages- Run 2: 5.4 Stages, Run 3: 4.8 Stages.

Both Runs feed rates are MIBK (LLI) = 551 mL/min, H2O (HLI) 755 mL/min (~80% of Flood capacity) at ~38C with standard extraction elements. Run 2 yielded 5.4 Stages and Run 3 yielded 4.8 Stages.

	mmol Acetic Acid / 1 L phase		mmol Acetic Acid / 1 L phase	
	MIBK Light	Water Heavy	MIBK Light	Water Heavy
Graph	Y_1	X_1	Y_2	X_2
Equilibrium Line	43.6	83.0	523.7	996.1

Operating Lines		755 mL/min	551 mL/min	
Graph Axis ID	Y_1	X_1	Y_2	X_2
McCabe Theile Designations	x_N (L_N) Raffinate	y_{N+1} (V_{N+1}) Solvent	x_0 (L_0) Feed	y_1 (V_1) Extract
	LLO	HLI	LLI	HLO
Run 2	25	0	626.5	778.5
Run 3	36.5	0	626.5	738.5

Figure 7 Composition Data for Figure 6

Similar extraction efficiency understanding can be developed for each process under consideration. B&P uses MIBK-Acetone- Water Ternary system for extraction efficiency expression since equilibrium data is readily available. Better understanding can be developed by the amount of testing conducted for an application. End users will often have better equilibrium data available specific to their process and can develop further separation/ extraction performance data as needed.

2.1.4 Effects of Viscosity

Increasing viscosity decreases the Pod Contactor capacity. When a new process without capacity test data is considered for anticipated capacity, it is de-rated by an amount approximately proportional to the inverse of viscosity to 1/3rd power. If pilot scale testing data becomes available, that data is preferred.

This viscosity capacity relationship can be illustrated with this chart and is based on demonstrated historical capacity from various process testing:

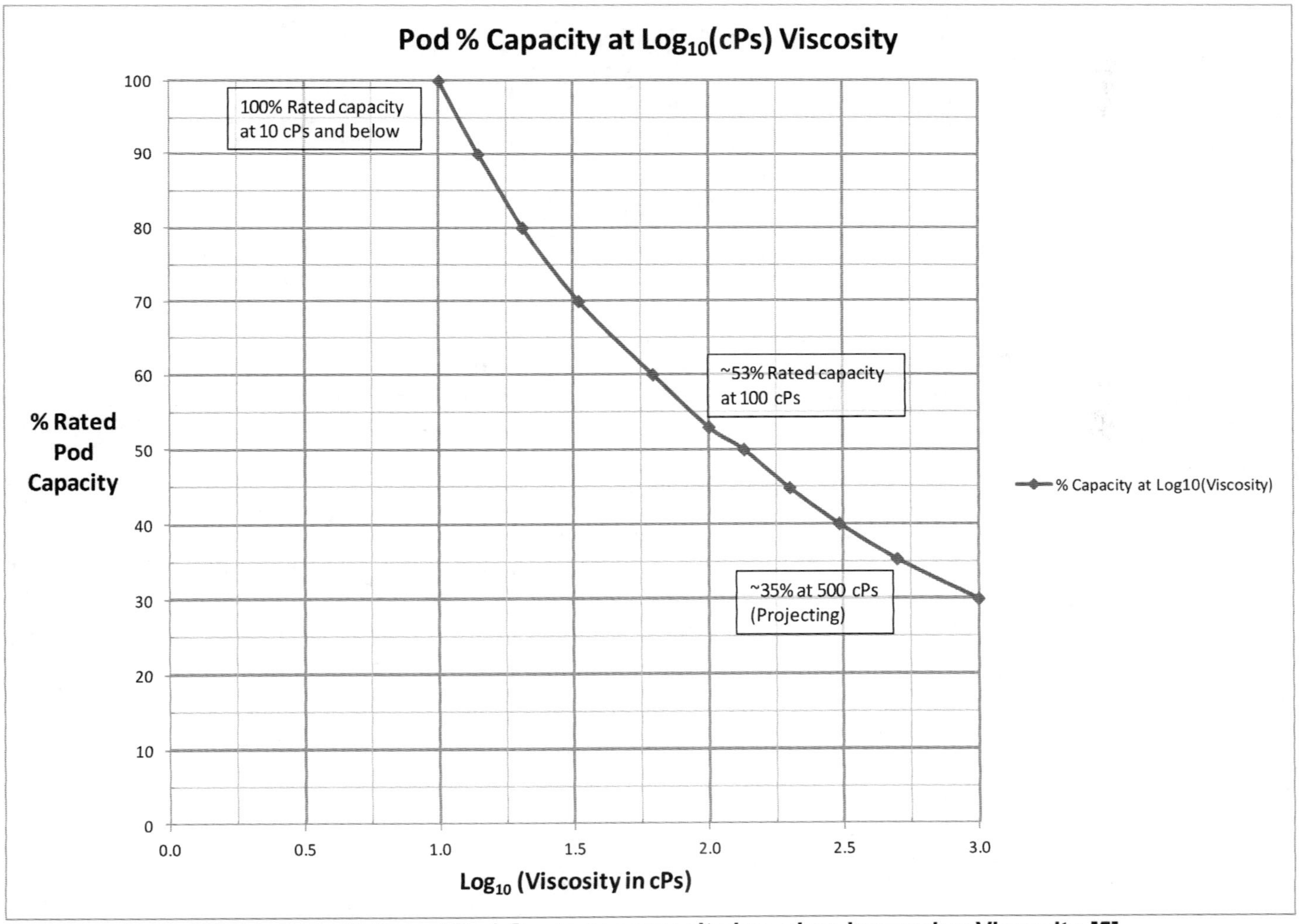

Figure 8 De-Rating of Pod Contactor capacity based on increasing Viscosity [5]

2.2 Operation of a Pod Contactor:

The internal pressure for the Centrifugal separator is treated as a Centrifugal manometer [1, 7, 8] - where the pressure is relative to the radial "height" of the column of liquid. The outer Rim equates to the bottom of the tower and the center shaft equates to the top of the tower.

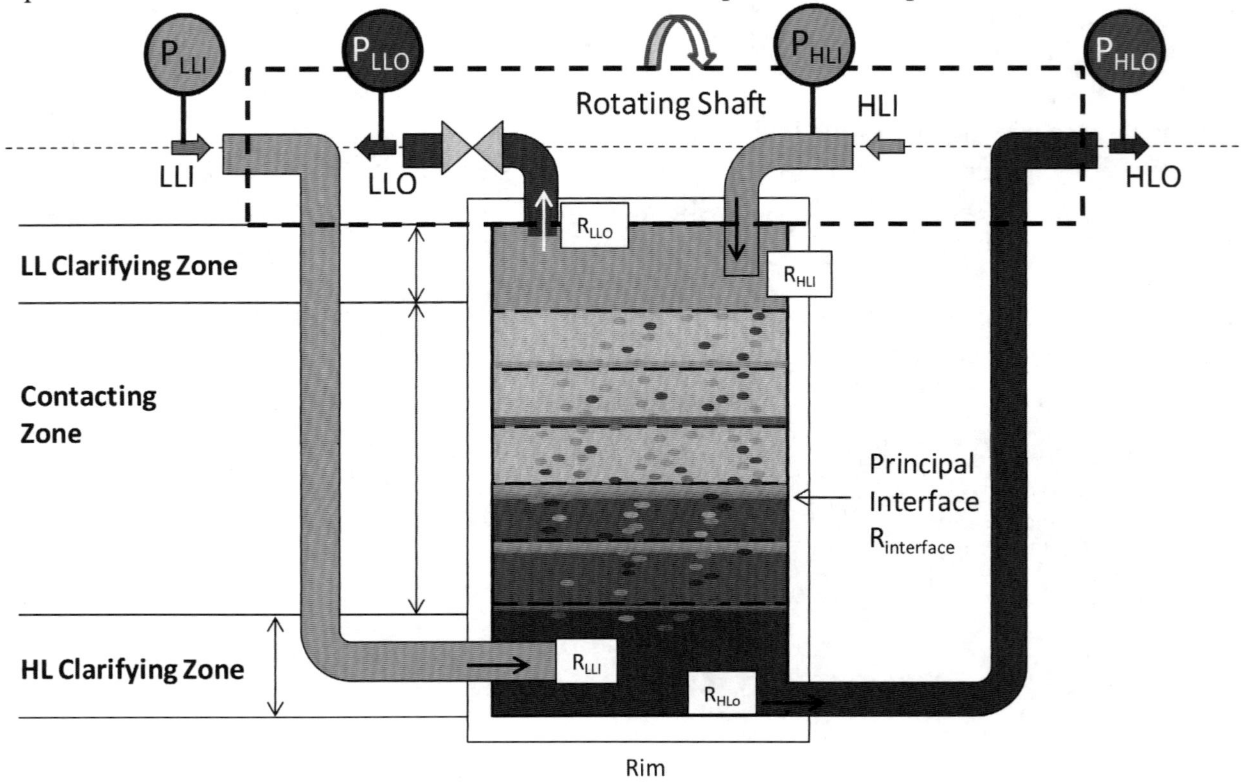

Figure 9 Illustration of Centrifugal Manometer

Streams and process conditions are identified below in Table II by a series of B&P often used abbreviations. Many will be seen in the upcoming operations and equation descriptions.

Variable	Definition
Di	Diameter position of "i" relative to shaft Center line (D_{Rim}, $D_{interface}$, D_{LLI}, D_{HLI}, D_{LLO}, D_{HLO}, etc)
HL	Heavy Liquid
HLI	Heavy Liquid In
HLO	Heavy Liquid Out
LL	Light Liquid
LLI	Light Liquid In
LLO	Light Liquid Out
Pi	Pressure of "i" (P_{LLI}, P_{HLI}, P_{LLO}, P_{HLO}, etc)
Pri	Pressure at Radial position of stream i (P_{RHLI}, P_{RLLI}, P_{RLLO}, P_{RHLO}, P_{RIM}, etc)
Ri	Radial Position of "i" relative to shaft center line
$R_{interface}$	Radial Position of principle interface boundary
SG_i	Specific Gravity of considered species (SG_{HLI}, SG_{LLI}, SG_{LLO}, SG_{HLO}, etc)

Table II Variables around Centrifugal Manometer

Pod Contactor operation is controlled by the differential pressure (P'LLO or ΔP'LLO) of the Light Liquid Out, Heavy Liquid Out. Increased pressure differential ΔP'LLO prevents the Heavy Liquid from overcoming the Light Liquid phase and permits the phases to separate for discharge from the machine. If ΔP'LLO is too high, the light liquid passes out with the heavy liquid. Likewise if it is too low, heavy liquid will pass out with the light liquid stream. Adjusting the ΔP'LLO also sets the radial position of the internal Principle (Major) Interface Boundary and can manage separation quality. The differential pressure (ΔP'LLO) is adjusted by altering the Light Liquid Out back pressure- P_{LLO}.

Back pressure adjustments made at the HLO position-P_{HLO}- do not help the differential pressure to change. Changing the P_{HLO} changes total system pressure; IE P_{LLI}, P_{HLI}, P_{LLO} all change by the same ΔP as the pressure change at the P_{HLO}. There may be process reasons to operate at a higher P_{HLO}, but generally it is not a separation quality control parameter.

The Principle Interface Boundary is a radial boundary where concentric element chambers transition from continuous Light Liquid into a continuous Heavy Liquid. (IE the Light Liquid is continuous for element bands inboard of the interface boundary and the Heavy Liquid is continuous outboard of the Interface):

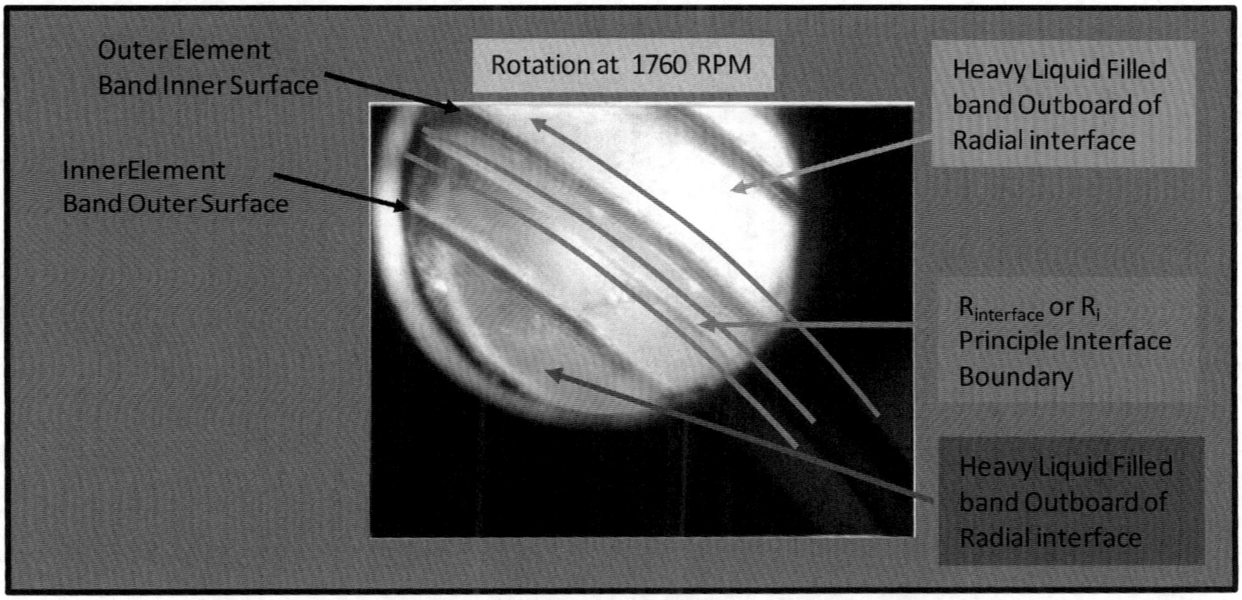

Figure 10 Interface boundary for Pod Contactor operation. [7]

Figure 11 shows the Radial liquid distribution images:

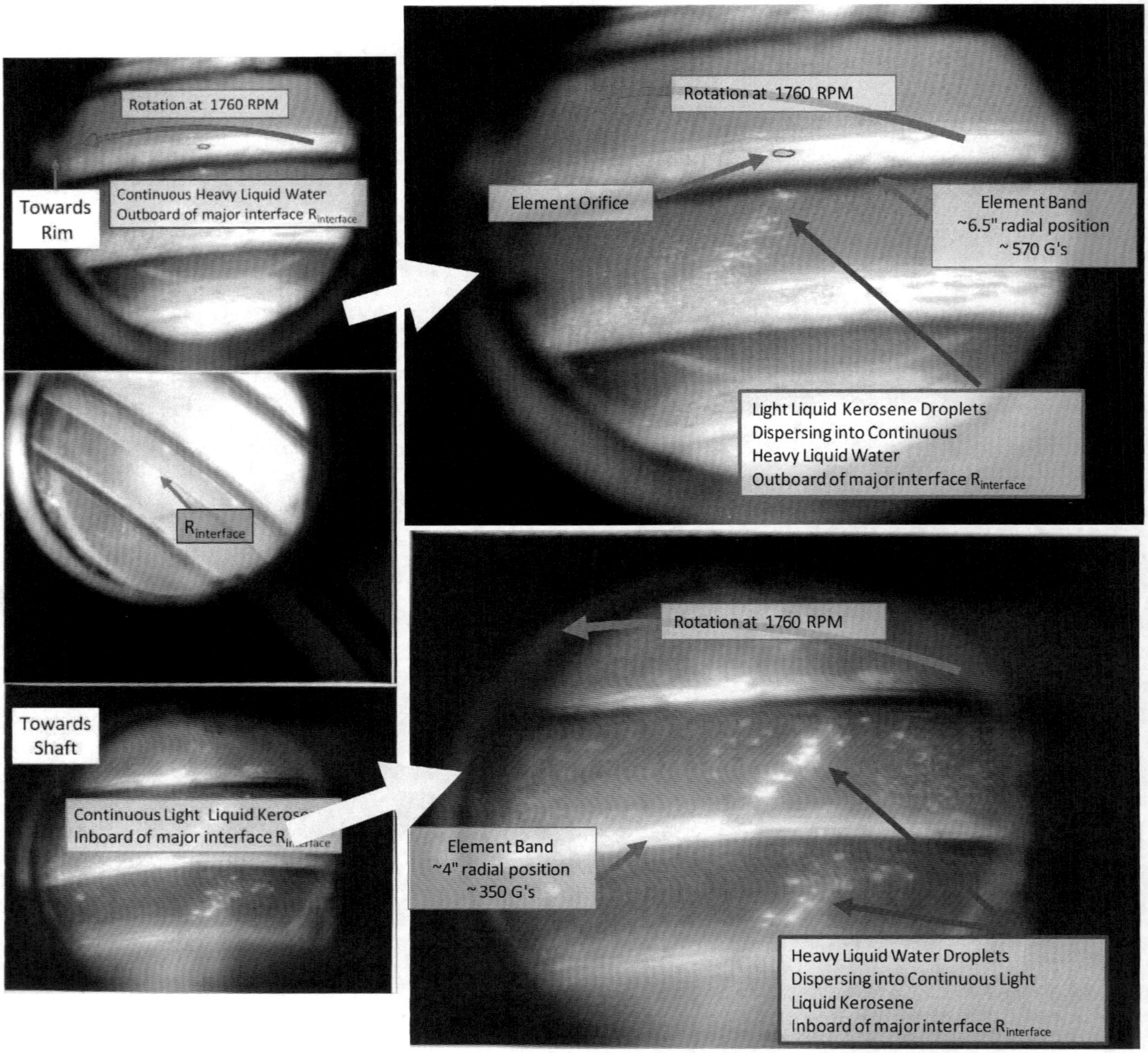

Figure 11 Clear window Pod Contactor images illustrating phase breakup and droplet formation at elements [7]

Pod Contactor rotor speed may also be used to manage Pod Contactor operation. Increase of extractor rotational speed increases the G-field the feed streams are exposed to. It can be used to vary separation quality or capacity- provided the rotor speed remains within the limits of the machine.

Two additional methods for separation / extraction management are the element design and the feed ratios. Element sets can have open area decreased to improve turbulence & extraction efficiency; or the area can be increased to accommodate problem emulsifications, high viscosity or –provided quality remains acceptable to the end user- allow higher feed rates. Feed ratios can be varied so that solvent use is optimized for extraction. Again though, both of these design parameters are limited by the required process quality requirements.

Temperature and operating pressure of the system can be manipulated, but these are generally set by the individual process needs. These can be evaluated during testing based on anticipated or current process conditions. Sometimes temperature increases due to seal friction needs to be managed with higher flow rates, upstream heat exchangers or buffer fluids. Operation adjustments can be made as process experience is gained. Only direct Pod Contactor operation variables are discussed here.

2.2.1 ΔP'LLO- Differential Pressure and Interface boundary ($R_{interface}$)

The ΔP'LLO is used to control where the principle interface boundary ($R_{interface}$) is positioned. The pressure values may also be used to determine that boundaries' radial position. This is relevant due to operating concerns. If the interface boundary is too close to the shaft LLO exit, there can be loss of separation at the LLO position. If it is too close to the HLO exit, there can be loss of separation at the HLO. Operation with $R_{interface}$ too close to one side or the other and the process can be much more susceptible to minor process upsets such as composition, pressure fluctuations, feed rates or other process variables. The best Operating ΔP'LLO (and hence $R_{interface}$ boundary) can be determined from which tests yield the best process quality acceptable to the end user. Additional refinement can be made with gains in process operation experience.

For extraction cases, sometimes there are large gains in extraction efficiency by adjusting the $R_{interface}$. Other times there is little or no gain. Typically the HLI & LLI feed diameters are positioned so the $R_{interface}$ ($D_{interface}$) is midway between the feed positions. Sometimes interface placement is best for just outboard of the LLI or just inboard of the HLI positions. Unfortunately, which operation is better remains an empirical understanding and relies on testing or process experience.

For separations, the best feed may be the inner radial heavy feed position if the feed is mostly heavy phase and low content of the light phase. The outer light feed position may be better if it is a majority of light phase components. But this may not always be the case and a fully intermediate position may be better for the given process. Testing a single stream from either the HLI or the LLI position can yield comparative data for what is an optimal feed location.

The outlet stream differential pressure is calculated as follows:

$$\Delta P'LLO_{TT} = P_{LLO} - P_{HLO}$$

Equation 5 Outlet Stream Differential Pressure

P_{LLO} = Pressure on Light Liquid Discharge

P_{HLO} = Pressure on Heavy Liquid Discharge

To get to pressures inside of the Pod Contactor Rotor, several equations from physics are used. Since pressure is Force/ Area, the first equation will be Force.

$$Force = m * a$$

m = mass

a = acceleration (g for a manometer pressure calculation)

For uniform circular motion, the magnitude of centripetal acceleration a_r, is:

$$a_r = \frac{|v|^2}{r} = r\omega^2$$

a_r = magnitude of centripetal acceleration

v = speed in distance/sec

r = radial position under consideration

ω = angular speed (s^{-1}) = (N rpm)*(2π/60)

As in a conventional manometer, the pressure (force/ area) is a product of the (mass * acceleration) over area. Extending this definition to find pressure P at any point R on a rotating cylinder of liquid and neglecting effects of gravity due to orientation,

$$P = \frac{Force}{Area} = \frac{m * a_r}{A} = \frac{(\rho * Vol) * a_r}{A} \implies \frac{(\rho*(\pi r^2 w)) * (r\omega^2)}{2\pi r w} \implies \frac{\rho r^2 \omega^2}{2}$$

$$\frac{dP}{dr} = \rho\omega^2 r \implies dP = \rho\omega^2 r\, dr$$

$$\int_0^P dP = \int_0^R \rho\omega^2 r\, dr \implies P = \frac{1}{2}\rho\omega^2 R^2$$

To express this equation in Specific Gravity, Rotational speed in RPMs and generic units of radial length, this becomes:

$$P_i = \left(\frac{2\pi^2 \rho_{water}}{3600}\right) SG_i N^2 R^2 \implies \left(\frac{\pi^2 \rho_{water}}{1800}\right) SG_i N^2 R^2$$

From this, the following expressions can be used for English customary units:

$$P_i = \left(0.513 * 10^{-6}\right) N^2 R_i^2 SG_i$$

Equation 6 Single Phase Liquid Leg Pressure at radial position

P_i	=	Liquid Leg Pressure in PSI
N	=	Rotor speed in Revolutions/ Minute (RPMs)
R_i	=	Liquid leg radial position (Inches)
SG_i	=	Specific Gravity of liquid leg under consideration (Dimensionless)

This expression can be used for SI units:

$$P_{i\,(N/m^2)} ==> \left(5.483\right) SG_i N^2 R_i^2$$

Equation 7 Single Phase Liquid Leg Pressure (SI)

P_i	=	Liquid Leg Pressure in N/m^2
N	=	Rotor speed in Revolutions/ Minute (RPMs)
R_i	=	Liquid leg radial position (meters)
SG_i	=	Specific Gravity of liquid leg under consideration (Dimensionless)

The interface boundary can also be obtained from the pressure balance of the heavy and light liquid. This equation can be used with English units:

$$\Delta P'LLO = \left(0.513 * 10^{-6}\right) N^2 R_{interface}^{\ 2} \Delta SG_{HL\text{-}LL}$$

Equation 8 ΔP'LLO at given major interface boundary $R_{interface}$

Rearrangement to solve for $R_{interface}$ becomes

$$R_{interface} = \sqrt{\frac{\Delta P'LLO}{\left(0.513 * 10^{-6}\right) N^2 \ \Delta SG_{HL\text{-}LL}}}$$

Equation 9 Major Interface boundary position calculation

$R_{interface}$	=	Interface position from shaft center in inches
N	=	Steady State rotational speed in RPM
ΔP'LLO	=	Outlet Differential Pressure in PSI
$\Delta SH_{HL\text{-}LL}$	=	SG Difference between Heavy and Light Liquid

Other equations appropriate to available units can be derived.

2.2.2 Effects of Rotor Speed on Capacity

Centrifugal extractors may be operated at different speeds depending on process needs. B&P design limits rim pressure are approximately at ~625 PSI differential pressure. Since it is a pressure limitation, the Specific Gravity of the heaviest phase will set the rotor speed limits for a given process. The rotation speed limit becomes reduced by the following relationship:

$$N_{Max} = N_{model\ RPM\ Limits} * \sqrt{\frac{1.0}{SG_{HL}}}$$

Equation 10 Rotational Speed De-rating due to SG>1.0

N_{max} = Model RPM limit for process ($SG_{HL} > 1$)

$N_{Model\ RPM\ limits}$ = RPM limits for machine under consideration

SG_{HL} = SG of heaviest condition of Heavy Liquid

(See Table VI In Appendix for values suitable for B&P machines)

The Model RPM limit is not increased if the involved process fluids have an SG < 1.0. This is due to various design limits and not specifically associated with the pressure limits.

Gains may be made for the process by running slower or faster. With respect to machine process capacity, the increase in RPMs expands the feed rate capacity as illustrated in Figure 12 and Figure 13.

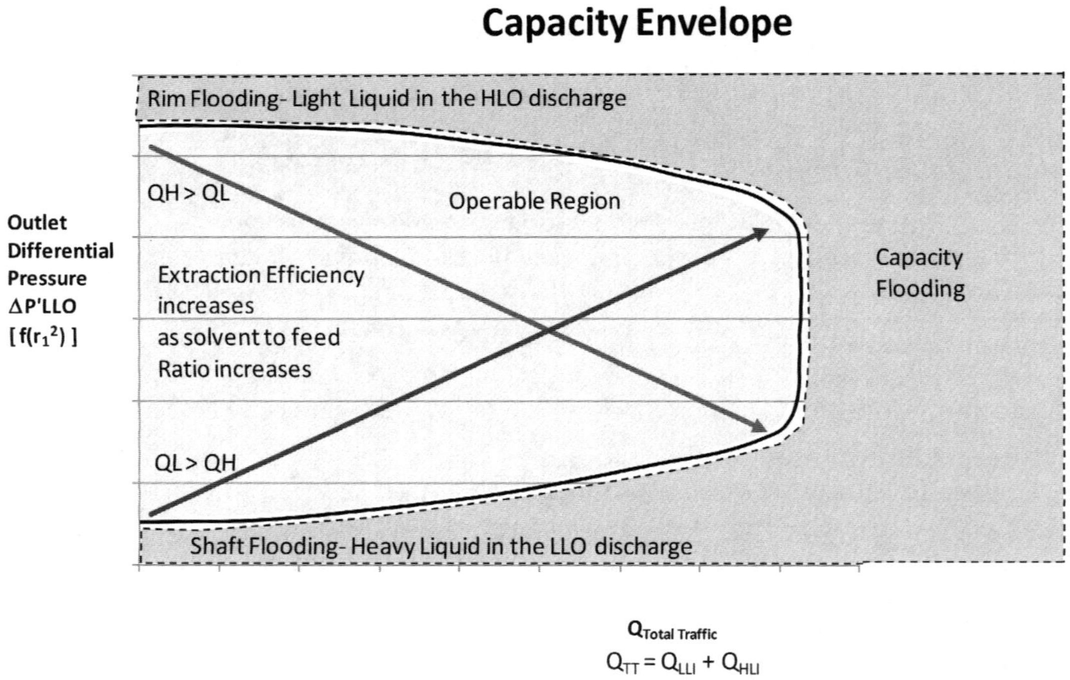

Figure 12 Qualitative Operable Capacity Envelope for Centrifugal Extractor [6]

When the RPM is increased or decreased, the operable region expands or contracts as illustrated in Figure 13

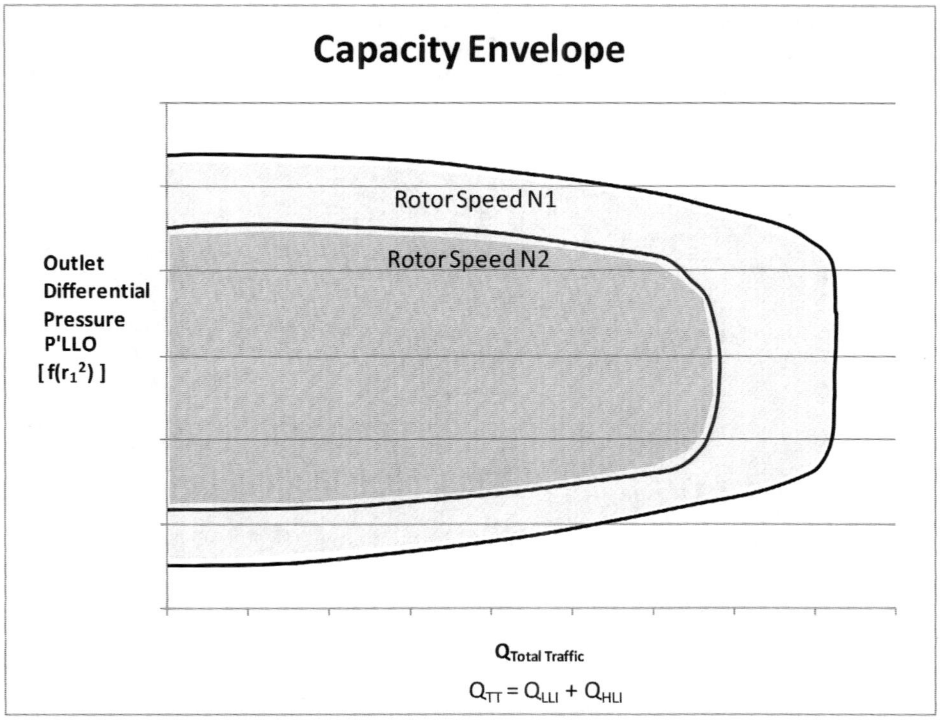

:

Figure 13 Effect of Rotor Speed Variation- N1> N2. [1, 6]
(Speed Variation must be within rotor speed limits for design and process)

2.2.3 Effects of Internal element design

Specific demonstrations of the above capacity envelop have been documented with the Kerosene & Water system. Similar behavior has been noted with other processes. Here is an illustration of the Kerosene & water capacity envelope:

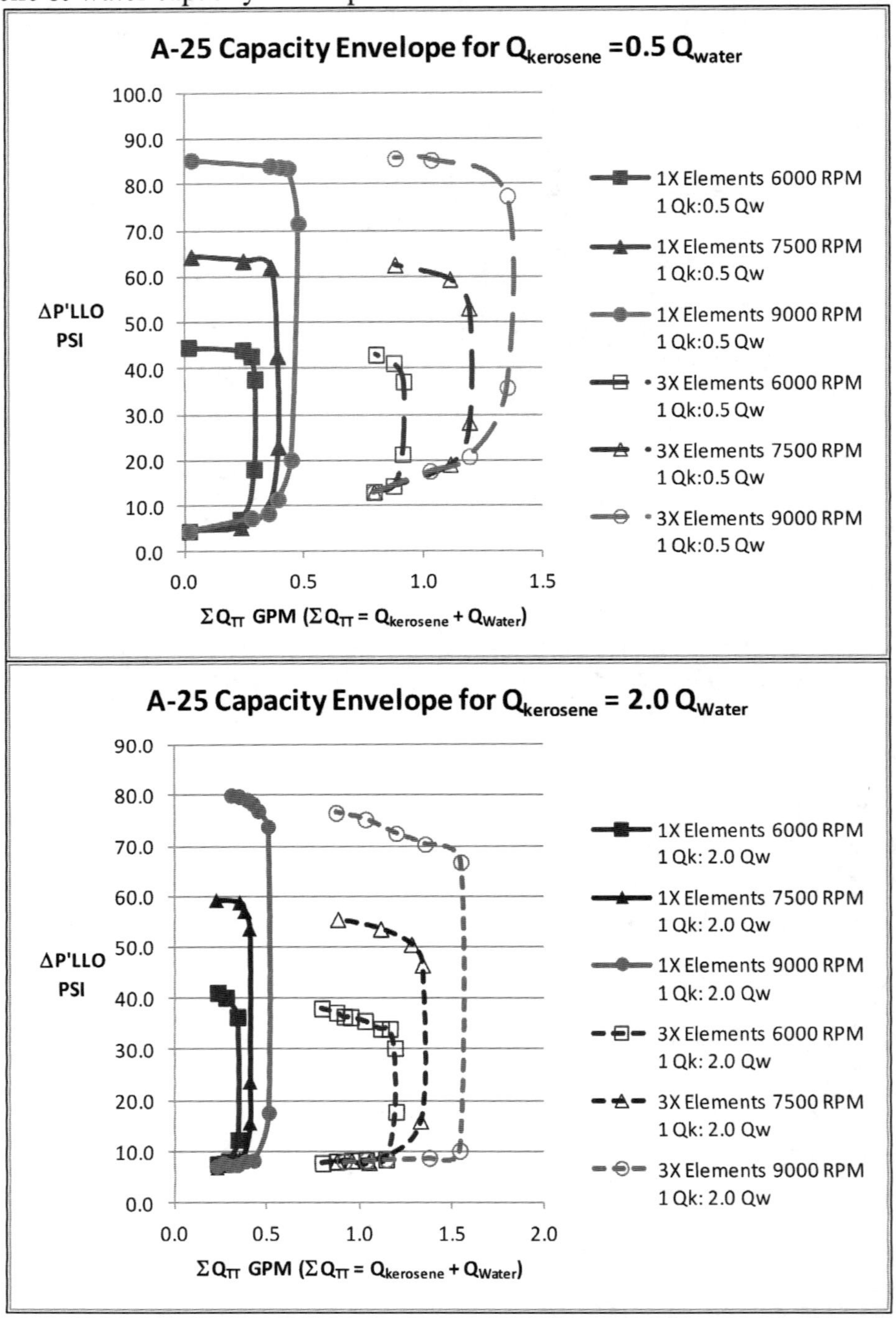

Figure 14 Flood capacity envelope for two feed ratios of Kerosene & water[1]

Two styles of elements were tested for the Flood capacity charts. The 1X element set is a "standard" A-25 extraction element set and is the set most often used with the A-25 demonstration unit. It features 4 to 6 holes per sheet metal contacting element band and there are ~10 of 14 bands in the contacting zone. The 2 inner most and outer most elements have more open area and allow coalescing of the light and heavy phases before they pass out of the machine. The 1X set is also used with the example MIBK- Acetic Acid- Water extraction and is selected due to its observed better extraction performance.

The "3X" set carries 3 times the number of holes/ single element and was a set used to explore possible capacity increases when increasing the Percent open area in the machine's element set. This increase in open area results in lower traffic velocities at the process streams. This can be good for easily emulsifying systems or if high viscosity is present. The process cost however can be a reduction in extraction efficiency; likely attributable to lower velocity & turbulence. Lower velocity & turbulence allow larger droplets, reducing exposed phase surface area. Conversely, higher velocity results in higher turbulence and more & smaller droplet formation, tending to promote mass transfer.

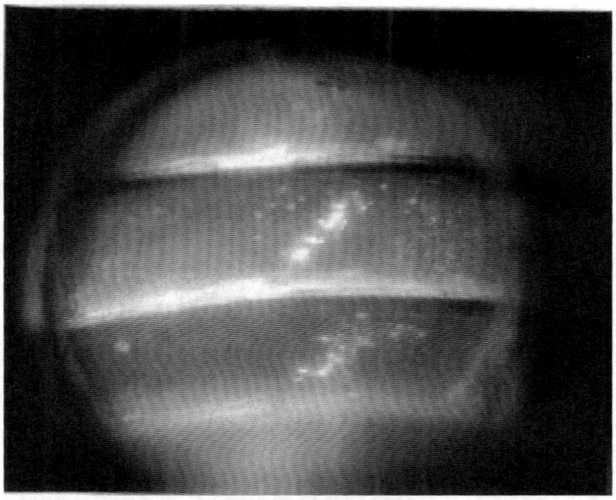

Figure 15 Example droplet dispersion Heavy Water droplet into Continuous Light phase

2.2.4 Effects of varying feed ratios

Figure 16 is a comparison of the 1X element with respect to variation in the Light ($Q_{Kerosene}$) & Heavy (Q_{water}) liquid feed ratios (Excerpt of same data in previous charts):

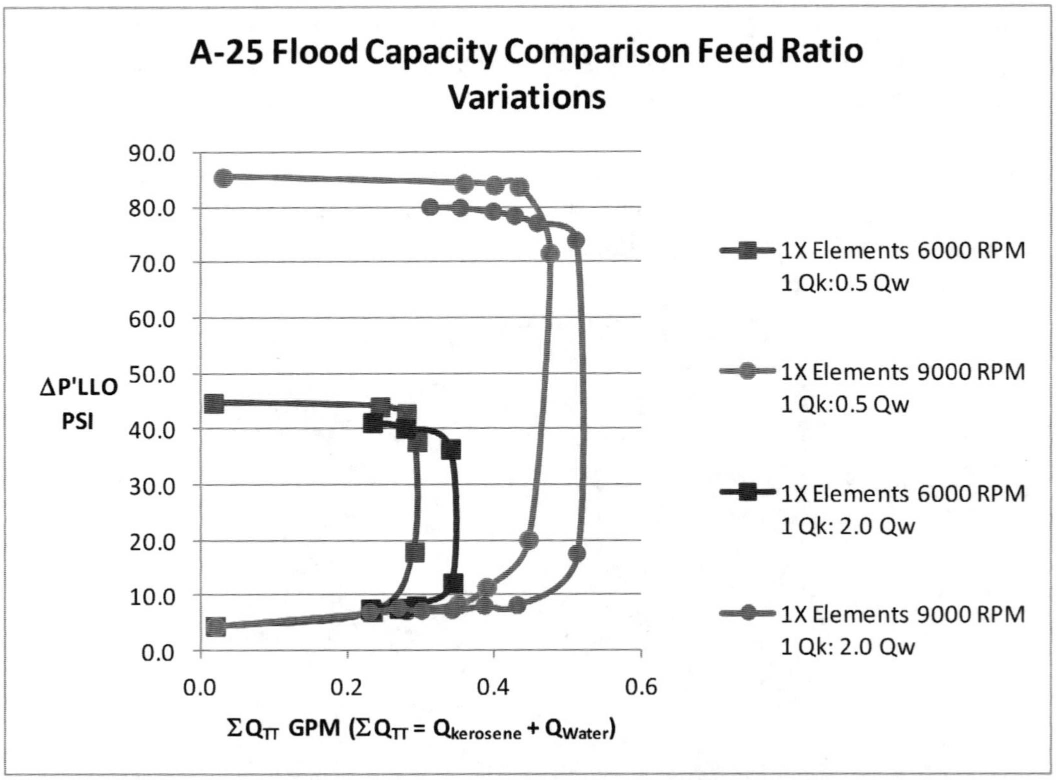

Figure 16 Comparison of Kerosene/ Water system Flood capacity when Feed ratio is changed

The ΔP'LLO envelope contracted somewhat, but for this particular system the flood the P'LLO values remain relatively close (at 9000 RPMS, ~85 PSI ΔP for 1 $Q_{kerosene}$: 0.5 Q_{water} versus ~79 PSI ΔP for 1 $Q_{kerosene}$: 2.0 Q_{water}) on the Rim Flooding (LLO in the HLO).

B&P observes and expects this same behavior in other Liquid- Liquid extractions. Full mapping of the capacity envelope can be plotted, but requires a large number of tests. Often for limited time and resource initial pilot tests, feed ratios are matched to existing process gravity separation equipment and full capacity is not fully charted. Some feed ratio variation can be made based on test performance and the amount of material available, but overall, the interest level of the process owner contributes to how well a capacity envelope is defined.

Some processes get an initial operation and the party is satisfied with quality without further process optimization. Other times, operational or historical experience offers a wider view of the capacity envelop and may offer better or alternative feed ratios to consider for the pilot scale testing. Because extensive material may be required, conducting a pilot plant Pod Contactor test at the customer's site can be more advantageous than shipping material for a remote site trial.

2.2.5 Existing process troubleshooting

As a final note concerning operation for those who may already be using these machines-
When doing process troubleshooting for centrifugal contactors consider these process troubleshooting tips:

- If there is Heavy Liquid in the Light Liquid Out (LLO) position, the Light Liquid Out pressure (P_{LLO}) may be too low
- If there is Light Liquid in the Heavy Liquid Out (HLO) position, the Light Liquid Out pressure (P_{LLO}) may too high
- If there is entrainment at both positions at previously demonstrated Differential pressure ($\Delta P'LLO$ / P_{LLO} / P_{HLO}), the feed rate may be too high.
- A seal leak may be suspect if entrainment is on one machine side or the other at a previously demonstrated successful feed rate (Q_{TT}) and operation differential pressure ($\Delta P'LLO$).

For further and more in-depth analysis, the cited sources carry more details and further background than can be covered in this discussion. Copies of the Podbielniak specific articles may also be obtained from B&P process Equipment upon request.

3. Conclusions:

Liquid-Liquid Contacting centrifuges such as Pod Contactors used for extractions and/ or separations remain a valuable tool in many commercial chemical processes. These may be used as a simple two phase separation, some distillation applications that could be preferentially accomplished by solvents (possibly due to temperature concerns for product), tower stripping/ rectification of immiscible phases or any gravity settling application that may benefit in process purity due to the higher G field. Their smaller size can also be advantageous where large process equipment hold-up can be a concern or where better product separation quality for less space is desirable.

References remain vague when discussing what is considered for scaling of continuous extraction centrifuges and this paper covered what B&P Process Equipment considers when evaluating and scaling an application. Other methods may be devised, but B&P has been using these practices with success for many years.

Factors that optimize process operation include

- Adjusting thermodynamic equilibriums available to the process during passage through the unit
  - Flows rates to adjust residence time τ
  - Operation pressures
  - Differential pressure
  - Temperatures.
- Operational speed adjustments
  - Adjust to optimize mass transfer and/ or separation quality
  - Bypassing emulsify regimes of operations

- Fluid velocity profile management on internal centrifuge design-
  - Internal design adjustment to accommodate viscosity or emulsification
  - Adjusted internal turbulence/ mixing to increase phase mass transfer surface area

The method B&P uses to scale from one process rate to another relies on the mathematical product of the maximum multiple of gravity and the residence time. Ideally, when a process is to be scaled, a pilot plant test is conducted. This provides operation data such as best feed ratios and outlet differential pressures that will be applicable to the full scale production units. Scale up may then be accomplished with the outlined equations. See Table VI in the Appendix for examples of design values suitable for B&P Centrifugal Extractors.

Historical test data can also provide reliable scale up for operation and capacity provided operating conditions are closely matched. For many existing Liquid-Liquid contacting & separation processes this data may be available. In the absence of historical data or pilot testing, chemically similar applications may have close operation behavior and conditions, providing a basis for initial test planning and or potential scale up rates.

Applications include major chemical process steps such as separations, extractions, rectification, ion exchanges, neutralization & washings. Examples are found industry wide:

- Large bulk and general chemical production
  - Fumaric Acid extraction, Polycarbonate & poly-alcohol washing & others
- Pharmaceutical applications
  - Penicillin, Erythromycin, Vitamin extractions & others
- Petroleum Refining
  - Acid treatment, "White" oil purifications (High purity refined mineral oils), contaminate stripping, neutralization & washing steps & others
- Liquid ion exchange
  - Citric acid, rare earth metal ion recovery & others
- Food industries
  - Flavor and essential oil extractions & others

4. Appendix Data

	1X perforation 6000 RPM 1 $Q_{Kerosene}$: 0.5 Q_{Water}		1X perforation 7500 RPM 1 $Q_{Kerosene}$: 0.5 Q_{Water}		1X perforation 9000 RPM 1 $Q_{Kerosene}$: 0.5 Q_{Water}		3X perforation 6000 RPM 1 $Q_{Kerosene}$: 0.5 Q_{Water}		3X perforation 7500 RPM 1 $Q_{Kerosene}$: 0.5 Q_{Water}		3X perforation 9000 RPM 1 $Q_{Kerosene}$: 0.5 Q_{Water}	
	□		□		□		□		□		□	
Data Point	□P'LLO	GPM	□P'LLO	GPM	□P'LLO	GPM	□P'LLO	GPM	□P'LLO	GPM	□P'LLO	GPM
1	4.5	0.02	4.5	0.02	4.5	0.02	13.0	0.79	13.0	0.79	13.0	0.79
2	7.0	0.23	5.3	0.24	7.3	0.28	14.3	0.87	19.1	1.11	17.6	1.03
3	18	0.29	9.9	0.35	8.4	0.35	21	0.91	28	1.19	21	1.19
4	38	0.29	23	0.39	11.5	0.39	37	0.92	53	1.19	36	1.35
5	43	0.28	43	0.39	20	0.45	41	0.87	59	1.11	78	1.35
6	44	0.24	62	0.36	72	0.47	43	0.80	63	0.88	85	1.03
7	45	0.02	63	0.25	84	0.43					86	0.88
8			64	0.03	84	0.40						
9					84	0.36						
10					85	0.03						

Table III 1 Flood Rate & Differential pressure for $Q_{Kerosene}$: 0.5 Q_{Water} Table [1]

	1X perforation 6000 RPM 1 $Q_{Kerosene}$: 2.0 Q_{Water}		1X perforation 7500 RPM 1 $Q_{Kerosene}$: 2.0 Q_{Water}		1X perforation 9000 RPM 1 $Q_{Kerosene}$: 2.0 Q_{Water}		3X perforation 6000 RPM 1 $Q_{Kerosene}$: 2.0 Q_{Water}		3X perforation 7500 RPM 1 $Q_{Kerosene}$: 2.0 Q_{Water}		3X perforation 9000 RPM 1 $Q_{Kerosene}$: 2.0 Q_{Water}	
	□		□		□		□		□		□	
Data Point	□P'LLO	GPM	□P'LLO	GPM	□P'LLO	GPM	□P'LLO	GPM	□P'LLO	GPM	□P'LLO	GPM
1	7.6	0.23	6.8	0.23	7.0	0.23	7.8	0.80	8.1	0.87	8.1	0.92
2	7.7	0.27	7.3	0.27	7.7	0.27	8.1	0.87	8.2	0.96	8.3	1.03
3	8.1	0.29	7.4	0.30	7.2	0.30	8.2	0.96	7.9	1.05	8.5	1.14
4	12	0.34	8.5	0.35	7.3	0.34	8.3	1.03	16	1.33	8.8	1.38
5	36	0.34	9.2	0.38	8.0	0.39	8.5	1.14	46	1.34	10.2	1.54
6	40	0.28	16	0.41	8.1 PSI	0.43	18	1.20	51	1.28	67	1.55
7	41	0.23	24	0.41	17	0.51	30	1.19	54	1.11	70	1.35
8			54	0.41	74	0.51	34	1.16	56	0.88	73	1.20
9			57	0.38	77	0.46	34	1.11			75	1.03
10			59	0.36	78	0.43	36	1.03			77	0.87
11			59	0.23	79	0.40	36	0.95				
12					80	0.35	36	0.92				
13					80	0.31	37	0.88				
14							38	0.79				

Table IV 1 Flood Rate & Differential pressure for $Q_{Kerosene}$: 2.0 Q_{Water} Table [1]

Run Number	P (psig) Q (ml/min) T (°C/°F)	Light Liquid Out	Heavy Liquid Out	Heavy Liquid In	Light Liquid In	Remarks
	P					
	Q					
	T					
	P					
	Q					
	T					
	P					
	Q					
	T					

Table V Example of typical Contacting Centrifuge data collection during tests

B&P Process Centrifugal Contactor design dimensions							
Model	Inside Rim Diameter	Camber width	Hold up Volume	RPM limits at SG = 1.0 ($N_{Model\ RPM\ limits}$)	G's at RPM Limit	Clean Process rated	Fermented Process
	Inches	Inches	US Gal	RPM	Multiples of g	US GPM	US GPM
A-25	7.0"	1.0"	0.16	10,000	10,030 G's	0.40	0.20
B-250	23.0"	10.0"	15.0	3,200	3,340 G's	40	25
D-450	34.0"	18.0"	58.0	2,100	2,130 G's	150	65
D-900	34.0"	36.0"	116.0	2,100	2,130 G's	300	130
E-1200	44.0"	48.0"	260.0	1,600	1,630 G's	600	260

Table VI Generally accepted design parameters for B&P Contactor centrifuges [1]

3. References

[1] Dr. Todd, David, "Performance of Centrifugal Extractors", Proceedings of the International Solvent Extraction Conference", Lyon, Vol 3, 2380-2398, September, 1974, Print

[2] Bach, A, Chan, T, "Pod Presentation-Jan2014.PPTX", B&P Process internal documentation, Tech Center\Library\Presentation, 2014, electronic

[3] Aweibel, L, "History Over View", *Introduction & History.doc* B&P Process Pod Contactor internal documentation, Marketing\Tech Papers\PODS\1999 Pod Sales Manual, 1998, electronic

[4] Bush, Lisa, "3- Pod Contactor Applications List.doc", B&P Marketing documentation, Marketing\Sales Resource Manual\b - II Pod Contactor\2 - Overview, 2009, electronic

[5] B&P Process internal process documentation sources, "De-rating Pods due to Viscocity.pdf", ENGINEER\Calculations\Pods\Pod Process Design Manual\, (Various Dates), electronic

[6] "Capacity Envelope", *Pod Contactor Process*, 22, B&P Process internal process documentation sources- Pod Process Design Manual\01 Podbielniak Centrifugal Contactor Process Design Rev 1.pdf, (Various Dates), electronic

[7] Todd, D.B, Davies, G.R, Lange, H.A, "Behavior of drops in a centrifugal field", B&P Process internal process documentation sources- Pod Process Design Manual\Pod clear window rotor flow pictures & presentation\Behavior of Drops in a Centrifugal Field.pdf, (1999), electronic

[8] Davies, G.R. "The History of the Podbielniak Contactor", Tech Center\Library\POD\The History of the Podbielniak Contactor- Gordon R. Davies.PDF

Additional references not cited:

[9] Todd, D.B & Podbielniak, W.J., "Advances in Centrifugal Extraction", Chemical Engineering Progress, 61, No. 5, 69, 1965

[10] Todd, D.B, "Improving Performance of Centrifugal Extractors", Chemical Engineering, 79, No. 16, 152, 1972

[11] *Pod Contactor Process*, B&P Process internal process documentation sources- ENGINEER\Calculations\Pods\Pod Process Design Manual\01 Podbielniak Centrifugal Contactor Process Design Rev 1.pdf, (Various Dates), electronic

443334 Expanding Applicability of Integrated Separation Systems

Abstract:

Methathesis of mixed butenes to propylene and hexene will be used as a case study. An integrated separation system will be explored as a potential process intensification. This approach expands the applicability of such systems to processes involving a combination of reaction and multicomponent separations

444320 Process Development for Synthesis of Bio-Based Plasticizers

Abstract:

E3Tec and MSU are pursuing the commercialization of an integrated, energy-efficient process technology for the production of bio-based plasticizers. Organic acids, typically produced by fermentation of carbohydrate feedstocks, constitute an important class of bio-renewable platform chemicals that can be further converted to useful products. Bio-based plasticizers are currently gaining significant attention because they can be used in place of petroleum-based phthalate plasticizers that have shown to have health concerns in certain consumer products. E3Tec received and recently completed US DOE SBIR Phase I and II projects. The objective of the project was to design and evaluate processes using heat integrated reactive distillation (HIRD) with side reactors and PerVap membrane for esterification of organic acids. The model chemical system studied in the SBIR project was esterification of citric acid with ethanol to produce tri-ethyl citrate (TEC). ASPENPlus™ design method was developed and validated with the MSU pilot-scale test unit. The process analysis is based on ASPENPlus™ integrated with validated rate-based models of side reactors. The primary focus has been to develop a design methodology for rapid scale-up of bio-based processes. The validated ASPENPlus™ design model is applied for synthesis of two leading bio-based plasticizer candidates, dioctyl adipate (DOA) and dioctyl succinate (DOSX). Techno-economic merit comparisons based on C-Footprint, CAPEX, and OPEX for both the new and current commercial process technologies will be presented.

444077 Energy Efficient Subcritical Oxidation of High COD Wastewaters

Abstract:

Supercritical wet oxidation is the holy grail for destruction of tough to biodegrade species in industrial wastewaters - ones with high COD and not necessarily high BOD. Subcritical approaches are nearly as effective but equally restrained by high energy costs. A process intensification approach to equipment design and energy integration can find a sweet spot for implementing an economic solution.

444336 Modeling and Optimization of Simultaneous Natural Gas Purification and Storage

Abstract:

Natural gas is the fastest growing energy source with applications in power generation, vehicular transportation and chemicals production, among others. It is projected to account for more than a quarter of the total primary energy consumption by 2035.[1] To meet the increased demand, we need to tap into both conventional and unconventional sources of methane, which is the key component of natural gas. Since many of the unconventional sources are stranded, transportation becomes a key issue.[2] Moreover, many methane sources including shale gas, coal bed methane and landfill gas have impurities (e.g. CO_2, N_2, H_2S) as high as 70%.To transport methane through pipelines, the impurity content has to be less than 3-4%.[3,4] Traditionally, methane is first separated from the impurities, then compressed and transported. Each of the steps in the methane supply chain is energy and capital intensive. For instance, the operating cost of absorption-based CO_2 removal from methane is about \$ 0.18-0.3/MMscf.[5] The cost of cryogenic distillation to remove N_2 from methane is even higher (\$ 0.3-1.0/MMscf).[6] Pressure swing adsorption (PSA)[7] using microporous materials (e.g., zeolites) is a cost-effective alternative with operating cost ranging from \$0.04-0.66/MMBtu.[8] The operating cost of natural gas transportation through pipelines ranges from \$0.3-0.5 /MMBtu per 1000 miles depending on the transportation pressure and flow rate.[9]

Due to the standalone development of separation and transportation technologies, excess energy is consumed. For instance, PSA processes often involve high pressure adsorption for purification, vacuum depressurization for column regeneration and then re-pressurization of the product gas for pipeline transport. Energy is consumed at each pressure change. There is a potential for energy savings if the purification and storage steps are combined and carried out in the same vessel. Thus, our objective is to streamline the current separate purification, storage and distribution entities in the supply chain into fewer steps resulting in reduced energy costs. We propose a combined separation and storage (CSS) technology to purify and store methane in the same vessel which can eliminate the need for separate purification and storage. Microporous materials such as zeolites and metal organic frameworks (MOFs) can be used as adsorbents due to their good preferential adsorption and storage capabilities.

In the CSS process, feed gas is sent to the column packed with adsorbent and the impurities are vented out from the other end while the product gas (CH_4) gets stored in the adsorbent column till saturation is reached at that pressure. We model the adsorption of CH_4 in the column using a non-linear algebraic partial differential equation model[8,11,12] which is discretized over space and time to a system of algebraic equations. The completely discretized model is then optimized to identify optimal operating conditions and unit specifications (e.g. adsorption pressure, adsorption time, length of column) for obtaining maximum storage capacity and purity of $CH_4$and minimum energy cost for a particular material. To further improve the CSS technology, we identify candidate materials which have high storage capacity while meeting the requisite purity specifications at reduced cost.

References:

1) Energy Outlook 2035, British Petroleum, February 2015.

2) Rufford, T., Smart, S., Watson, G.C.Y., Graham, B.F., Boxall, J., Diniz da Costa, J.C. and May, E.F. "The removal of CO_2 and N_2 from natural gas: A review of conventional and emerging process technologies", *Journal of Petroleum Science and Engineering* 94-95 (2012): 123-154.

3) Tagliabue M., Farrusseng D., Valencia S., Aguado S., Ravon U., Rizzo C., Corma A. and Mirodatos. C. "Natural Gas Treating by Selective Adsorption: Material Science and Chemical Engineering Interplay" *Chemical Engineering Journal*155(3) (2009): 553–566.

4) Baker, R.W. and Lokhandwala, K.A. "Natural gas processing with membranes: an overview". *Industrial and Engineering Chemistry Research*47(7) (2008):2109-2121.

5) Peters, L., Hussain, A., Follmann, M., Melin, T. and Hägg, M.B. "CO_2 removal from natural gas by employing amine absorption and membrane technology - A technical and economical analysis" *Chemical Engineering Science* 172(2-3) (2011): 952-960.

6) "Nitrogen Removal from Natural Gas". Membrane Technology and Research, Inc., Menlo Park, CA, United States (1997). http://www.osti.gov/scitech/servlets/purl/493341.

7) Ruthven D.M., *Principles of Adsorption and Adsorption Processes,* John Wiley & Sons, New York, NY, USA (1984).

8) First, E.L., Hasan, M.M.F. and Floudas, C.A., "Discovery of Novel Zeolites for Natural Gas Purification through Combined Material Screening and Process optimization." *Journal of American Institute of Chemical Engineers*. 60 (2014): 1767–1785.

9) Comot-Gandolphe, S., Appert O., Dickel, R., Chabrelie, M-F. and Rojey, A., "The challenges of further cost reductions for new supply options (Pipeline,LNG,GTL)", *22nd World Gas Conference*, Tokyo, Japan, (2003).

10) Ko, D., Siriwardene, R. and Biegler L., "Optimization of pressure swing adsorption and fractionated vacuum pressure swing adsorption processes for CO_2 capture", *Industrial and Engineering Chemistry Research*44(2005): 8084-8094.

11) Hasan, M.M.F**.,** Baliban, R.C., Elia, J.A. and Floudas, C.A. "Modeling, Simulation and Optimization of Post-combustion CO_2 Capture for Variable Feed CO_2 Concentration and Feed Flow. Part 2: Chemical Absorption and Membrane Processes" *Industrial & Engineering Chemistry Research*51(48) (2012): 15665–15682.

12) Hasan, M.M.F., First, E.L. and Floudas, C.A. "Cost-Effective CO_2 capture based on *in silico* screening of zeolites and process optimization", *Physical Chemistry Chemical Physics* 15(2013): 17601-17618.

444034 Enhancing the Resolution of Process Design: Incorporating High Fidelity Process Unit Models in Flowsheet Optimization

Abstract:

Enhancing the resolution of process design: Incorporating high fidelity process unit models in flowsheet optimization

R.C. Pattison and M. Baldea

McKetta Department of Chemical Engineering, The University of Texas at Austin, Austin, TX 78712

Process design is a vital activity to the chemical industry. Given the massive and long-term investments required to carry out a green-field project or a major retrofit, there's a significant and justifiable interest in devising optimized designs with a high level of detail around key process units.

Research progress in equation-based simulation has been steady, and some commercial simulators offer equation-oriented options as well. In this presentation, we discuss the possibility of further enhancing the resolution of equation-oriented process flowsheet models by embedding detailed models of unit operations such as reactors, multi-stream heat exchangers, etc. Such models capture physical characteristics over multiple length scales, including the (smaller) dimensions that are typically not captured at the design stage.

We begin from our previously developed concept of pseudo-transient flowsheet simulation [1-3], and show that detailed unit models described by partial-differential material and energy balance equations lend themselves naturally to a pseudo-transient reformulation. This, in turn, facilitates incorporating such models in simulation and optimization calculations at the flowsheet level.

We conclude our presentation with several design case studies pertaining to industry-relevant problems, including production of alternative fuels and natural gas processing.

References

[1] Pattison, R.P.; Baldea, M. Equation-oriented flowsheet simulation and optimization using pseudo-transient models. *AIChE J.*, **2014**, 60, 4104-4123.

[2] Pattison, R.P.; Baldea, M. Multistream heat exchangers: Equation-oriented modeling and flowsheet optimization. *AIChE J.*, **2015**, 61, 1856-1866.

[3] Pattison, R.P.; Gupta, A.M.; Baldea, M. Equation-oriented optimization of process flowsheets with dividing-wall columns. *AIChE J.*, **2015**, doi: 10.1002/aic.15060.

444134 Planning Optimal River Transport of Petrochemicals Concerning Uncertainties of Water Levels by Two Stage P-Graph

Abstract:

A novel algorithmic method is proposed for synthesizing the optimal plan of transporting petrochemicals while taking into account the uncertainties of different modes of transportation, especially from refineries of the MOL Group to its major consumers in Central Europe. The preferred mode of transportation is barge, but the availability of river transport highly depends on the uncertain water levels of river Danube connecting ten countries as the second longest river of Europe.

The current work reveals a methodology that provides an adequate basis to model supply chains mathematically and formally as well as to synthesize optimal and alternative supply scenarios algorithmically. The proposed methodology is based on the combinatorial foundations of algorithmic process synthesis or more specifically on the P-graph framework and its recent adaptations for stochastic optimization.

The supply chain in the focus of actual examinations involves two refineries, a storage site, and target locations of the consumers. Decisions on reserving barge and rail capacities are supported by optimization algorithm and software taking into consideration probabilities of different water levels limiting the capacity of river transport. The software generates the mathematical model, the optimal transport plan, and alternative suboptimal plans.

445225 Integration of Divided Wall Columns in a Shale Oil Processing Unit

Abstract:

Use of Divided Wall Column (DWC) in petroleum processing has significantly increased in the past decade. Savings of 20-30% in capital and energy have been reported. This paper discusses the application of DWC in a Shale Oil Processing unit. Impact of utility constraints on the economics for application of DWC is discussed. The paper will also discuss how Pinch Technology and Process Integration techniques can be used to integrate the DWC with the rest of the process and the insights that can be developed to optimize the process configuration.

443937 Dividing-Wall Column Screening Guidelines and Applications

Abstract:

A dividing-wall column (DWC) is a distillation column with a vertical partition that can be implemented as a potentially attractive alternative to conventional multi-column or sidecut column arrangements; use of a DWC may enable capital cost and energy savings of around 30% or even higher to be achieved. Rules-of-thumb involving process conditions like feed composition and relative volatility are often cited for consideration when trying to determine if a DWC might make sense (or not) in a given flowscheme. The intent of this work is to present and review these guidelines in the context of refining and petrochemical processes and obtain an updated perspective that will allow for more detailed guidance to be provided for future screening evaluations.

444354 New Superstructure-Based Optimization for Process Intensification

Abstract:

Process intensification enables "multi-tasking" by combining reaction, separation and other operations in a single unit, and leads to a substantially smaller, cleaner, safer, and more energy-efficient technology [1-3]. However, there is a lack of a systematic method for the identification of intensification opportunities and the optimal synthesis of intensified processes at the flowsheet level. Traditional process synthesis and integration methods have focused on classical unit operations, where units are integrated but they are not merged to perform multiple operations *in situ*. Though research on process intensification has brought about many promising technologies including reactive distillation, divided wall columns, membrane reactors and task integrated columns, current methods are often suboptimal and consider only a few of all plausible alternatives. While recent phenomena-based design methods [4-6] incorporate process intensification, they are difficult to solve and are based on heuristics and/or decomposition approaches leading to suboptimal solution. Instead of considering classical unit-operation based superstructure representation of a flowsheet, we propose a new flowsheet based on building blocks in which each block represents a unit use of materials performing a specific function (reaction, separation, storage, etc.). An assembly of blocks of the same material would obtain a classical unit operation, while blocks with different materials or functionalities would result in intensified unit operation. This allows a systematic identification, representation and generation of intensification alternatives at the flowsheet level without exhaustive enumeration. We pose the optimal process intensification problem as a mixed-integer nonlinear optimization (MINLP) problem which aims to synthesize a process with intensified units by minimizing or maximizing a process metric given the feed and product specifications, feed and product prices, material properties and bounds on flow rates. In this talk, we will present the overall process intensification method with specific applications.

References

[1] Reay, D.; Ramshaw, C.; Harvey, A. Process intensification: engineering for efficiency, sustainability and flexibility. Waltham, MA: Butterworth-Heinemann, 2013.

[2] Stankiwicz, A.I. , Moulijn, J.A.,. Process intensification: transforming chemical engineering. Chem. Eng. Process, January, 2000, 22-34.

[3] Baldea, M. From process integration to process intensification. *Computers & Chemical Engineering* 2015, Doi.10.1016/j.compchemeng.2015.03.011.

[4] Arizmendi-Sanchez, J. A., and Sharratt, P. Phenomena-based modularisation of chemical process models to approach intensive options. *Chemical Engineering Journal* 2008,135, 83-94.

[5] Lutze P, Babi DK, Woodley JM, Gani R. Phenomena based methodology forprocess synthesis incorporating process intensification. *Ind Eng Chem Res* 2013, 52(22):7127–44.

[6] Babi, D. K., Holtbruegge, J., Lutze, P., G_orak, A., Woodley, J. M., and Gani, R. Sustainable process Synthesis–intensification. *Comput. Chem. Eng.* 2015, .doi:10.1016/j. compchemeng.2015.04.030.

445008 A Novel Pressure Swing CO_2 Adsorption Process Based on Self-Heat Recuperation

Abstract:

A novel pressure swing adsorption (PSA) process based on self-heat recuperation technology is proposed. By using pressure swing and heat circulation modules the adsorption heat and pressure can be recuperated and recycled to reduce the energy consumption. The process simulation results indicate that the energy consumption of proposed PSA process for the CO2 capture from the flue gas is reduced to more than 40% that of the conventional PSA process.

445047 Extendible Multi-Objective and Multi-Period Optimization Framework with Monte Carlo Simulation to Manage Industrial Flares during Normal and Abnormal Process Operations

Abstract:

In this work, an extendible multi-objective and multi-period optimization framework is presented for the systematic synthesis of energy alternative tools to manage industrial flares during normal and abnormal operations. Flaring is a very common practice across all industrial plants during abnormal situation management. Industrial flaring and its negative impacts on environment, ecosystem and society have gained the attention of researchers, environmentalists and decision makers. It does not only waste potentially valuable source of energy, it also adds significant carbon emissions and other toxic materials to the atmosphere that have been linked to cause diseases in human health and contribute to global warming or several natural disasters. Although, governments and companies have had success in reducing flare gas with significant investments; global gas flaring has remained largely stable the past fifteen years in the range of 140-170 billion cubic meters (BCM) which is 5% of global gas production (Kazi et al.,

2015). Therefore, much attention is still needed to mitigate and manage industrial flares by innovating green process engineering for sustainable energy and environment.

Target of this work is to utilize the surplus energy from industrial flaring to manage waste water treatment facility. In addition, the uncertainty of flaring incidents should be taken into consideration. Thus, it is essential to develop a robust methodology to integrate flare and water management tools with process industries considering the risk factor form uncertain flaring incidents.

The objective of the proposed framework is to reduce the environmental footprint of abnormal flares by enumerating and assessing possible process configurations in order to manage flares from uncertain sources and to utilize unused energy resources for waste water treatment. Following the concept of multi- period optimization approach, the core of this optimization framework is developed using genetic algorithm and Monte Carlo simulation results. Its objective function is aimed at minimizing the total annualized cost which accounts the fixed and operating costs of the system, the value of produced by co-products (i.e., power, waste water treatment savings, income from permeate), and taxes/credits associated with GHGs. An ethylene process plant is used to demonstrate the applicability of the developed framework. The results of different alternative configurations demonstrate the economic, energetic and/or environmental trade-offs of integrating thermal membrane distillation (TMD) and cogeneration (COGEN) unit with the process plant both for flare mitigation and during normal operation.

Kazi, M.-K., Mohammed, F., AlNouss, A. M. N., & Eljack, F. (2015). Multi-objective optimization methodology to size cogeneration systems for managing flares from uncertain sources during abnormal process operations. Computers & Chemical Engineering, 76, 76-86.

443840 A Systems Framework for Shale Gas Monetization

Abstract:

The aim of this paper is to provide a systems perspective on shale gas monetization to fuels and chemicals. The properties of shale reservoirs and shale gas production processes will be discussed briefly before focusing on four main building blocks of the chemicals industry: methanol, ethylene, propylene, and butadiene. For each of these building blocks, the main derivatives and their chemical processes are discussed as well as the changes incurred on their markets because of the shale gas boom. In addition, the chemistry of gas-to-liquids (GTL) technology and existing commercial applications of it are detailed. Also, an overview of existing and proposed plants for each of the five monetization pathways will be given. Based on this survey, an optimization formulation is developed and solved to determine the optimal pathways of a given shale gas resource. The objective of the optimization formulation is to maximize profit based on capital and operating costs of the given processes, feedstock prices, the sales prices of the produced chemicals while accounting for supply, demand, technical, and environmental factors (e.g. C). A case study is solved for the Barnett and Eagle Ford shale formations.

450663 Enabling Process Innovation through Computational Optimization of the Chemical Engineering Process Plant Design of Stationary, High Temperature Fuel Cell Systems that Co-Produce Hydrogen

Abstract:

U.S. automobiles consume about 16.17 Exajoules (EJ) of primary energy, which translates into about 2.59 EJ of useful motive energy and about 13.58 EJ of energy dissipated to the environment. On a well-to-wheels basis, the U.S. automotive transportation supply chain is, on average, only about 16% efficient (2.59 EJ/13.58 EJ). Studies indicate that an automotive supply chain based on hydrogen-fueled fuel cell vehicles (FCVs) may be ~two times more efficient on a well-to-wheels basis. Substitution of gasoline fuel with hydrogen fuel derived from local sources of natural gas and renewables can help address this energy efficiency bottleneck, while also increasing security of fuel supply and reducing air pollution and greenhouse gas emissions.

One of the most efficient ways to produce hydrogen fuel for FCVs is with stationary, high temperature fuel cell systems (FCSs) that co-produce hydrogen. These hydrogen co-production systems (H_2-FCSs) can be fueled by renewable methane (such as biogas) or natural gas, and can be designed to produce electricity, heat, and hydrogen simultaneously. This research investigates the chemical engineering process plant design of H_2-FCSs. Process models are developed and deployed to optimize H_2-FCS design for maximum hydrogen yield per unit of methane input, and other parameters. Special attention is paid to scenarios in which H_2-FCSs are designed to be more energy efficient by recovering electrochemical heat from the high temperature fuel cell stack to heat the endothermic steam reforming process to generate additional hydrogen for vehicles. Process plant models are used to explore the impact of key input parameters (including (A) stack/reformer operating temperature, (B) steam-to-carbon ratio for the fuel processing sub-system, (C) heat transfer effectiveness between inlet and outlet gases, etc.) on important design parameters, including, but not limited to, (1) hydrogen output, (2) overall energy conversion efficiency, and (3) net water balance. Model results indicate that, compared with stand-alone natural gas steam methane reforming (SMR) hydrogen generators, H_2-FCSs can reduce fuel consumption by about 20%.

443692 Group Contribution Modelling for the Prediction of Safety-Related and Environmental Properties

Abstract:

We present a new set of property prediction models based on group contributions to predict major safety-related and environmental properties for organic compounds. The predicted list of properties includes lower and upper flammability limits, heat of combustion, auto ignition temperature, global warming potential and ozone depletion potential.

Process safety studies and environmental assessments rely on accurate property data. Safety data such as flammability limits, heat of combustion or auto ignition temperature play an important role in quantifying the risk of fire and explosions among others. Global warming potential and ozone depletion potential became a standard to analyze the environmental impact of processes and products.

In the early stage of process development and analysis, experimental values are often not available due to cost or time constraints. In this case property prediction models like group contribution (GC) models can estimate data. However, the estimation needs to be accurate, reliable and as little time-consuming as possible so that the models can be used on the fly.

In this study the Marrero and Gani group contribution (MR GC) method has been used to develop the models for safety-related and environmental properties. The method considers the group contribution in three levels: The contributions from a specific functional group (1st order parameters), from poly-functional (2nd order parameters) as well as from structural groups (3rd order parameters). The latter two classes of GC factors provide additional structural information beside the functional group. The contributions of all three factors are then summed up. For the database, DIPPR are used together with the reported measurement accuracy. For parameter estimation, a comprehensive statistical methodology was followed to improve prediction accuracy and reliability (95% confidence).

The method is simple and easy to apply. Taking into account higher order groups increases the accuracy. Furthermore, the application range is high due to the high number of considered functional and structural contributions.

A thorough uncertainty analysis provides information about the prediction error, which is important for communicating the reliability of the predicted values for the user for potential applications in process safety studies and environmental assessments.

Example application of the models are shown for a selected class of chemicals to highlight the application range of models.

442533 Mechanism-Free Chemical Kinetics. Part 3. The Phenolphthalein/Base Reaction

Abstract:

Einstein is quoted as saying that Thermodynamics is the only scientific discipline that will remain unchanged through the millennia. Why so for Thermodynamics and not for the other scientific disciplines? The answer is simple: Thermodynamics does not depend on mechanism. All the other scientific disciplines depend upon hypothesizing a mechanism then comparing experimental results to the results produced by the proposed mechanism. With time, the proposed mechanisms change: they are either discarded for more relevant ones or they are modified to explain new data or more precise data. Thermodynamics presents mathematical methods for describing changes of state for materials. In other words, Thermodynamics is mechanism-free.

The same cannot be said for chemical kinetics. Currently, we guess a mechanism for a chemical reaction, then compare experimental data to the datum generated by our assumed mechanism. If the correlation is "good", then we assume the mechanism is correct and we use it to design a chemical plant. However, there is no proof that the assumed mechanism is correct.

Today's computer power and user-friendly software provide us with the tools to possibly remove mechanism from chemical kinetics. This paper extends the material presented during the AIChE Spring 2014 and Spring 2015 meetings. This discussion presents new experimental data and expands the discussion for adopting a mechanism-free kinetics in chemical engineering.

442542 Mechanism-Free Chemical Kinetics: Part 4. Design of Inherently Safe Chemical Plants

Abstract:

The design of chemical processing reactors is curently based on a set of assumptions about the chemical kinetics occurring in them. A generally stated assumption is that the chemical reaction follows a first-order or second-order mechanism. The most common unstated assumption is that we can prove the first assumption graphically. Unfortunately, graphical coincidence does not constitute logical proof. Thus, many chemical reactors are designed via a flawed procedure.

For the past several years, we have been investigating whether it is possible to formulate a mechanism-free chemical kinetics. This concept improves the inherent safety of chemical plants since it removes the many assumptions required by our current reactor design procedure.

We have studied several common chemical reactions and demonstrated that they can be fully quantified without first assuming a mechanism for the chemical reaction. We have reported our results in the last several Spring AIChE meetings.

To extend the scope of our concept, we have been applying it to published kinetic studies. We discuss in this presentation the results of our literature investigation with regard to developing a mechanism-free chemical kinetics.

443466 Estimating Maximum Pond Water Temperature in an Industrial Facility

Abstract:

Estimating water temperature in an outdoor pond or reservoir is a challenging issue for domestic and industrial purposes. The temperature estimation requires the simultaneous application of radiative, convective and conductive heat transfer along with evaporative heat loss. There are many industrial and domestic applications where water temperature estimation is critical. As an example, estimating water temperature can help outdoor swimming pool owners to have a proper plan for heating the water or covering the pool in different seasons. In pisciculture, it is significant to know the maximum daily water temperature rise to avoid losing fish due to extreme low or high water temperatures. Many industrial plants in tropical areas use outdoor water reservoirs for firefighting purposes. It is critical to estimate highest water temperature in such reservoirs to avoid malfunction of firefighting equipment and piping during fire incidents.

This paper uses a real industrial application as a case study to estimate the maximum pond water temperature. Accurate temperature estimation is required to choose the proper piping material for fire service piping for facility in this case study. This study proposes a simple but accurate method to estimate maximum water temperature in outdoor ponds or reservoirs. The study uses correlations available in literature with minor modifications to calculate interfacial heat transfer which is a combination of radiation heat transfer, convection heat transfer, evaporative heat loss, and conduction heat transfer. The results of proposed method are compared with real water temperature measurements and results of numerical simulation of the reservoir by ANSYS Fluent software. The results show a reasonable accuracy for the proposed method to calculate the maximum water temperature in open outdoor reservoirs.

(151b) One Year Operation of a Salinity Gradient Solar Pond in Northern Cyprus- Experimental Investigations and CFD Simulation

Obaidullah, M.,

Kadyrov, S., Chemical Engineering Program, Middle East Technical University Northern Cyprus Campus

Gorjinezhad, S., Civil Engineering Department, School of Engineering, Nazarbayev University

Taylan, O.,

Amouei Torkmahalleh, M., Chemical Engineering Department, School of Engineering, Nazarbayev University, 53 Kabanbay batyr ave,

Ahmadi, G., Department of Mechanical Engineering, Clarkson University

Originally presented on: 4/13/2016 8:30:00 - 9:00:00

Long Term Operation of a Salinity Gradient Solar Pond in Northern Cyprus- Experimental Investigations and CFD Simulation

[1]Madina Obaidullah, [2]Sultan Kadyrov, [3]Soudabeh Gorjinezhad, [4]Onur Taylan, [5]Mehdi Amouei Torkmahalleh, [6]Goodarz Ahmadi

1- *Sustainable Environment & Energy Systems, Middle East Technical University ? Northern Cyprus Campus, Guzelyurt, Kalkanli, Mersin 10, Turkey*

2- *Chemical Engineering Program, Middle East Technical University Northern Cyprus Campus, Guzelyurt, Mersin 10, Turkey*

3- *Civil Engineering Department, School of Engineering, Nazarbayev University, 53 Kabanbay batyr ave., Astana, 010000, Kazakhstan*

4- *Mechanical Engineering Program, Middle East Technical University Northern Cyprus Campus, Guzelyurt, Mersin 10, Turkey*

5- *Chemical Engineering Department, School of Engineering, Nazarbayev University, 53 Kabanbay batyr ave., Astana, 010000, Kazakhstan*

6- *Mechanical Engineering Department, Clarkson University, Potsdam, NY, 13699- 5725*

1. Introduction

Today, renewable energy sources gain importance day by day. It is crucial to develop devices and processes to supply energy from non-polluting and renewable energy sources for sustainable development of the world. Solar pond is an example of such devices that basically collects solar energy and stores it as thermal energy for a long period. Studies indicate that the temperature in solar ponds may in general reach up to 70-80°C implying that the thermal energy from solar ponds can be useful to applications with low grade energy demand [1]. One of the most important advantages of solar pond over other renewable energy sources, such as solar collectors, is its lower cost of investment [1]. Solar pond is environmentally friendly in particular when it is used for electricity generation. The heat obtained from solar pond can be converted into electric power even at low temperatures [2]. In this regard, organic Rankine cycle engines are generally run using the temperature difference in a solar pond. For applications where the organic fluid fails to operate due to low temperature difference, thermoelectric generators can be a good candidate to replace organic Rankine cycle engines for power generation [3].

Solar Ponds normally consist of three different salinity layers. The first layer, known as the upper convective zone (UCZ), is located at the top of the pond, and contains the least salinity level. The second layer, whose salinity level increases with depth, is called non-convective zone (NCZ). This layer is responsible to act as an insulator to prevent heat from escaping to the UCZ, maintaining higher temperature at deeper zones. The last layer made of a saturated salt solution, is responsible for energy storage, and is known as the lower convective zone (LCZ) [5]. The working principles of solar ponds are quite simple. In the absence of salinity gradient, the solar radiations reaching a pond are mostly absorbed by the lower water levels and will cause the water to heat up. The heated water is then risen up to the pond surface due to buoyancy effect and loses its thermal energy to the atmosphere. Thus, the main reason behind imposing salinity gradient to solar ponds is to create a density gradient restricting such buoyancy driven natural convection and as a result to trap the thermal energy at the bottom of the pond [4].

Northern Cyprus is enriched in solar energy in particular during summer. The performance of a solar pond in this region has not been evaluated yet. To take advantage of solar energy in Northern Cyprus, a salinity gradient solar pond has been constructed and operated at Middle East Technical University Northern Cyprus Campus (METU NCC) located at Guzelyurt, Northern Cyprus (***Figure 1***). This study is primarily concerned with describing the development, validation and use of CFD for computer modeling of flow and thermal analysis of the experimental salinity gradient solar pond at METU NCC. The pond has started operating since Oct. 2014 and experimental results are being recorded accordingly. The CFD commercial software considered to be utilized is ANSYS FLUENT and the challenge is to visualize the transient effect of temperature distribution, salinity concentration, velocity field and the degree of influence of wind speed on the overall stability and performance of the solar pond over time.

Figure 1. Constructed Solar Pond at Middle East Technical University (Northern Cyprus)

2. Materials and Experiment

A cylindrical salinity gradient solar pond of 61 cm diameter, 55 cm height and 1.2 cm thickness installed and operated since October 8th, 2014 at Middle East Technical University Northern Cyprus Campus (METUNCC) located at Guzelyurt, Northern Cyprus. Three inlet ports were installed to carefully add the solutions to the pond to create the three salinity layers. Also, these inlets were used to add proper amount of solutions to each zone during the experiments to compensate for evaporation and sampling losses as well as surface washing. The pond has been divided in three zones. The bottom zone (LCZ) consisted of 75 liters of saturated salt solution (concentration C). The middle zone (NCZ) consisted of three equally divided sub-layers, each 15 liters in volume, with 0.75C, 0.5C and 0.25C concentrations. The top zone (UCZ) consisted of 15 liters of fresh water. The rest of the pond was left empty. The pond was equipped with three sampling valves to withdraw samples from each layer to monitor the salt concentrations using conductivity measurements. Six thermometers (2 in each zone) were installed to monitor the temperature variations at 9 a.m., 1 p.m., 5 p.m., and 10 p.m. every day. Temperature recording at other time of the day was performed as needed. The bottom and sides of the pond were insulated with a 2.5cm thick thermal insulator. The inner surfaces of the pond were painted black. Excess (undissolved) salt was added to the bottom of the pond, to ensure saturation of the bottom layer. The ambient temperature and solar radiation were monitored using a sun tracker (Kipp and Zonen) placed next to the solar pond. Wind velocity and relative humidity were also monitored throughout the experiment.

3. Computational Scheme and Modeling

The geometry and meshing of the solar pond were constructed by ANSYS Workbench Design Modeler and Meshing environment. FLUENT version 14.5 was used as the CFD solver. A two-dimensional computational domain is considered for investigating the transient behavior of the pond representing a cross-section of the solar pond. Accordingly, the rectangular domain of consideration is that of the existing pond dimensions with 61 (cm) in width and 50 (cm) in height. The lower wall of the rectangle is assumed to be the ground and upper boundary to be the top water level open to atmosphere. Vertical walls and the bottom surface of the pond are set to be impermeable and thermally insulated with the corresponding thermal conductivity value of the insulation material. Also the top surface, due to contact with ambient air, is taken to be at ambient temperature with an average wind speed according to local meteorological data. The salinity concentration of the LCZ is kept at saturated level, while the top surface is set to zero concentration, initially.

The modeling of the pond is divided into two phases. Firstly, a simple model is going to be developed in which it is assumed that the concentration level remains the same due to the present regular injection of brine solution to the system. Accordingly, the thermofluid properties will be assumed to be primarily a function of temperature and will be evaluated for each layer at its initial concentration. In the second phase, the model will be enhanced by defining the concentration as a varying parameter to the system. In this regard, the variation of properties with concentration can be taken into account and the significance of salt diffusion and its influence on the stability of the pond can be investigated.

4. Results and Discussion

Figure 2 shows the average monthly solar pond temperature as well as the insolation variations from October 2014 to April 2015. As can be seen, the pond has been functioning as the temperature differences among the three layers have been established. The pond temperature variations are well correlated with the ambient temperature and insolation variations. The temperature at all layers decreased from October 2014 to end of December 2015, but then beginning January 2015, it increased.

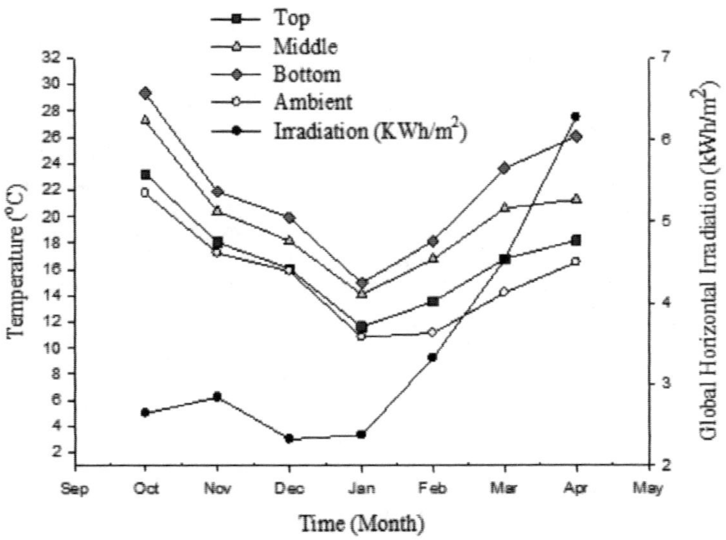

Figure 1. Average monthly temperature and insolation variations of the solar pond

References

[1.] Sukhatme, S., & Nayak, J. (2008). *Solar energy: Principles of thermal collection and storage* (3rd ed.). New Delhi: Tata McGraw-Hill.

[2.] Akbarzadeh, A., Bernad, F., Casas, S., Gibert, O., Cortina, J. L., & Valderrama, C. (2013). Salinity gradient solar pond: Validation and simulation model. *Solar Energy*, *98*, 366-374.

[3.] Singh, B., Gomes, J., Tan, L., Date, A., & Akbarzadeh, A. (2012). Small Scale Power Generation using Low Grade Heat from Solar Pond. *Procedia Engineering,* 50-56

[4.] Mehdizadeh, M., & Ahmadi, G. (2014). Two-dimensional Computer Simulation of Salinity Gradient Solar Pond Operation. *Proceedings of ASME 2014 Fluids Engineering Summer Meeting, FEDSM2014.*

[5.] Jaefarzadeh, M. (2005). Thermal behavior of a large salinity-gradient solar pond in the city of mashhad. *Salinity Gradient Solar Pond.* Retrieved October 14, 2014, from http://profdoc.um.ac.ir/articles /a/203017.pdf

438111 Biological Activities of Cyclometalated Half-Sandwich Complexes of Iridium(III) with Heterocyclic Scaffolds

Abstract:

Dihalobridged binuclear complex [{(h^5-C_5Me_5)Ir(m-Cl)Cl}$_2$] undergo halide bridge cleavage reactions with multidentate N, N heterocycles namely 5'-chloro-3'-methyl-1',2-diphenyl-5-(pyridin-2-yl)-3,4-dihydro-1'H,2H-3,4'-bipyrazole (L_1) / 4-(5-chloro-3-methyl-1-phenyl-1H-pyrazol-4-yl)-6-(pyridin-2-yl)pyrimidin-2-amine (L_2) / 7-(5-chloro-3-methyl-1-phenyl-1H-pyrazol-4-yl)-5-(pyridin-2-yl)-[1,2,4]triazolo[1,5-a]pyrimidine (L_3) forming a new half sandwich cationic mononuclear iridium(III) complex [(h^5-C_5Me_5)Ir(k^2-L)Cl]Cl. The complexes were fully characterized by various physicochemical techniques, namely, elemental analyses, spectral (ESI-MS, IR, ^{1}H, ^{13}C NMR, UV/vis). The *in vitro* antibacterial activity against three gram negative and two gram positive microorganisms and *in vitro* antituberculosis activity against *Mycobacterium tuberculosis* H37Rv strain were studied for all compounds and all the complexes shows good inhibition as compared to ligands. The *in vitro* cytotoxicity was determined using brine shrimp bioassay and LD_{50} value obtained in the range of 7–12 mg mL^{-1}. Cellular level cytotoxicity was performed using bioassay of *S. pombe* and it shows inevitable viability as concentration increases. The DNA binding activity was studied by absorption titration and viscosity measurements. The complexes bind to CT-DNA through intercalation mode (kb = 10^5). DNA nuclease activity was studied by agarose gel electrophoresis method and the compounds show 70–80% cleavage.

Study on Dynamic Risk and Flexible Protection Layer Theory of Process Safety

Yong Zhao

Sinopec Safety Engineering Institute, Qingdao, China, 266071

(eddile@foxmail.com)

Yong Zhao, Shanjun Mu, Shucai Zhang, Wei Xu, Cuihua An, Xiaogang Ding and Xue Jiang

Sinopec Safety Engineering Institute, Qingdao, China, 266071

Prepared for Presentation at
American Institute of Chemical Engineers
2016 Spring Meeting and 12th Global Congress on Process Safety
Houston, TX
April 13, 2016

Study on Dynamic Risk and Flexible Protection Layer Theory of Process Safety

Yong Zhao

Sinopec Safety Engineering Institute, Qingdao, China, 266071

(eddile@foxmail.com)

Yong Zhao[1], Shanjun Mu[1], Shucai Zhang[1], Wei Xu [1], Cuihua An[2] ， Xiaogang Ding[1] and Xue Jiang[1]

(1)Sinopec Safety Engineering Institute, Qingdao, China, 266071

(2) Qingdao Star Vocational and Technical College， Qingdao, China， 266100

Key Words: Dynamic Risk; Flexible Protection Layer Theory; Process Safety

Abstract

The research re-defined the process safety risk from risk's essential characteristics and set the scope of new definition. Positive and Negative Effects and timeliness of risk were introduced for the description more reasonable than before. The definitions laid basis during the essential characteristics of dynamic risk description. Risk and protection layer characterization of dynamic changes and related dynamic risk model gives a new definition to the interpretation of the dynamic nature of the safety risk of property, and added to the safety management related human factor, rich past of static risk management and prevention and control methods. Meanwhile, since the protection layer itself changing nature of risks in addition to its own internal dynamic changes caused by the protection layer is also closely related to the changes, which form the meaning of comprehensive dynamic safety risks. Through the study of changes in the law and the interaction between the respective mechanisms, new safety risks characterization methods can be formed. Based on the risk and protection layer characteristics of analysis and mathematical characterization of the essential characteristics of risk research, optimization and upgrading of the current risk management and control measures, and supplement the current countermeasures defect, forming a comprehensive response.

1.0 Introduction

On November 22, 2013, Shandong Qingdao Donghuang Crude Oil Pipeline had leakage of oil and gas going into the municipal drainage culvert, and accumulated in confined space, where spark triggered the explosion, resulting in 62 people killed and 136 people injured. In addition, the accident has also blocked traffic near the scene of the accident, and the supply of heat, gas, water and power was shut down in some areas, with 5.5 km roads and storm water culvert damaged, many residential buildings destroyed, and 10,000 square kilometers sea polluted with oil leakage, resulting in more than 750 million yuan as direct economic losses. On August 12, 2015， an explosion at a container terminal in Tianjin Binhai New Area caused fire and mushroom clouds. Strong tremors were felt in neighboring areas. The explosion killed 165 people (including 110 people of public security firefighters, 55 people of accident enterprises, surrounding businesses employees and surrounding residents), 8 people missing (including 5 people of the Tianjin Port firefighters, 3 people of surrounding employees and families of firefighters Tianjin Port), 798 people were injured (including 58 people injured heavier or weight wounded, 740 people injured slightly). According to the accident investigation and accident responsibility confirmation result, the investigation team suggested deal with 123 people in the explosion.

As can be seen from the above analysis, it's not incidental at all to have the occurrence of major accident like 11 • 22 or 8•12 Accident; the evolution and occurrence of accident can be traced to the day-to-day operation and maintenance work of the enterprise, as well as the local government's planning layout, emergency program actions, accident investigation approach, and lessons learning process; only the materialization of uncontrollable risks associated with the parties in relation to the accident shall inevitably lead to accident of such serious consequences. Investigation of the deep-seated reasons behind the accident suggests that the nature of risk and the corresponding management strategies is misunderstanding and shortcomings, so systematic improvements are required for safety management.

2.0 Dynamic Risk

Risk (R) is a combination of the possibility (P) and consequences (C) of a particular dangerous situation that may occur, with the combined effect of both. It's measured by general risk assessment of the possibility of (P) and consequences (C), in the form of product [1,3] as per the following equation:

$$R=(P,C)=P\times C\text{.........................} \quad (1)$$

For engineering applications, the likelihood of risk generally refers to the occurrence probability of incident. Under certain conditions, an incident A occurs m times in n times of repeated test, then the frequency (F) of the incident occurrence is expressed as m/n A. If n is infinite, then F will tend to stabilize at a value near p, and the value of p is the probability of the incident A under this condition, where F is approximately equal to P. Nevertheless, given the limited sample and time in engineering practice, it is impossible to obtain a probability value in absolute sense, so the risk assessment process often substitutes F for P in the risk calculation, where the most common value of F is the frequency of the incident in a year [4,5], namely as below:

$$R=(P,C)\rightarrow F\times C\text{...........................} \quad (2)$$

This technique features great uncertainty as to the risk calculation results. The figure above shows the most common histogram of frequency normal distribution. You can see the random characteristics of the incident frequency when the number of sample is small, so the true distribution of the probability of an incident can not be described; only when the number of sample is large enough, can the incident exhibits the general characteristics of a normal distribution. Therefore, the risk value as calculated based on equation (2) can not equal to the actual risk value.

This leads to that the enterprise in actual business process shall focus the limited resources and safety management concerns on the prevention and control of the incident with frequent occurrence in the past, and pay little attention to the incident with rare occurrence in the past. Hereunder, we will focus on the review and discussion on changes in risk, combined with equation (1) and (2).

2.1 Risk Dynamic Changes

Due to the foregoing equation (2) does not introduce time as a parameter, the traditional risk characterization ignores the risk dynamic characteristic caused by the time generated. The size of the risk can be added in the timeline to characterize changes in value:

$$R=f\ (F,C,t) \quad \text{.........................} \quad (3)$$

其中 C 为事件 The concept of risk often comes with strong timing feature, as expressed as follows:

$$\mathrm{R} = f\ (\mathrm{C}, F, t) = \varphi\ (f_1(C,t), f_2(F,t)) \quad \text{................} \quad (4)$$

Where C is the consequences of the incident A under a certain condition, and F is the probability of A occurrence under this condition, and t is time.

If a longer time dimension is taken to observe the occurrence of incident A, then the risk of A incident can be described as follows:

$$R_A = \int_{t=1}^{\infty} \varphi\ (f_1(C,t), f_2(F,t)) \quad \text{...................} \quad (5)$$

Because the value of $f_1(C,t)$ and $f_2(F,t)$ as above are not less than 0, RA must be greater than 0 as shown by the integrating calculations, and the value is not necessarily small; $f_1(C,t)$ integration value must be very big on the timeline; $f_2(F,t)$ integration value is also not necessarily small on the timeline. So you can get the following conclusion: No matter how small the probability of an incident, it is bound to occur, provided that the time is long enough. This is consistent with the well-known expression of the Murphy's Law. The above conclusion is based on statistical theory, and, more often than not, this kind of accidents may occur suddenly, not necessarily in a very long time in the future. Meanwhile, the risks facing the enterprise is dynamic and static methods of management can not cope with the changing environment of risk, so we can not take chance without prevention of low-probability and high-consequence Black Swan Event; instead, we shall take regular assessment of risks, and make timely adjustment of the risk management and control strategies.

2.2 Dynamic Risk Spatial Modelling

2.2.1 Specific Examples of R = f (F, C)

It has been elaborated In 2.0 that in engineering we commonly used formula (1) for risk calculation, therefore, we'd make a deeper analysis for the formula (1) in order to analyze the time-varying characteristics of risk. Regard R = PC as a function, and use function Z = XY to a preliminary study, the spatial expression of this function is to form a saddle surface in space, which is shown as below.

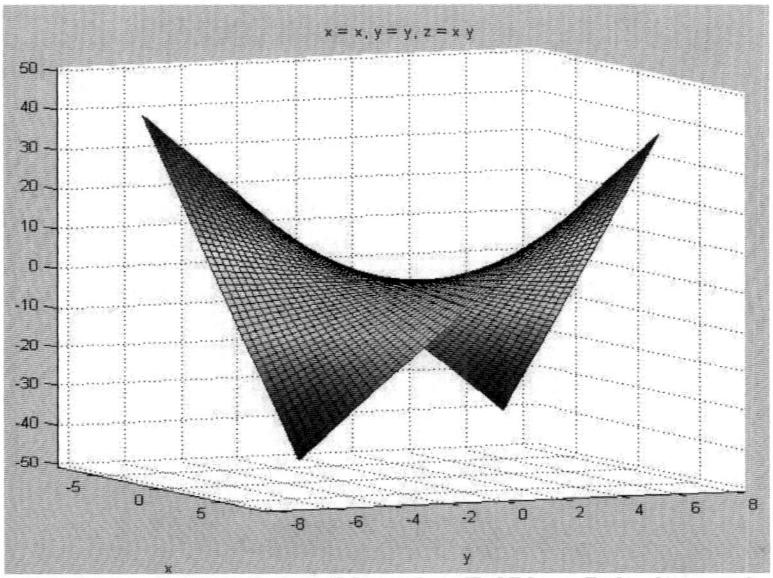

Fig.1. The saddle face of function Z=XY on finite interval

2.2.2 Dynamic variation of function R = f (F, C)

Since F∈(0,1), C∈[0, +), it is not easy to express the variation of function R = FC in the first quadrant, we make grading for the calculation results, with the upper limit to 1 and C∈[0,1], which leading to the risk function, R = FC, R∈[0,1], F∈[0,1], C∈[0,1]. And the function spatial expression is shown in Figure2.

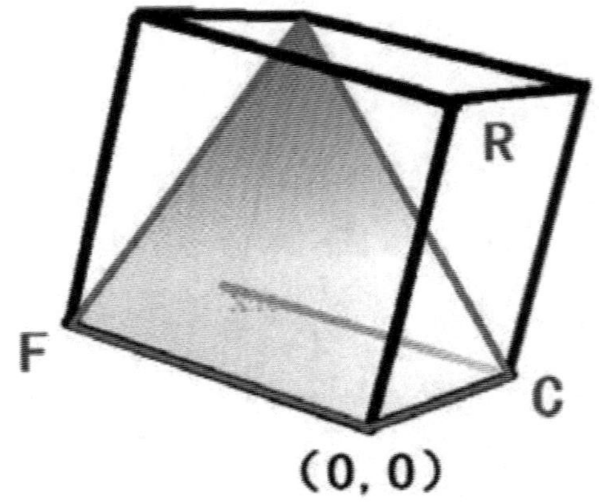

Fig.2.The spatial curved surface among (0,0,0) to (1,1,1)

Assume that C hardly changes over time(0 <C <1), and it ranges from C1 to C2 to analyze the accident probability.

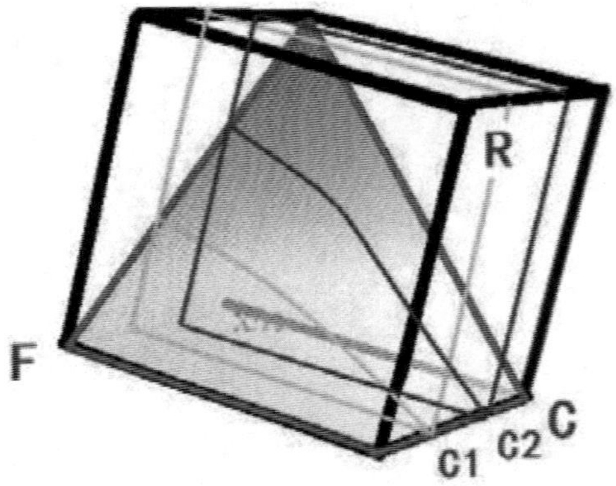

Fig.3. The probability of plane line among C1 to C2

Changes in the probability has been concluded that the possibility characterizes the probability of an event, not a certainty, but uncertainty, and the next change events is unpredictable at any point for $F \in (0,1]$. The probability change is a discrete function of time.

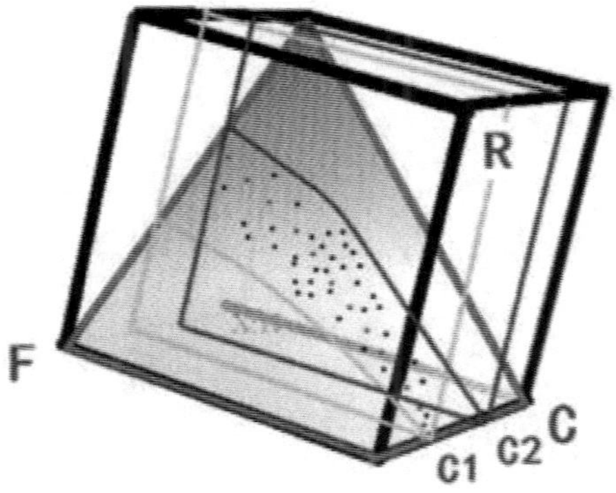

Fig.4. The discrete probability among C1 to C2

3.0 Flexible Protection layer

3.1 The protection layer's flexibility

The so-called "flexible" (Flexiblity) refers to flexible variability, meaning that the body can quickly respond to external uncertainty and stimuli. According to Malecki's view of the flexibility, it can enhance the fitness to environment, and increase its ability to resist risks. The traditional view has not taken into account the flexibility, insisting that the protection layer is rigid.

3.2 The protection layer's porosity

The protection layer itself is not entirely equally thick for each layer, there are various types of hollow, empty holes arisen by hardware or software defect in different parts of each layer. These hollow, empty holes exist like Swiss cheese model, and these default protection layers have a tremendous impact on building the flexibility model, shown in Figure 5:

Fig.5.The hollow modeling for each protection layer

3.3 The Break Rules for Dynamic Risk

Because of these empty holes (defects) in protection layers, which's been discussed in Chapter 2.0, 2.1 and 2.2, the risk is always in a dynamic change. The risk has been as well become a thorn-like objects ,for the core portion of flexible protection layers as shown in Fig.6, and will continue to look for holes or weak spots to break the protection layers, and finally, accidents may will happen. This situation is based on neglect of the common cause failure, assuming that each protection layer is independent. If we consider the common cause failure, we will find that there may be some correlation between these holes in these protection layers.

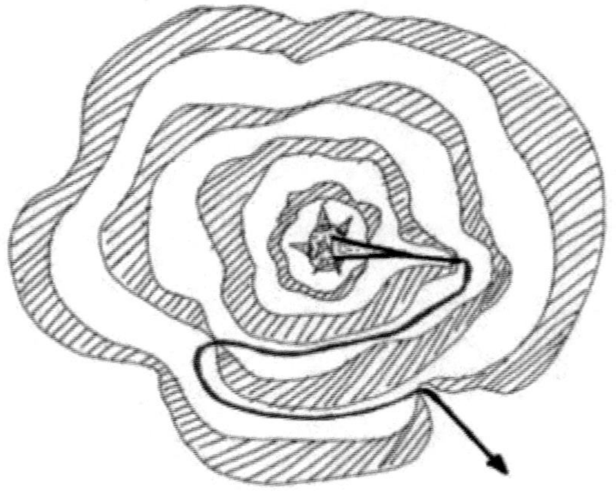

Fig.6. The break rule for protection layer

4.0 Conclusions

4.1 The conclusion of dynamic risk analysis

（1）Why is the probability of the accident can not be zero

If the density functions are normal distribution, then x integral from negative infinity to zero is greater than zero, so the probability will not be zero, the accident is inevitable, according to Murphy's Law, that all accidents can be prevented, but not avoided .

（2） Uncertainty of probabilistic characterization

To characterize the probability of uncertainty, its value has a discrete jump, similar to the quantum mechanics of particles. Which results in the uncertainty of risk.

（3） The default of probability

The error can't be avoided if we use frequency instead of probability, because we has taken the unit of time as one year (indirectly reducing sample size), especially for small probability event, the deviation will be great.

(4) The role in risk control of value's stability for the consequences

If we can stabilize the size of the consequences of C, making the [C1, C2] section as small as we can, we will make the curve F tends to a spatial one, which will be a certain significance for the risk study.

(5) The positive significance of introduction of risk timeline

It showed the dynamic characteristics of risks, and led to the dynamic of protection layer.

4.2 Flexible Protection Layer Analysis Conclusion

(1) Dynamic Risk Characterization Technology plays an important role in guiding on the setup of flexible modelling of inherently safe design, basic process control system (BPCS), key alarm and intervention, Safety Instrumented Functions (SIF), physical protection (safety valves, bursting discs, etc.), protective equipment after the release, plant emergency response and community response.

(2) Flexible protective is greatly helpful to establish the complete protection layer model, and to promote the protection layer's design and failure mechanism research, aiming to improve the protective effect of protection layer.

5 References

[1] M.F. Milazzo, T. Aven, An extended risk assessment approach for chemical plants applied to a study related to pipe ruptures, Reliability Engineering & System Safety, 99(2012) , 183-192.

[2] Q.M. Cao, Q.M. Yu, D.W. Connell, Health risk characterisation for environmental pollutants with a new concept of overall risk probability, Journal of Hazardous Materials, 187(2011), 480-487.

[3] F. Gabsi, M. Hammers-Wirtz, V. Grimm, A. Schäffer, T. G. Preuss,Coupling different mechanistic effect models for capturing individual- and population-level effects of chemicals: Lessons from a case where standard risk assessment failed, Ecological Modelling, 280(2004), 18-29.

[4] S. Busura, F.Khan, K. Hawboldt, A. Iliyas, Quantitative risk-based ranking of chemicals considering hazardous thermal reactions, Journal of Chemical Health and Safety, 21(2014), 27-38

[5] A. F. Egorov, T. V.Savitskaya, Methods and models for the risk analysis and safety management of chemical plants, Theoretical Foundations of Chemical Engineering 44 (2010), 326-338.

16AIChE
Spring Meeting, Houston, TX

METHOD AND DEVICE FOR REAL-TIME MONITORING AND PREDICTION OF HAZARDOUS GAS LEAK FLOW IN PETROCHEMICAL PROCESS

Ilyas Uyanik
Halliburton
10200 Bellaire Blvd, Houston Texas 77072
Ilyas.Uyanik@halliburton

Avinash Wesley,
Jianxin Lu,
Andrey Filippov
Halliburton
Avinash.Wesley@halliburton.com
Jessie.Lu@halliburton.com
Andrey.Filippov@halliburton.com

Prepared for Presentation at
American Institute of Chemical Engineers
2016 Spring Meeting
11th Global Congress on Process Safety
Houston, Texas
April 10-14, 2016

UNPUBLISHED

METHOD AND DEVICE FOR REAL-TIME MONITORING AND PREDICTION OF HAZARDOUS GAS LEAK FLOW IN PETROCHEMICAL PROCESS

Ilyas Uyanik
Halliburton
10200 Bellaire Blvd, Houston Texas 77072
Ilyas.Uyanik@halliburton

Avinash Wesley,
Jianxin Lu,
Andrey Filippov
Halliburton

Keywords: gas leaks; operational safety; CO_2 emission; notification; mobile application; HSE

Abstract

During upstream drilling/production operations or downstream refining processes, poisonous, highly flammable hazardous gases can be released into the environment. This paper presents an initial experience in developing a real-time monitoring and surveillance algorithm and system, consisting of desktop and smart-phone applications with mobile and stationary sensors. The system monitors and reports the spatiotemporal evolution of hazardous gases, such as H_2S, CH_4, and CO_2 clouds, in upstream/downstream operation facilities in real time and provides the necessary actions for a safe operation. This article will introduce a method and system for gas-leak detection to determine the most likely distribution of a leak and notify nearby workers of the risk. Real-time data obtained from multiple sensors throughout a facility are sent to a server for data aggregation and cleaning. A predicted gas distribution is obtained through the mathematical simulation of real-time gas-leak data in combination with wind velocity and facility geometry. Predetermined risk profiles can be used to evaluate the danger of the gas leak, and an alert can be created based on the risk profiles and predicted gas flow and sent to nearby workers. The system and method described in this article can be implemented on both land- and sea-based upstream or downstream facilities.

1. Introduction

It is increasingly important to monitor and control gaseous contaminants during chemical processes, upstream drilling/production, or downstream petrochemical processes, both onshore and offshore. Poisonous, highly flammable, hazardous gases can be released into the environment and pose serious health risks to those working in and around the facility. Facilities typically use gas detectors to alert workers that a leak is present within the facility, which sound an alarm when a certain gas is detected.

Computational fluid dynamics (CFD) simulations for gas dispersion can help to understand how wind direction, wind speed, and the local facility layout affect the distribution of pollutants. CFD simulation is a widely applied method for the study of fluid behavior in many fields (i.e., weather forecasting). In this work, a similar approach is used to predict the dispersion of gas pollutants. CFD simulations are carried out in the background, while the result contours are illustrated using a mobile device application.

2. Methods, Procedures, Process

The method monitors harmful gas leaks using on-site sensors and predicts the distribution of leaked gas using computational fluid dynamics (CFD) simulations. Through real-time data acquisition from on-site sensors, the system provides a visualization of sensor readings and locations. The sensor readings are provided as an input to the simulation along with other environmental conditions to simulate the leak in the facility layout. In the simulations, the turbulent flow model is used for the multicomponent gas flow, while the species-diffusion equation is solved for the gas-dispersion calculation [5].

Figure 1 illustrates a system for detecting and notifying workers of a gas leak within a facility. Multiple sensors can be placed throughout a facility such that a gas leak in any part of the facility can be quickly detected. The sensors are configured to measure characteristics of the environment, including the presence and concentration of various gases, such as carbon dioxide (CO_2), hydrogen sulfide (H_2S), and weather. Most sensors are commercially available (e.g., the Model 1000-H_2S-CO_2 hydrogen sulfide/carbon dioxide analyzer). These sensors can be placed throughout the facility at the height at which an average person would be breathing (e.g., 4–6 ft above ground level). Alternatively, the sensors can be placed at a height at which workers may be sitting or standing during wellbore operations. Weather sensors configured to measure wind speed and direction can also be purchased commercially, such as the 5-in-1 weather sensor.

Environmental characteristic data are transmitted over a network to a server. The server includes a processor, which is communicatively coupled with a memory and configured to store instructions and a database, including data, such as facility geometry. When a gas leak is detected within the facility, a notification is sent from the server over a network to each of the subscribed mobile devices, apprising workers of the leak. The notification can be generalized, showing the gas distribution across the entire facility. Furthermore, for mobile devices that are equipped with a global positioning system (GPS), the notification can be customized based on the proximity of the mobile device to the leak. Although the use of a mobile device is described above, the proposed monitoring system is equally well-suited for use on laptops, smart phones, or similar devices capable of accessing a web application.

The server includes a data aggregator configured to receive data transmitted from each sensor and is communicatively coupled with a data cleaner. The data-cleaning module removes noise from the data, including, but not limited to, sensor noise, spikes, and white noise. The data cleaner sends the cleaned data to a data simulator, which is configured to run a CFD simulation on the data to provide multiphase, multiphysics flow behavior. The simulation is achieved through the use of a steady-state solution of a realizable k-ε model [6] for turbulent gas flow and species diffusion using the convective diffusion equation.

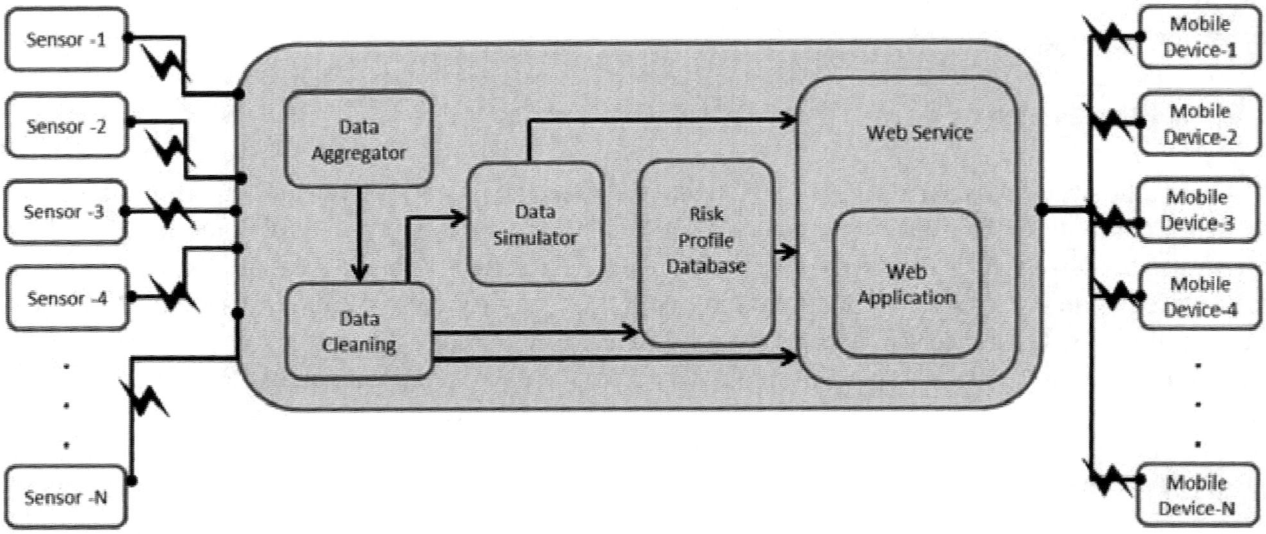

Figure 1 − Depicts system architecture. The system includes multiple sensors, a server, and mobile devices.

The CFD simulation is created using information, including facility geometry (obtained from the database, as shown in Figure 1), leak location, leak rates, gas components, gas concentration, and wind speed and direction. The completed simulation output is time stamped (as shown in Figure 3), and the above processes can be repeated indefinitely in order to create a continuous stream of current gas-leak information. The data simulator communicates with the risk-profile database containing statistical risk data, which correspond to the measured gas leak. The risk profile database obtains environmental characteristic data from the data simulator, as well as directly from the data cleaner. The risk data can be obtained from government agencies, including the Environmental Protection Agency (EPA), the Occupational Safety and Health Administration (OSHA), etc. Examples of such risk data are shown in Tables 1 and 2 below. Risk profiles, such as low risk, medium risk, and high risk, are created based on the risk data, and alarms or notifications can be configured for each risk profile.

Table 1: Toxicity of CO_2 at various concentrations.

CO_2 Concentration (%)	Time	Effects
17–30	1 minute	Loss of controlled and purposeful activity, unconsciousness, convulsions, coma, death
>10–15	1 to several minutes	Dizziness, drowsiness, severe muscle twitching, unconsciousness
7–10	Few minutes 1.5 minutes to 1 hour	Unconsciousness/near unconsciousness, headache, increased heart rate, shortness of breath, dizziness, sweating, rapid breathing
6	1–2 minutes 16 minutes Several hours	Hearing and visual disturbances Headache, dyspnea Tremors
4-5	Within a few minutes	Headache, dizziness, increased blood pressure, uncomfortable dyspnea
3	1 hour	Mild headache, sweating, and dyspnea at rest
2	Several hours	Headache, dyspnea upon mild exertion

A web service is configured to receive data from the data cleaner, data simulator, and risk-profile database. This service combines the simulation, environmental characteristic data, and alarms such that the gas leak can be displayed on a web application within the web service. The web application displays the predicted gas-flow distribution obtained from the simulation on a rendering of the facility, as well as the risk profiles associated with the different exposure levels throughout the facility. The simulation rendering is in the form of a still photograph, which shows the current location of the gas leak or where the gas leak will expand to in the near future. Alternatively, the simulation can also be rendered in the form of a video, showing the viewer the leak point and the expected gas distribution throughout the facility. The web application is accessible to mobile devices connected to a network such that workers throughout the facility can make informed decisions in case of emergency.

Table 2: Toxicity of H_2S at various concentrations.

H_2S Concentration (ppm)	Effects
0.1–3	Odor threshold
3–10	Offensive odor
10–50	Headache, nausea, throat and eye irritation
50–100	Eye injury
100–300	Conjunctivitis, respiratory tract irritation, olfactory paralysis
300–500	Pulmonary edema, imminent threat to life
500–1,000	Strong nervous system stimulation, apnea
1,000–2,000	Immediate collapse with respiratory paralysis, risk of death

The method for creating and sending a notification of a hazardous gas leak within a facility follows the flow diagram depicted in Figure. 2. Real-time data are collected at multiple locations where sensors have been placed throughout the facility. The real-time data are transmitted from each sensor via a network and received at a data aggregator communicatively coupled with a server. The data are then accumulated on the data aggregator and transmitted to a data cleaner. The data cleaner removes noise from the real-time data, including spikes, white noise, and sensors noise. It is then determined whether an increased gas level has been detected at any point throughout the facility. If no increased gas level is detected, the process does not proceed. However, data are continuously collected at all times such that a gas leak can be detected as soon as it occurs.

Alternatively, if increased gas levels are detected at a point within the facility, the data are then transferred to the data simulator, as shown in flow chart. A simulation to determine the potential gas-leak distribution is conducted, and the information is transmitted to a risk-profile database. The above-described process can be repeated on a continuous loop such that the simulation is constantly updated with real-time data acquired from each sensor, thus producing an accurate simulation of the movement of the gas plume. A risk profile is created by applying predetermined parameters to the simulation showing the potential gas-leak distribution. The risk profile is then transferred to a web service, which is communicatively coupled with the risk-profile database. A notification is configured describing the potential gas-leak distribution and risk level in different areas throughout the facility. A web application is used to display the simulation on a rendering of the facility. The web application can be accessed by workers via a mobile device connected to a network such that they can receive a notification that a gas leak has been detected and can view the potential path of the gas leak.

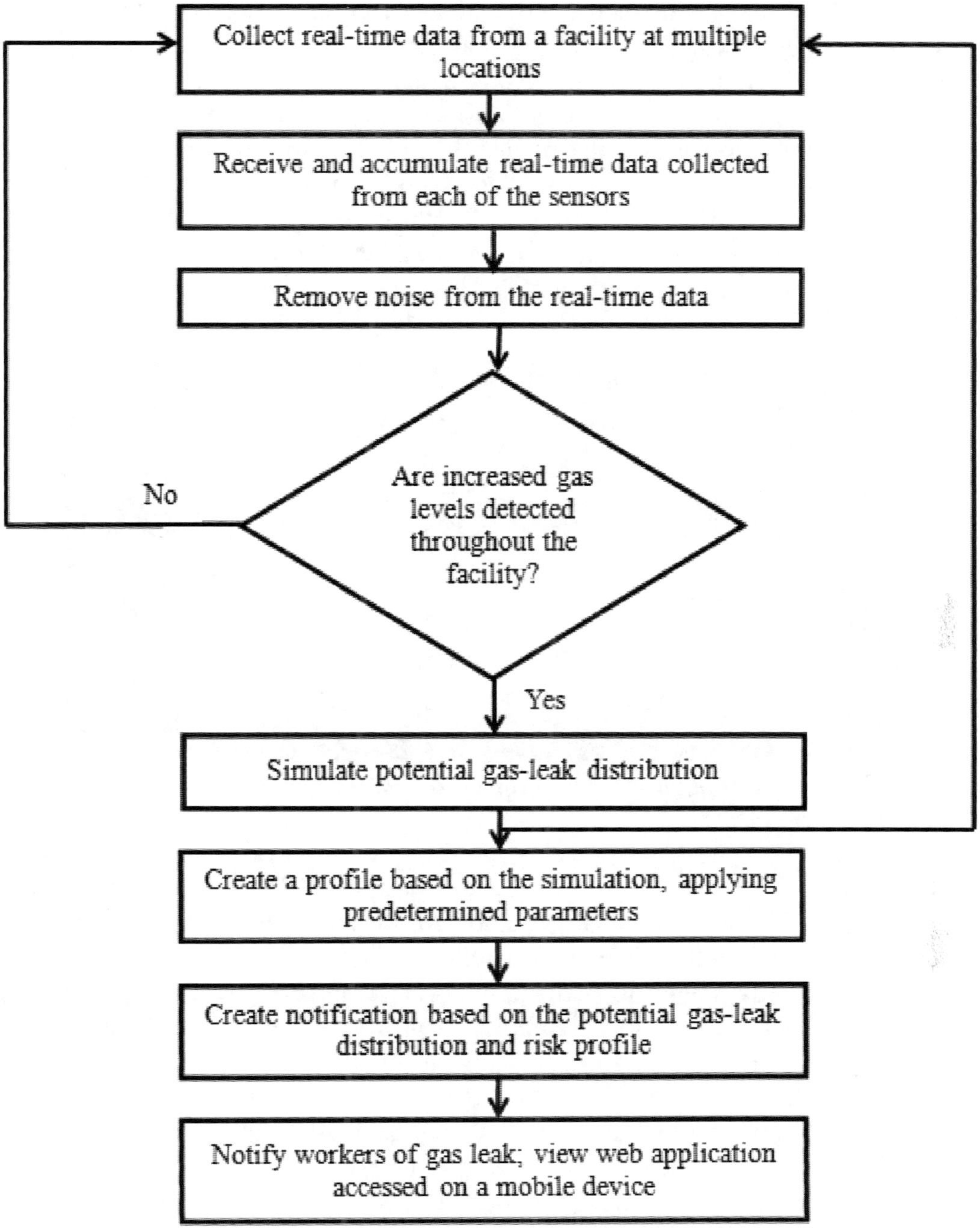

Figure 2 - System architecture flow chart.

Multiple sensors are shown to be spaced throughout a rendering of the facility (see Figure 4). The rendering shows the floor plan of the facility, as well as the location of doors and other objects that can affect the dispersal of a gas. Real-time wind speed and direction, as well as an alert notifying a viewer that an increased gas concentration has been detected at one of the sensors, are also shown. The potential gas-leak distribution is shown starting from the sensor where the increased gas level was detected. The web application can show a picture of the potential gas distribution at a specific time in the future. Alternatively, the web application can show a video of the gas plume dispersing from the point of origin throughout the facility (see Figure 3). Workers within the facility can access the web application at all times to determine whether an increased gas level has been detected within the facility. Additionally, the web application can show the location of the mobile device, as shown in Figure 4.

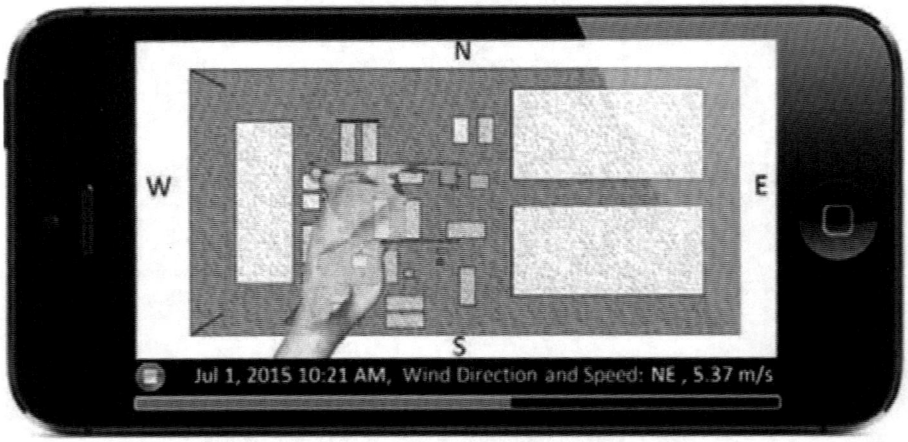

Figure 3 – Time-stamped simulation on mobile device rendered as a short video clip.

An annotation icon (Figure 4) displays a near-real-time reading of the hazardous gases, as measured via sensors that transmit the gas readings through, for example, a Bluetooth or other connection, to one or more mobile devices. This allows the user to know his or her potential exposure and can prompt the user to take precautionary measures, such as using respiratory protection equipment or performing adjustments to operating equipment, to manage exposure if needed or desired. The system includes indicators, such as bars, icons, or other symbols, to provide information about a worker's current exposure level. The indicators can be color-, shape-, size-, or audio-coded to indicate a range of exposure levels and where a user's existing exposure level fits within a specified range. The screen provides a near-instantaneous notification of the gas levels. It also displays other relevant information, including the cumulative or total exposure levels received by a user over a certain period of time, such as a workday or shift, and the rate of change in exposure levels. The screen can also provide a display showing the predicted exposure over a period of time based on current estimated conditions. The sensors can transmit gas-level information to a base station, separately or in combination with additional information, such as a time stamp to indicate when each measurement was taken and/or a global positioning system (GPS) coordinate to indicate the location of a sensor when the measurements were taken. This use is similar to a mobile application called OzoneMap [7, 8].

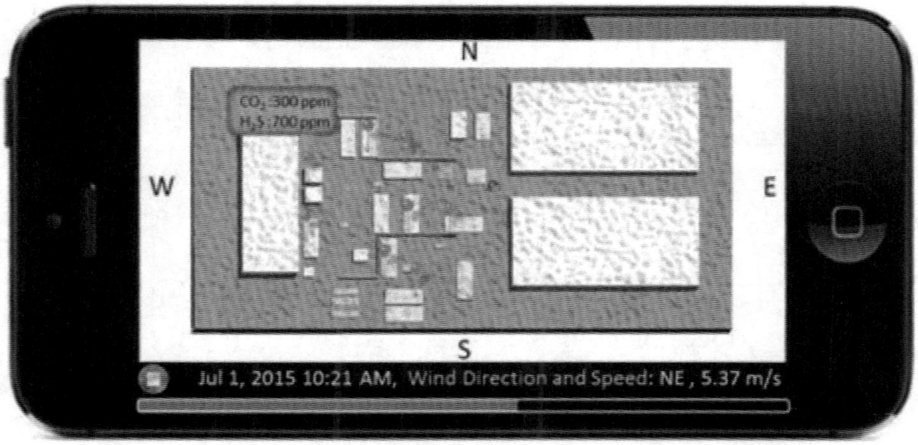

Figure 4 – Real-time sensor readings on the mobile device. Blue point indicates the location of the mobile device.

3. Conclusion

The system provides visualization of sensor readings and locations. It also includes the numerical simulation of the environment, predicting where the gas cloud will float in the next few minutes. All such visualizations are high resolution and are overlaid on the facility layout. Each worker can keep track of his environment and record how much he or she has been exposed to the hazardous gases. Because all the raw and computed data reside in the online data server, all pertaining information will be sent to the onshore headquarters office as well. Risk profiles and alarms can be configured and designed by the health, safety, and environment (HSE) management workflows. The exposure data for each worker is stored on the data servers for further analysis. A unique feature of the system is the functionality of interpolation and aggregation among sensor stations, which yields a continuous measurement plane superimposed on the corresponding geospatial map. The application monitors exposure levels in real time and analyzes their history.

This system can be used by HSE management to determine the risk of moderate and non-tolerable hazardous gas-level profiles. Remedial actions and exit strategies can be guided by the numerical simulation and sensor readings, which are provided by the system. To the best of our knowledge, this is the first time hazardous gas leaks will be monitored at such a fine resolution in real time.

4. References

[1] Fontaine, D. and Hall, M. 1991. Dispersion Modeling of Gas Releases on Offshore Platforms. Presented at the First Internal Conference on Health, Safety and Environmental, The Hague, The Netherlands.

[2] U. S. Environmental Protection Agency, Plume-in-Grid Treatment of Major Point Source Emissions, Available: http://www.epa.gov/amad/CMAQ/Documents/ch09.pdf. Accessed 18th January 2015.

[3] U. S. Environmental Protection Agency, Acute Health Effects of Carbon Dioxide. Available: http://www.epa.gov/ozone/snap/fire/co2/appendixb.pdf. Accessed 19th January 2015.

[4] ANSYS Fluent User Guide, ANSYS.

[5] Lu, J., Filippov, A., Marcancola, F., and Alcocer, V. 2015. CFD Analysis of Gaseous Leaks with

80% CO_2 on Offshore Facilities. Presented at the Proceedings of the 11th Global Congress on Process Safety, Austin, Texas, U.S.

[6] Zhu, H., Lin, Y., Zeng, D., Zhou, Y., Xie, J., and Wu, Y. 2012. Numerical analysis of flow erosion on drill pipe in gas drilling. *Engineering Failure Analysis*, **22**: 83–91.

[7] Uyanik, I., Price, D., Tsiamyrtzis, P., and Pavlidis, I. 2013. Interfacing Real-Time Ozone Information. Presented at the Proceedings of the 1st ACM SIGSPATIAL International Workshop on MapInteraction ACM. 20–23.

[8] Uyanik, I., Khatri, A., Tsiamyrtzis, P., and Pavlidis, I. 2014. Design and Usage of an Ozone Mapping App. Presented at the Proceedings of the Wireless Health 2014 on National Institutes of Health ACM. 1–7

DIMENSIONLESS FLOW REGIME MAPPING AND MODELING FOR PARTICLE TRANSPORT IN FLUID

Feifei Zhang, Andrey Filippov, and Jianxin Lu, Halliburton, Houston, TX

Abstract

All the flow regimes for particle transport in fluid flow are classified into three major flow patterns: flow with a stationary particle bed, flow with moving particle bed or dunes, and flow of fully dispersed particle-liquid mixture. In this work, analysis of experimental data showed the flow regimes can be defined by two dimensionless parameters: the ratios of the flow velocity and turbulent shear velocity to the particle settling velocity.

A flow regime map was developed using these two dimensionless parameters. This flow regime map is applicable to a wide range of conditions. In the considered experiments, the size of the solid particles varied from several micrometers to 1 cm, the solids fraction varied from less than 1% to approximately 40%, and the solid phase was formed by different particles, such as glass spheres (1.24 sg) and sand (2.5 to 3.0 sg). The inclination angle of the flow measured from the vertical changed from 0 to 180°.

This mapping method has important applications in sand-related flow assurance problems, such as sand cleaning during heavy oil production. It can also be used in other operations involving solids transport in fluid flow, such as cuttings transport in drilling, proppants transport during hydraulic fracturing, and hydraulic transport in the mining industry.

Introduction

Solid particles transport in fluid flow can be encountered in various industries, such as mining, civil, and oil and gas. The configurations of solid particles in fluid flow, which are called flow patterns, vary depending on the flow parameters. Flow patterns are critical for the operation processes. High solids fraction or significant solids deposition in the flow could cause severe pressure loss and block the flow conduit, which will damage the system and even lead to accidents. Accurate predictions and monitoring of the flow pattern are extremely important during the design and operation of the flow system.

There are a number of different terms to describe the various flow patterns for solid-liquid mixtures (Vacaldo and Charles, 1972; Parzonka et al., 1981; Shamlou, 1987; Doron et al., 1997; Kelessidis and Mpandelis, 2003; Zhang et al., 2015). In general, these flow patterns can be summarized into four categories: flow with a stationary solid bed, flow with moving bed or dunes, flow with moving dunes, and saltation flow, as shown in Figure 1 (Govier and Aziz, 1972). The second and third flow patterns can be considered as the transition from pack solids to dispersed solids. In this study, these two flow patterns are combined and named as transitional flow. The stationary bed flow has a stationary packed solids bed at the bottom of the flow conduit. Above the packed bed is the flow of solids and liquid mixtures. The dispersed flow means all the solids are fully dispersed in the flow and there are no packed solids in the conduit. The solids distribution can be homogeneous or heterogeneous. The transitional flow is the transition between stationary bed flow and dispersed flow.

In this flow pattern, the packed solids bed shows some wavy features or the continuous bed turns into dunes. Additionally, these wavy beds and dunes are not stationary in the flow.

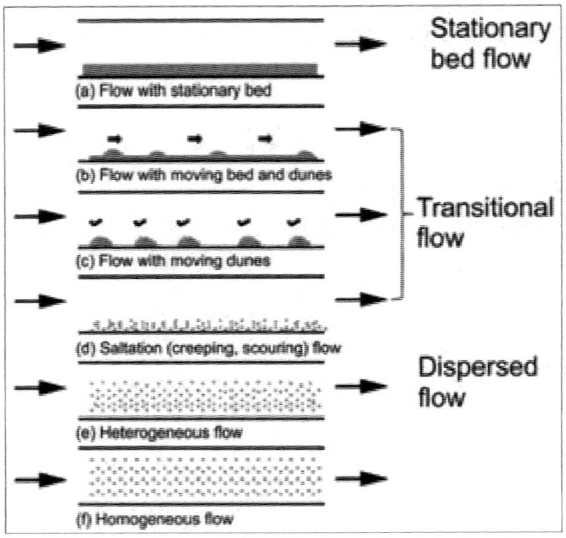

Figure 1. Schematic configuration of the flow patterns (Govier and Aziz, 1972)

Flow Pattern Maps

Previous Flow Pattern Maps

The traditional approach to predict the flow pattern is to conduct a series of experimental tests and plot the flow patterns using superficial solids velocity and fluid velocity, as shown in Figure 2 (Doron et al., 1997). The flow pattern maps developed in this approach are only valid within their test range. The flow patterns depend on various parameters, such as solid particle shape and size, solid density, solid fraction in the flow, flow conduit shape and size, flow conduit inclination angle, fluid velocity, and fluid viscosity and density. The size of the transported particles varies from tens of microns to several centimeters. The fluid properties (viscosity, density, Newtonian, non-Newtonian) and the flow conduit properties (shape, size, and inclination angle) also change significantly for different operations. It is unpractical to conduct experimental tests to cover all the parameters.

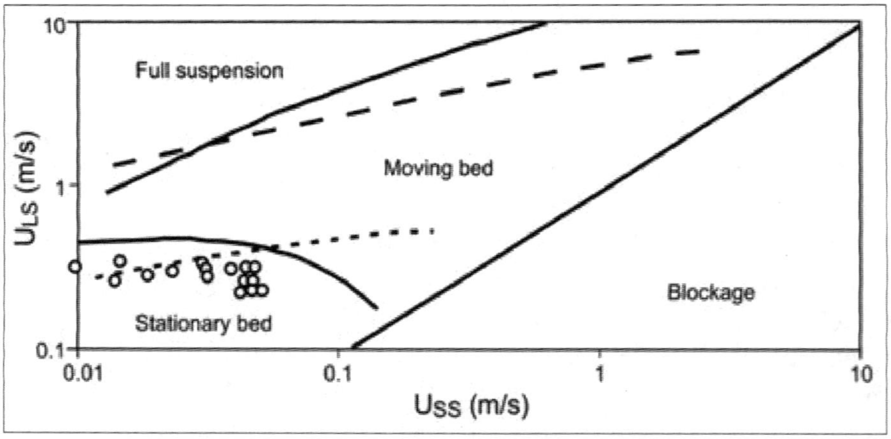

Figure 2. Example of the flow pattern map plotted by superficial velocities (Doron et.al, 1997)

Another approach to plot the flow pattern map is to use the pressure gradient of the flow and the mixture velocity, as shown in Figure 3 (Govier and Aziz, 1972). At low flow velocity conditions, the solids tend to pack in the conduit, which forms a stationary or moving solids bed. The formation of the bed or dunes decreases the flow area and leads to high pressure loss. As the flow velocity increases, the pressure gradient decreases initially, then increases after it reaches a minimum point, as shown in Figure 3. Below the minimum point, the pressure gradient is dominated by solids; beyond the minimum point, the pressure gradient is dominated by friction loss of the flow. This flow pattern plotting approach has broader application than Doron's approach, but it is still restricted by the experimental range since the two axes are dimensional.

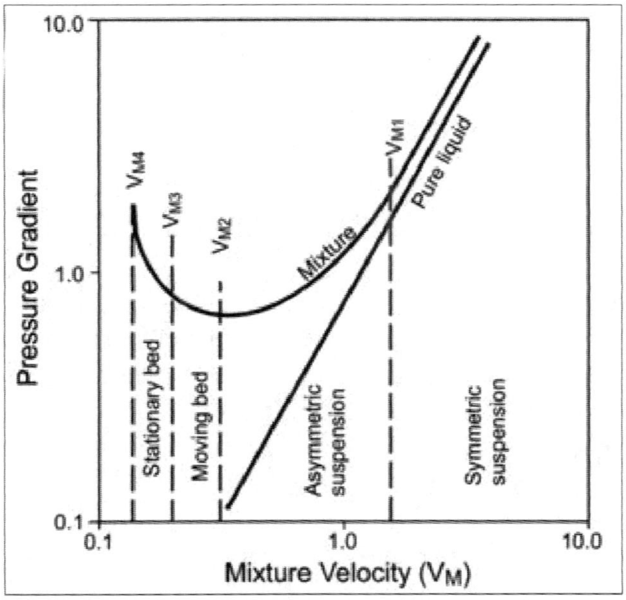

Figure 3. Schematic representation of flow patterns and pressure gradient (Govier and Aziz, 1972)

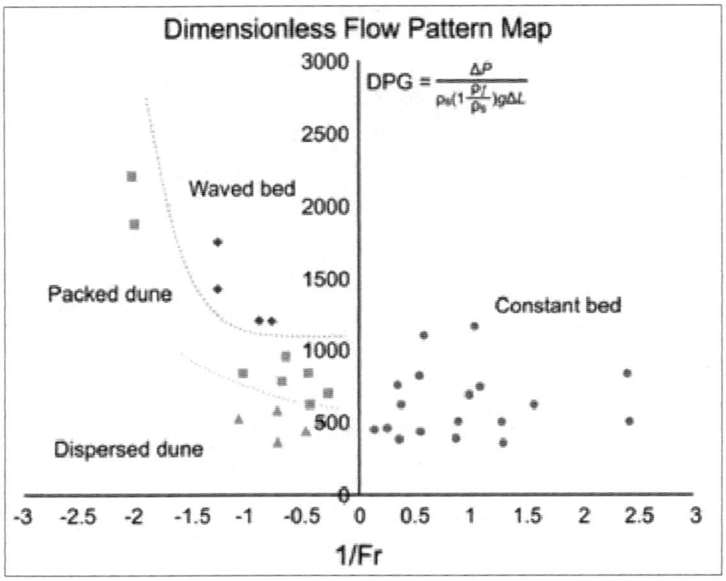

Figure 4. Dimensionless flow regime map (Zhang, 2015)

Recently, Zhang (2015) presented a dimensionless flow regime map, as shown in Figure 4. The two parameters used are dimensionless pressure gradient and a modified Froude number (Fr). This flow pattern map breaks the restrictions from the experimental data since the two axes do not have any dimension. It is also valid for inclined conduits since the modified Froude number including the conduit inclination angle (refer to Zhang, 2015 for details). However, this flow regime map is hard to understand because of the negative Froude number.

Generalized Dimensionless Flow Regime Map

In this paper, two dimensionless parameters, which consider most of the important factors in solid phase transport, are used to plot a generalized flow regime map. In this way, the flow regime map is valid for different solids, different conduits, and different fluids, which significantly reduces the number of experimental tests required to predict flow regime in practical applications.

The following two dimensionless parameters are found to govern the flow regimes of solids in fluid flow:

$$\Omega_h = \frac{u^*}{u_{settling} \cdot sin\theta} \tag{1}$$

$$\Omega_v = \frac{u_f}{u_{settling} \cdot sin\theta} \tag{2}$$

where u* is the friction velocity of the flow, which can be defined as Equation (3), $u_{settling}$ is the settling velocity of the particles in the fluid, θ is the inclination angle of the conduit deviated from vertical, and u_f is the fluid velocity. The dimensionless number Ω_h is used as the horizontal axis in the flow regime map and Ω_v is used as the vertical axis in the flow regime map. The solids-liquid, two-phase flow is classified into three different flow regimes: flow with a stationary solids bed, flow with moving solids bed or dunes, and solids fully dispersed in the flow.

$$u^* = \sqrt{\frac{\tau}{\rho_f}} \tag{3}$$

In Equation 3, τ is the shear stress, which is related to pressure gradient, and ρ_f is the fluid density.

The available experimental data from several different studies (Shook and Roco, 1991; Admiraal, 2003; Balbino, 2010; Lee et al., 2013; Vlasák et al., 2006) are plotted using these two dimensionless parameters, and the flow regime map for pipe flow is shown in Figure 5. The data used to plot this mapping method is from studies with significant different experimental conditions. The size of the solid particles varies from several micrometers to one centimeter, the solids fraction varies from less than one percent to about forty percent, and the solids include glass spheres (1.24 sg) and sand particles (2.5~3.0 sg). The inclination angle of the flow conduits changes from 0 to 180° (0~90° mean upward flow, and 90~180° mean downward flow).

Both the horizontal and vertical axes of the flow regime map are dimensionless. The two dimensionless parameters used to plot the flow regime map represent the three physical factors that control the solid-liquid flow regime. The vertical axis represents the inertia of the flow, which is the source for moving solid particles forward; the horizontal axis represents the turbulent degree of the flow, which is the source to suspend solid particles. The settling velocity $u_{settling}$ and sin θ represent the gravity of the solid particles. Thus, these two dimensionless parameters are a combination of the major parameters that affect the two phase flow regime. The solid particle properties (size, shape, and density) are included in the settling velocity, $u_{settling}$, and the fluid properties are (density and viscosity) included in the settling velocity and the friction velocity, u*. The solids concentration and conduit geometry can also be covered by the friction velocity. The conduit inclination angle is also considered plotting. Thus, the applicable range of this map is not limited by the range of the experimental data used to plot the map.

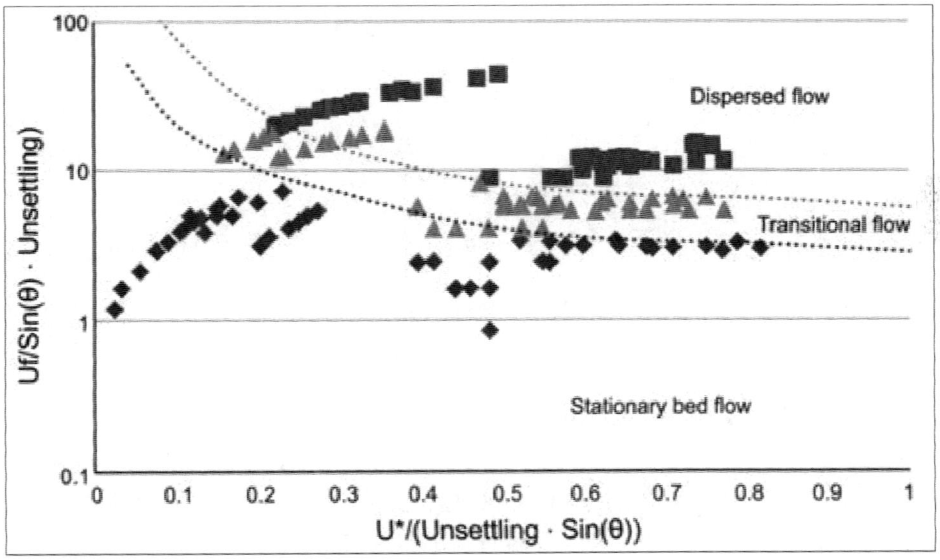

Figure 5. Proposed dimensional flow regime map for pipe flow

The Mechanistic Models

For each flow regime, mechanistic models are developed to calculate the pressure gradient and solids concentration in the wellbore. These mechanistic models are based on the configurations of the flow regimes, mass balance, and momentum balance.

The layer models (Wilson, 1970; Doron et al., 1997) are used for the stationary bed flow and transitional flow. Improvements are made to increase the accuracy of the models. The model is based on the mass and momentum balance of each section (packed solids, solids-liquid mixture, and liquid) in the flow regime. Figure 6 shows a schematic of the geometry for the most complex regime, which has three sections: pure liquid layer, mixture layer, and solids bed layer. The formulations of the mechanistic model are shown in the following.

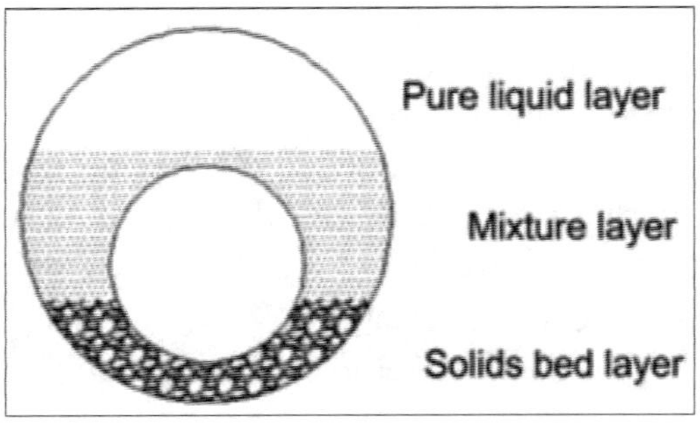

Figure 6. Layer model geometry

The mass conservation for solid phase is

$$\rho_s A_{sb} \emptyset U_{sb} + \rho_s A_m C_m U_{ms} = \rho_s A U_{ss}\,. \quad (4)$$

The mass conservation for liquid phase is

$$\rho_f A_u U_u + \rho_L A_m (1 - C_m) U_{mf} + \rho_L A_{sb} (1 - \emptyset)(U_{sb} + \mathrm{U_r}) = \rho_L Q. \quad (5)$$

The momentum conservations for the three sections in the geometry are

Upper liquid layer: $-\mathrm{A_u}\left(\frac{\Delta P}{L}\right) = -\tau_u S_u - \tau_{u\_m} S_{u\_m} - \rho_u g A_u \cos\theta$ (6)

Mixture layer: $-\mathrm{A_m}\left(\frac{\Delta P}{L}\right) = -\tau_m S_m + \tau_{u\_m} S_{u\_m} - \tau_{m\_sb} S_{m\_sb} - \rho_m g A_m \cos\rho$ (7)

Solids bed layer: $-\mathrm{A_{sb}}\left(\frac{\Delta P}{L}\right) = -\tau_{sb} S_{sb} + \tau_{m\_sb} S_{m\_sb} - \rho_{sb} g A_{sb} \cos\theta$ (8)

where A is the cross-sectional area, $\frac{\Delta P}{L}$ is the pressure gradient, τ is the shear stress, S is the perimeter of each section, ρ is the density, and θ is the conduit inclination angle. The subscript u represents the upper pure fluid layer, m represents the mixture layer, sb represents the solid bed layer, u_m is the interface between upper layer and mixture layer, and m_sb represents the interface between the mixture layer and solid bed layer.

For the stationary bed flow regime, the velocity of the solids bed layer is zero, so it can be eliminated from the calculation; for the transitional flow regime, the movement of the solids bed layer must be included in the calculation; for dispersed regime, the upper layer and solids bed layer can be eliminated from the calculation.

Flow Regime and Pressure Prediction Examples

By combining mechanistic models and a flow regime map, the flow regime, pressure gradient and solids concentration can be predicted automatically for any given flow parameters. The procedure for the flow regime prediction is the following:

1) Run all three mechanistic models with the given input parameters.
2) Find the flow regime for each model by applying the outputs from that model to the flow regime map.
3) Compare the predicted flow regimes with the original flow regime for that mechanistic model. If the predicted flow regime is the same as the original flow regime, then the workflow is converged.
4) If there is more than one converged flow regime, apply physical boundaries for solids transport to remove the multiple converged values to one converged value.

The example in Figure 7 shows the flow regime prediction using the proposed workflow in a 2-in. pipe. The fluid phase is water, and the solid phase is sand. The density of the sand is 2650 kg/m^3, and the diameter of the sand particle is 1 mm. The flow rate changes from 25 to 70 gpm and the inclination angle of the pipe changes from 10° to 100° (0° means upward vertical, 90° means horizontal, and 180° means downward vertical). In this figure, the blue color represents the stationary bed flow regime, the green color represents the transitional flow regime, and the yellow color represents the dispersed flow regime. This figure shows that the solids tend to pack at near horizontal and low flow rate conditions, and tend to disperse at near vertical and high flow rate conditions.

Figure 8 is the comparison between the predicted pressure gradients and Matousek's (2002) data. This figure shows that the proposed workflow verifies the general trend for solid-fluid flow regime transitions and it is able to calculate the pressure gradient accurately for any flow regime conditions.

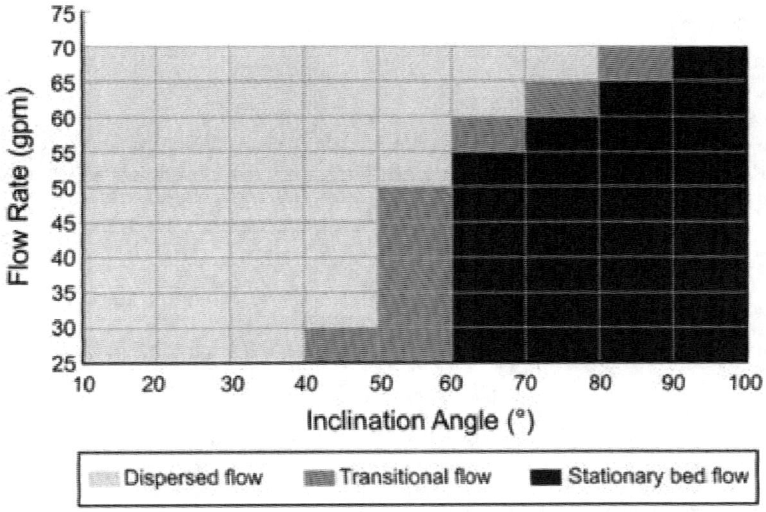

Figure 7. Flow regime prediction examples

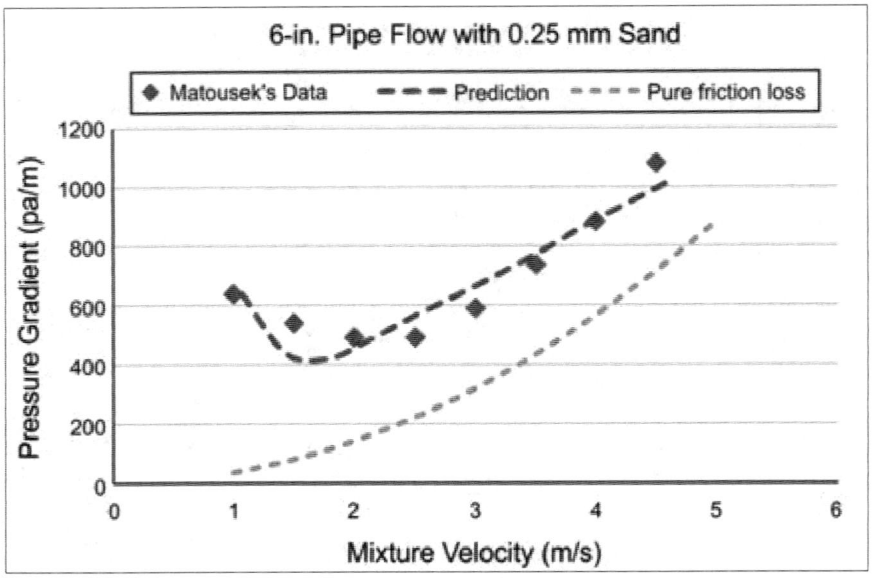

Figure 8. Pressure prediction verifications

Conclusions

The following conclusions are obtained through this study:

1. The solids transport flow regimes in fluid flow are summarized into three flow regimes: stationary bed flow, transitional flow, and dispersed flow for modeling purposes.
2. Two dimensionless numbers, which include major impacting factors for multiphase flow, are proposed to represent the change of flow regimes.
3. A generalized flow regime map is presented using the proposed dimensionless parameters and the available literature data. This flow regime map is not restricted by the experimental data.
4. By combining the mechanistic models and flow regime map, a workflow is developed to predict the flow regime, solid concentration, and pressure gradient automatically for any given flow conditions.

References

1. Admiraal, D. (2003), Influence of Pipe Angle on Bedload Transport in an Inclined Pipe. International Journal of Sediment Research, Vol. 18, No. 2, pp. 122-129
2. Balbino, G. (2010), Experimental Investigation of Critical Velocities for Sand Transport in Horizontal Single and Two-phase Flows at Low Sand Concentrations and Comparisons with Existing Models, MS thesis, University of Tulsa.
3. Govier, G.W. and Aziz, K. (1972), The Flow of Complex Mixtures in Pipes. Malabar, Florida, Robert E. Krieger Publishing Company.
4. Kelessidis, V.C. and Mpandelis, G.E. (2003), Flow patterns and minimum suspension velocity for efficient cuttings transport in horizontal and deviated wells in coiled-tubing drilling. SPE/ICoTA Coiled Tubing Conference and Exhibition, Houston, TX. SPE 81746

5. Lee, M.S., Matousek, V., Chung, C.K., and Lee, Y.N. (2013), Pipe Size Effect on Hydraulic Transport of Jumoonjin Sand –Experiments in a Dredging Test Loop. Journal of Fluid Dynamics, Vol. 3, pp. 152-162.
6. Matousek, V. (2002), Pressure Drops and Flow Patterns in Sand-Mixture Pipes, Experimental Thermal and Fluid Science, Vol. 26, I6–7, pp. 693–702.
7. Doron, P., Simkhis, M., and Barnea, D. (1997), "Flow of Solid-Liquid Mixture in Inclined Pipes," Int. J. Multiphase Flow Vol. 23, No. 2, pp. 313-323.
8. Parzonka, W., Kenchington, J.M., and Charles, M.E. (1981), "Hydrotransport of solids in horizontal pipes: effects of solids concentration and particle size on the deposit velocity," Can. J. Chem. Engng, Vol.59, pp 291-296.
9. Vlasák, P. Chára, Z., Krupička, J., and Konfršt, J. (2006), "Experimental Investigation of Coarse Particles-Water Mixture Flow in Horizontal and Inclined Pipes," J. Hydrol. Hydromech., Vol. 62, 2014, 3, pp. 241–247.
10. Vacaldo, J.J. and Charles, M.E. (1972), "Prediction of pressure gradient for the horizontal turbulent flow of slurries," In Proc. 2th Int. Conf. on the Hydraulic Transport of Solids in Pipes, Conventry, England, C1, pp.1-12.
11. Shamlou, A.P. (1987), "Hydraulic transport of particulate solids," Chem. Engng Commun., Vol. 62, pp. 233-249.
12. Shook, C. A. and Roco, M. C. (1991), "Slurry Flow: Principles and Practice," Oxford, UK, Butterworth-Heinemann.
13. Wilson, K.C. (1970), "Slip Point of Beds in Solids-liquid Pipeline Flow," Proc. ASME, J. Hyd.Div., Vol. 96, pp. 1-12.
14. Zhang, F. (2015), "Numerical Simulation and Experimental Study of Cuttings Transport in Intermediate Inclined Wells," Ph.D. Dissertation, The University of Tulsa, Tulsa, OK, USA.
15. Zhang, F., Miska, S., Yu, M., Ozbayoglu, E.M., and Takach, N. (2015), "Pressure Profile in Annulus: Solids Play a Significant Role," J. Energy Resour. Technol Vol. 137(6), 064502.

444189 Effective Design of the Inlet Flow Distributer That Significantly Improved the Dehydration Efficiency of a Horizontal Heater Treater Vessel

Abstract:

The heater treater vessels are used to efficiently dehydrate the crude oil emulsions. These vessels are divided into a heating section and a separation section. The heating section heats the emulsion fluid effectively using the U-tube design of the firetube. Effective contact between the firetube and the crude oil ensures efficient and even heating of the crude. As a result the associated gas and free water are effectively released from the emulsion prior to entering the separation section. The study presented here was conducted to improve the contact between the firetubes and the crude oil for effective heating. ANSYS FLUENT Computational Fluid Dynamics (CFD) software was used to understand the flow and temperature distribution of the crude oil entering into the vessel through an inlet device. As a result of this study a new inlet distributer was designed and evaluated using the CFD technique. CFD results showed significantly improved heat transfer between the crude oil and firetubes when compared to the original design. The improved design was implemented and tested in an actual heater treater vessel and the results showed significant improvements in performance.

444079 Spatially Resolved Fixed-Bed Versus Pseudo-Homogeneous Reactor Models Including a Detailed Reaction Mechanism

Abstract:

Fixed bed or packed bed reactors are widely used in the chemical and process industry amongst others for highly exothermic or endothermic catalytic surface reactions. A small tube to particle diameter ratio (D/d-ratio) facilitates a safe thermal performance of these reactors. Many correlations reported in literature for pressure drop and heat transfer are limited in their predictive capability for such packed bed reactors due to a dominant influence of the confining wall. It affects the porosity distribution and as a result the velocity field as well as the species and temperature distribution within the bed. Accurate measurements for development of new correlations can be cumbersome, expensive and many times inaccurate due to the complexity of the geometry. The increasing power of computational hardware in the last decade has enabled virtual experiments: Computational Fluid Dynamics (CFD) can now be used even for three dimensional spatially resolved investigations of those reactor types including detailed surface reaction mechanisms as recently shown by [Wehinger et al., 2015] for different catalytic particle shapes and by [Eppinger et. al., 2015] for an optimization study on operating conditions. A complete numerical workflow was shown where the packing generation with DEM (discrete element method), the meshing and the simulation were done with the finite volume code STAR-CCM+ by CD-adapco and for the optimization Optimate+/HEEDS by Red Cedar Technology were used.

The current contribution is an extension of our study which was presented at the AIChE annual meeting 2015 ([Eppinger et. al., 2015]) with a focus on a comparison of the spatially resolved packed bed results with the results achieved by a less computational intensive pseudo-homogeneous approach. The significant reduction in calculation time is very attractive for industrial applications, but on the other hand this simplification can lead to inaccuracies. The sensitivity of the yield and conversion on mass transfer limitation which are typically occur in such reactors will be shown and quantified as well as the effect of typically used model enhancements like including local porosity profiles.

References

N. E. McGuire, N. P. Sullivan, O. Deutschmann, H. Zhu, R. J. Kee *Dry reforming of methane in a stagnation-flow reactor using Rh supported on strontium-substituted hexaaluminate*, Applied Catalysis A: General. (2011), 257-265

G. D. Wehinger, T. Eppinger, M. Kraume. *Evaluating Catalytic Fixed-Bed Reactors for Dry Reforming of Methane with Detailed CFD*, Chem. Ing. Tech. (2015), 734-745

T. Eppinger, N. Jurtz, R. Aglave, Gregor D. Wehinger, M. Kraume. *A Numerical Optimization Study on the Catalytic Dry Reforming of Methane in a Spatially Resolved Fixed-Bed Reactor, AIChE Annual Meeting 2015, Salt Lake City, USA*

Computational Fluid Dynamic design of steam cracking reactors: extrusion method for simulation of dynamic coke layer growth

Laurien A. Vandewalle
Ghent University, Laboratory for Chemical Technology
Technologiepark 918, 9052 Gent, Belgium

Jens N. Dedeyne, David J. Van Cauwenberge, Kevin M. Van Geem*, Guy B. Marin
Ghent University, Laboratory for Chemical Technology
Technologiepark 918, 9052 Gent, Belgium
Corresponding author: Kevin.VanGeem@UGent.be

Prepared for Presentation at
American Institute of Chemical Engineers
2016 Spring Meeting
11th Global Congress on Process Safety
Houston, Texas
April 10-14, 2016

UNPUBLISHED

Computational Fluid Dynamic design of steam cracking reactors: extrusion method for simulation of dynamic coke layer growth

Laurien A. Vandewalle
Ghent University, Laboratory for Chemical Technology
Technologiepark 918, 9052 Gent, Belgium

Jens N. Dedeyne, David J. Van Cauwenberge, Kevin M. Van Geem*, Guy B. Marin
Ghent University, Laboratory for Chemical Technology

Keywords: Computational fluid dynamics, steam cracking, coke formation, enhanced heat transfer

Abstract

Computational fluid dynamics (CFD) gas been successfully applied for the evaluation of the start-of-run performance of three-dimensional (3D) coil geometries in steam cracking reactors. However, as the most attractive characteristic of these 3D coil geometries is the potential extension of the run length, determining the full economic potential of a coil involves tracking its performance throughout the run and not only at start-of-run. In the case of 3D tubular geometries, the growth of the coke layer will generally not be uniform. Because of this, the reactor geometry will change in time, which will in turn influence the fluid dynamics, product yields and successive coke formation. To take this into account, the coke layer growth needs to be incorporated in the CFD simulations. In this work an algorithm based on dynamic meshing will be discussed for simulating coke formation on the 3D reactor geometry and tracking the geometry deformation caused by the growing coke layer. As a proof-of-concept, a Millisecond propane cracker was simulated over the first days of its run length for a bare tube and a classically applied finned reactor design. Future work also includes comparison with other industrially relevant 3D reactor geometries.

1. Introduction

Steam cracking of hydrocarbons is the predominant industrial process for the production of many commercially important base chemicals such as light olefins (ethene, propene, butadiene) and aromatics (benzene, toluene, xylenes). These base chemicals are the key building blocks for large-volume polymers and other high-value chemicals. Steam cracking is one of the most energy-intensive processes in the chemical industry, representing approximately 8% of the sector's primary energy consumption [1]. A major factor for the process energy efficiency is the formation of coke on the inner wall of the tubular cracking reactors. This growing carbonaceous layer has two negative effects. First of all, the reactor pressure drop increases as the cross-sectional flow area decreases, resulting in a loss of selectivity to ethylene, the process' main product [2]. Secondly, the highly insulating coke layer blocks the heat transfer from the furnace to the process gas. To maintain the same cracking severity, the heat input is increased to counteract the increased heat transfer resistance, leading to higher tube metal temperatures (TMT). Eventually, either due to an excessive pressure drop over the reactor or metallurgical constraints of the reactor tube alloy, production needs to be halted to decoke the reactor. Obviously, these periodic production interruptions have a negative effect on the process economics. Furthermore, the reactor material deteriorates with successive coking-decoking cycles by tube corrosion, carburization and erosion [3-6].

In response to this economic and energetic drawback, several technologies to reduce coke formation have been investigated [7-10]. One such technology is the use of three-dimensional (3D) reactor geometries to enhance radial mixing, resulting in lower coking rates and longer furnace run lengths [11]. Previously, computational fluid dynamics (CFD) has been applied to evaluate the effect of these 3D reactor geometries on pressure drop, coking rates and product yields [11, 12]. These studies however focused on start-of-run performance, whereas the most attractive characteristic of the enhanced reactor designs is the extension of the furnace run length. Determining the full economic potential of a coil hence involves tracking its performance throughout the run, i.e., as function of the time on stream. In the case of 3D tubular geometries or reactors with a strongly non-uniform heat flux profile (e.g. due to shadow effects), the growth of the coke layer will generally not be uniform. Because of this, the reactor geometry will change in time, which will in turn influence the fluid dynamics, product yields and successive coke formation. To take this into account, the coke layer growth needs to be incorporated in the CFD simulations. Therefore in this work a methodology is developed and demonstrated for simulating a run length of a steam cracking reactor and tracking the geometry deformation caused by the growing coke layer. The method is based on dynamic meshing and mesh generation, and can be applied for a variety of reactor tube geometries (bare, longitudinally finned, helically finned, MERT, slit-MERT). As a proof-of-concept, a Millisecond propane cracker is simulated over the first days of its run length for a bare and a classically applied finned reactor design. However, substantially more work is needed to simulate the complete run length and make the algorithm more flexible.

2. CFD model

2.1. Governing equations

The steady-state governing equations for a compressible, reactive, single-phase fluid flow are the following:

Continuity
$$\frac{\partial \rho u_i}{\partial x_i} = 0 \tag{1}$$

Momentum
$$\frac{\partial \rho u_i u_j}{\partial x_i} = -\frac{\partial p}{\partial x_j} + \frac{\partial \tau_{ij}}{\partial x_i} \tag{2}$$

Gas-phase energy
$$\frac{\partial \rho u_i h}{\partial x_i} = \frac{\partial q_i}{\partial x_i} + S_h \tag{3}$$

Species transport
$$\frac{\partial \rho Y_k u_i}{\partial x_i} = -\frac{\partial J_{k,i}}{\partial x_i} + R_k \quad (\forall k = 1, n_k\text{-}1) \tag{4}$$

To be able to explicitly account for heat transfer from the metal reactor outer wall to the process gas (conjugate heat transfer), the Laplace equations for thermal conduction in the coke layer and metal tube wall surrounding the process gas are solved simultaneously with the governing equations for the fluid flow.

Solid-phase energy
$$\nabla \cdot (\alpha_s \nabla T) = 0 \tag{5}$$

In the above equations, τ_{ij} is the viscous shear stress tensor, which is evaluated using the Boussinesq eddy viscosity concept. The sensible enthalpy flux q_i and the diffusion flux $J_{k,i}$ are both modeled using an eddy diffusivity approach.

$$\tau_{ij} = (\mu + \mu_t)\left(\frac{\partial u_i}{\partial x_j} + \frac{\partial u_j}{\partial x_i}\right) - \frac{2}{3}k\delta_{ij} \tag{6}$$

$$q_i = c_p\left(\frac{\mu}{Pr} + \frac{\mu_t}{Pr_t}\right)\frac{\partial T}{\partial x_i} \tag{7}$$

$$J_{k,i} = \left(\frac{\mu}{Sc} + \frac{\mu_t}{Sc_t}\right)\frac{\partial Y_k}{\partial x_i} \tag{8}$$

The turbulent Prandtl number Pr_t and turbulent Schmidt number Sc_t are defined as the ratio of the momentum eddy diffusivity to the heat and mass eddy diffusivity respectively. A value of 0.85 was chosen for these parameters, in accordance to the typical recommendations for gaseous flows [13-15].

The turbulent eddy viscosity μ_t appearing in equations (6)-(8) is calculated using the shear stress transport (SST) $k - \omega$ turbulence model developed by Menter [16]. This two-equation turbulence model blends the robust and accurate formulation of the $k - \omega$ model in the near-wall region with the freestream independence of the $k - \epsilon$ model in the far field. These features make the SST $k - \omega$ model more accurate and reliable for a wider class of flows than the standard $k - \omega$ model.

2.2. Chemistry model

Steam cracking of hydrocarbons mainly proceeds through a free-radical mechanism, which is inherently characterized by a huge number of species and reactions [17-19]. The incorporation of such detailed kinetic networks is computationally very expensive. Therefore, to limit the computational cost, the full single-event microkinetic CRACKSIM model [20-24], developed in-house, was reduced to its relevant core for propane cracking. The final network consists of 151 reactions between 29 species, of which 13 radical species. The pseudo-steady state assumption was applied to the radical reaction intermediates, transforming the ODEs for these short-lived species into algebraic equations. In this way the fastest transients are removed from the system of differential equations, resulting in an overall reduction of the stiffness of the kinetic model [25]. This procedure is roughly identical to that adopted by Reyniers et al. [26], who also confirmed the validity of the approach.

The Reaction Mechanism Generator (RMG) Transport estimator [27] was used for the estimation of the Lennard-Jones (L-J) model parameters. These were then used in the kinetic theory method for calculation of the species' individual thermal conductivities and viscosities. Using the same L-J parameters, the Chapman Enskog formula was used to quantify the diffusion coefficients. Properties of the multicomponent mixture were calculated by ideal gas mixing laws. To simplify the diffusion terms a unity Lewis number ($Le = Sc/Pr$) was assumed.

2.3. Computational grid and boundary conditions

The simulation procedure is validated for a bare reactor tube, which serves as the base case, and then compared with a classically applied longitudinally finned reactor design. The geometric parameters of this tubular reactor geometry are depicted in Figure 1. The 3D computational grids were generated by extruding a specified number of layers from a cylindrical surface to the surface of the 3D tube geometry, as explained in §2.5. By symmetry considerations, the computational domain for the finned tube can be limited to one fin to reduce simulation time. Also for the bare tube, the computational domain comprises only 1/8th of the cylinder's cross section, while periodic boundary conditions are applied at the azimuthal domain boundaries. Near-wall grid resolution was fixed to 10^{-5} m for all cases, satisfying the $y^+ < 1$ condition and ensuring the first cell lies within the viscous sublayer.

The inner reactor surface was treated as a no-slip, no-species flux solid wall. At the inlet boundary, the process gas temperature, mass flow rate, turbulence parameters, and composition of the process gas were imposed. At the reactor outlet, a constant pressure boundary condition was set. A heat flux profile was imposed at the outer surface of the reactor, i.e., the metal tube wall. Thermal coupling between the solid regions and the process gas was established through the use of the *turbulentTemperatureCoupledBaffleMixed* boundary condition at the metal/cokes and cokes/gas interfaces.

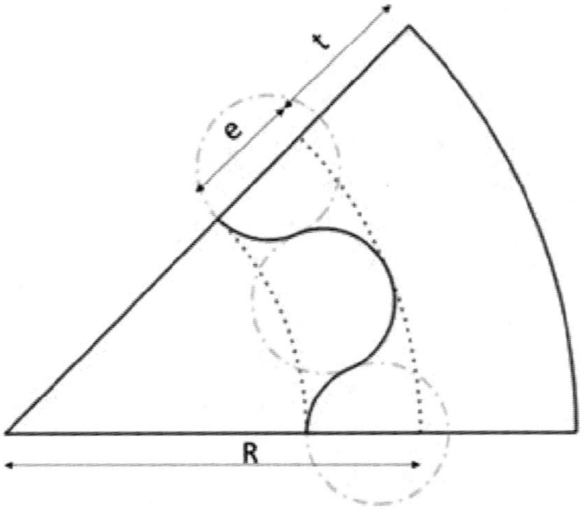

Figure 1. Geometric parameters of a longitudinally finned reactor tube. R: inner radius, e: fin height, t: minimum metal wall thickness.

2.4. Numerical model

The governing equations were solved using the open source CFD package OpenFOAM® [28].

The Semi-Implicit Method for Pressure-Linked Equations (SIMPLE) was used to couple the momentum and continuity equation, with one additional non-orthogonal correction step where necessary.

A second order central differencing spatial discretization scheme was used. The cell face values of the Gaussian surface integral were obtained through linear interpolation.

2.5. Dynamic modeling of coke formation

The overall methodology for the run length simulations can be summarized as follows. The reactor mesh is updated on a regular basis as coke deposits on the reactor wall until an end-of-run constraint is met, indicating that decoking is required. In industry, two constraints are typically used to end a furnace run and commence decoking: pressure drop and tube metal temperature (TMT). If one of these surpasses a critical value at the end of a simulation run, the algorithm is ended.

To perform the mesh update step, the open source CFD package OpenFOAM® [28] has been extended with a new post-processing utility for dynamically simulating the growing coke layer. The utility starts with reading the non-uniform temperature and concentration fields on the gas-coke interface. The corresponding coking rate is then calculated using the semi empirical coking model of Plehiers [29]. From the knowledge of the coking rate and a duration over which coke formation is assumed to be constant, the growth of the coke layer (in meters) is determined. With this information a new mesh is generated. The procedure can be repeated multiple times for different stages of the coking process.

The mesh generation itself starts from a structured cylindrical mesh. Layers are added to the mesh by extruding them in the radial direction starting from the core cylinder wall. Using a newly developed library of extrusion models it is possible to generate in this way structured meshes of high quality on several enhanced tube geometries. The library of extrusion models contains the surface definitions of several enhanced tube geometries. These surface definitions are used by the model to calculate the required extrusion distance in each point on the core cylindrical surface. The mesh of the 3D tube geometry is then created by radial extrusion of the core cylindrical surface over the extrusion distance calculated by the geometry-specific model. This is shown in Figure 2 for a finned tube. Initially, when no coke layer is present, the mesh for the gas region is extruded all the way to the specified surface. This means that the gas-coke interface coincides with the coke-metal interface (black line). However, when the utility calculates a non-zero thickness of the coke layer, the calculated coke thickness is subtracted from the initially calculated extrusion distance and the gas-coke interface 'moves' inward (red line).

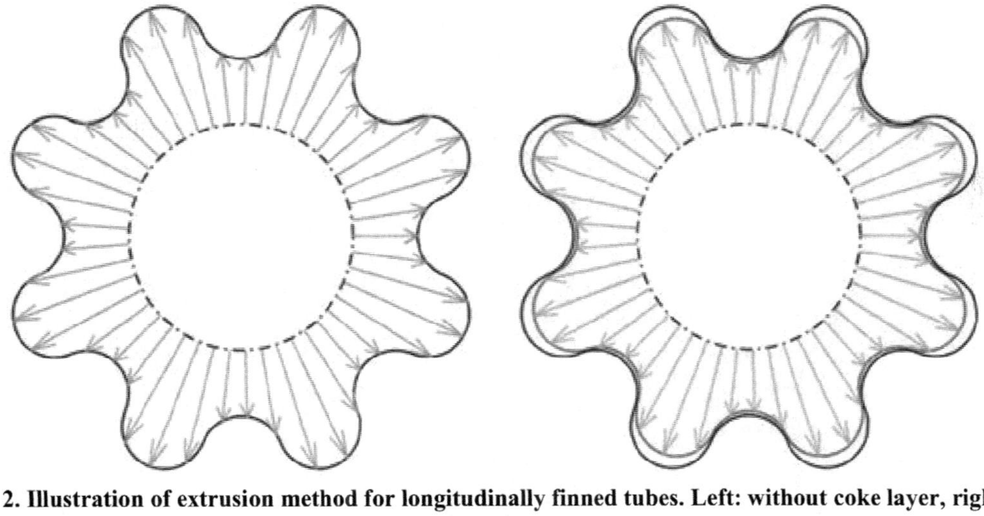

Figure 2. Illustration of extrusion method for longitudinally finned tubes. Left: without coke layer, right: with coke layer.

3. Run length simulation of an industrial propane cracker

Reactive simulations of one of the reactors in an industrial Millisecond furnace were performed to evaluate the effect of the growing coke layer on the fluid dynamics, product yields and successive coke formation. The simulated reactor has been previously described by Schietekat et al. [11] and is known to achieve very high selectivity towards ethene due to the short residence times in order of milliseconds. Disadvantages of this design are the very high operating temperatures and very short run lengths, sometimes as short as a week.

3.1. Process conditions and reactor configuration

The feedstock to the reactor consisted of 100 % pure propane. The hydrocarbon and steam flow rate were set to 0.033 and 0.011 kg s^{-1}, respectively, corresponding to a steam dilution of 0.33. The temperature at the inlet of the reactor was equal to 904 K, whereas the reactor outlet pressure, that is, upstream of the transfer line exchanger, was set to 170 kPa. Because a coupled reactor-furnace simulation is deemed too computationally expensive, a heat flux profile was imposed on the outer wall of the reactor. For the bare tube simulations, this heat flux profile is identical to that used by Reyniers et al. [26] and was taken from a furnace simulation where the boundary condition applied to the reactor tubes was the industrially measured outer wall temperature profile. For the simulations with the 3D tube geometry, the original heat flux profile was scaled in such way to obtain the same propane conversion in each simulation case. The same is also done after each simulation of coke layer growth, as the industry's response to coking is usually also an increase of fuel input to the burners to counteract the decreasing conversion and hence product yields.
The dimensions of the two reactor configurations simulated in this work are summarized in Table 1. First, a bare tube was simulated as a reference case. Second, an industrially applied finned reactor was considered, identical to the one simulated by Schietekat et al. [11]. An adiabatic entrance zone with a length of 0.4 m was added upstream both reactors to ensure fully developed profiles for velocity and turbulence parameters at the reactor inlet. An initial coke layer of 0.1 mm thickness was assumed.

Table 1. Reactor dimensions

Reactor ID	Bare	Finned
Reactor length [m]	10.556	10.556
Adiabatic inlet section [m]	0.444	0.444
Maximum inner diameter [10^{-3} m]	30.2	34.8
Initial coke layer thickness [10^{-3} m]	0.1	0.1
Number of fins [-]	-	8
Pitch [10^{-3} m]	-	-
Outer diameter [10^{-3} m]	43.7	48.3
Metal thickness [10^{-3} m]	6.75	6.75
Fin height [10^{-3} m]	-	4.8
Cross-sectional surface area [10^{-6} m^2]	707	707

3.2. Results and discussion

Table 2 summarizes the most important process conditions and product yields at start-of-run. For validation, data of an industrial bare tube Millisecond reactor running with a crude propane feedstock (i.e. 95 % pure propane or higher) under similar, but not identical conditions have been added. The agreement between the simulated bare reactor and the industrial data is reasonable, considering the accuracy of industrial yield measurements and the purity of the feed. The results are in line with those of Schietekat et al. [11].

Table 2. Start-of-run reactor conditions and product yields.

Reactor ID	Bare	Finned
Coil outlet temperature [K]	1165.6	1164.6
Pressure drop [kPa]	28.20	36.86
Pressure drop ratio [-]	1.00	1.31
Propane conversion [%]	80.00	79.99
P/E ratio [wt% $(wt\%)^{-1}$]	0.523	0.515
Product yields [wt%]		
H_2	1.50	1.48
C_2H_4	33.68	33.69
C_3H_6	17.61	17.37
C_3H_8	20.00	20.01

[a]Detailed feed composition not available.

Figure 3 shows the mixing-cup averaged temperature and pressure profile as function of the axial position at start-of-run conditions for the two reactor configurations. Little differences between the process gas temperatures are simulated as the same reactor volume and mass flow rate, and similar heat input were applied. However, the pressure drop varies drastically.

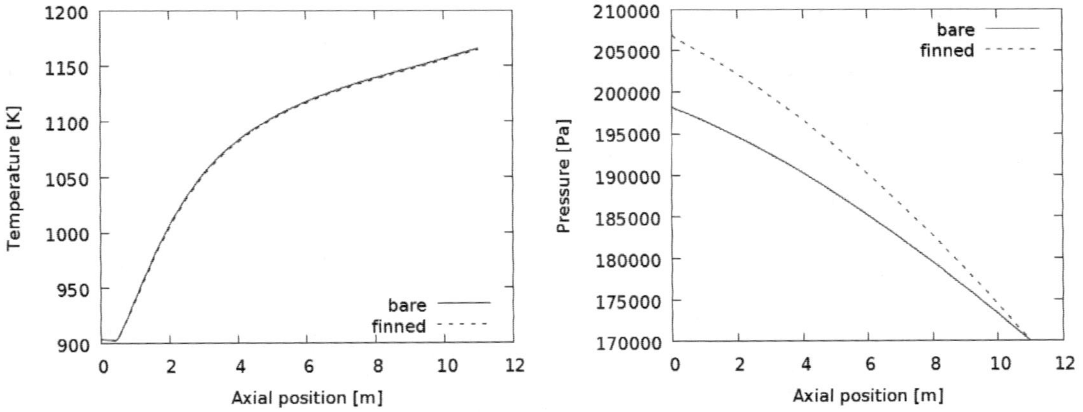

Figure 3. Start-of-run axial process gas temperature (left) and pressure (right) mixing-cup averaged over a cross section as function of axial position.

The most attractive characteristic of the finned reactor designs is the extension of the furnace run length, hence, one should not focus on start-of-run performance. Using the methodology presented in the previous section, the influence of the growing coke layer was investigated. Two mesh updates were performed, each corresponding with a duration of coke layer growth equal to 48 hrs, so a total run length of 96 hrs was simulated. As an illustration, the growing coke layer is shown in Figure 4 for the bare tube.

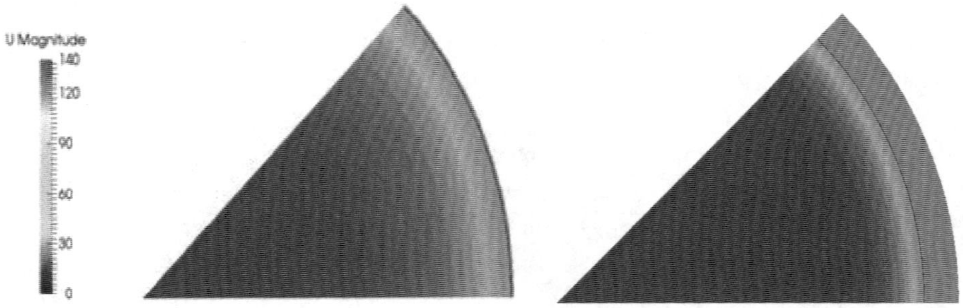

Figure 4. Contours of velocity at axial position z = 6 m: start-of-run (left) and after 48 hrs of coke layer growth (right).

4. Conclusions and future work

In this work a methodology was presented for simulating the run length of an industrial steam cracking reactor while tracking the geometry deformation caused by a growing coke layer. The method was tested for a bare tube and a classically applied finned tube reactor. The influence of the growing coke layer on product yields and selectivities still has to be studied in more detail. Future work also includes simulation of other types of 3D reactor geometries (MERT, helically finned).

List of symbols

c_p	Heat capacity	$\mathrm{J\,kg^{-1}\,K^{-1}}$
COT	Coil Outlet Temperature	K
COP	Coil Outlet Pressure	Pa
h	Specific enthalpy	$\mathrm{J\,kg^{-1}}$
J	Mass diffusion flux	$\mathrm{kg\,m^{-2}\,s^{-1}}$
k	Turbulent kinetic energy	$\mathrm{m^2\,s^{-2}}$
p	Pressure	Pa
Pr	Prandtl number	-
q	Sensible enthalpy flux	$\mathrm{W\,m^{-2}}$
R_i	Specific rate of formation of species i	$\mathrm{kg\,m^{-3}\,s^{-1}}$
Sc	Schmidt number	-
T	Temperature	K
TMT	Tube Metal Temperature	K
u	Velocity	$\mathrm{m\,s^{-1}}$
Y_i	Mass fraction of species i	-
α	Thermal diffusivity	$\mathrm{m^2\,s^{-1}}$
ρ	Density	$\mathrm{kg\,m^{-3}}$
τ_{ji}	Shear stress	$\mathrm{kg\,m^{-1}\,s^{-2}}$
μ	Dynamic viscosity	$\mathrm{kg\,m^{-1}s^{-1}}$

Acknowledgements

LAV acknowledges financial support from the Fund for Scientific Research Flanders (FWO), in association with the PI-FLOW project. The authors also acknowledge the financial support from the Long Term Structural Methusalem Funding by the Flemish Government – grant number BOF09/01M00409. The computational work was carried out using the STEVIN Supercomputer Infrastructure at Ghent University, funded by Ghent University, the Flemish Supercomputer Center (VSC), the Hercules Foundation and the Flemish Government – department EWI.

References

[1] T. Ren, M. Patel, et al. *Energy*. **2006**, 31, 425-451.
[2] K. M. Van Geem *Industrial & Engineering Chemistry Research*. **2005**, 44, 3402-3411.
[3] G. Zimmermann, W. Zychlinski, et al. *Industrial & Engineering Chemistry Research*. **1998**, 37, 4302-4305.
[4] W. Zychlinski, K. A. Wynns, et al. *Materials and Corrosion*. **2002**, 53, 30-36.
[5] Y. Nishiyama, N. Otsuka *Corrosion*. **2005**, 61, 84-93.
[6] A. E. Muñoz Gandarillas, K. M. Van Geem, et al. *Industrial & Engineering Chemistry Research*. **2014**, 53, 6358-6371.
[7] J. Wang, M. F. Reyniers, et al. *Industrial & Engineering Chemistry Research*. **2007**, 46, 4134-4148.
[8] J. Wang, M. F. Reyniers, et al. *Industrial & Engineering Chemistry Research*. **2008**, 47, 1468-1482.
[9] M.-F. S. G. Reyniers, G. F. Froment *Industrial & Engineering Chemistry Research*. **1995**, 34, 773-785.
[10] C. M. Schietekat, S. A. Sarris, et al. *Industrial & Engineering Chemistry Research*. **2015**, 54, 9525-9535.
[11] C. M. Schietekat, D. J. Van Cauwenberge, et al. *AIChE Journal*. **2014**, 60, 794-808.
[12] P. A. Reyniers, C. M. Schietekat, et al. *Industrial & Engineering Chemistry Research*. **2015**.
[13] P. Moin, K. Squires, et al. *Physics of Fluids A*. **1991**, 3, 2746-2757.
[14] M. C. William Kays, Bernhard Weigand, Convective Heat and Mass Transfer, Mcgraw-Hill (Tx), **2004**.
[15] S. B. Pope, Turbulent Flows, Cambridge University Press, **2000**.
[16] F. R. Menter *AIAA Journal*. **1994**, 32, 1598-1605.
[17] K. M. Van Geem, D. Hudebine, et al. *Computers & Chemical Engineering*. **2007**, 31, 1020-1034.
[18] E. Ranzi, A. Frassoldati, et al. *Industrial & Engineering Chemistry Research*. **2005**, 44, 5170-5183.
[19] T. Dijkmans, S. P. Pyl, et al. *Green Chemistry*. **2013**, 15, 3064.
[20] P. J. Clymans, G. F. Froment *Computers & Chemical Engineering*. **1984**, 8, 137-142.
[21] L. P. Hillewaert, J. L. Dierickx, et al. *AIChE Journal*. **1988**, 34, 17-24.
[22] K. M. Van Geem, M. F. Reyniers, et al. *Oil & Gas Science and Technology - Revue de l'IFP*. **2008**, 63, 79-94.
[23] P. A. Willems, G. F. Froment *Industrial & Engineering Chemistry Research*. **1988**, 27, 1959-1966.
[24] P. A. Willems, G. F. Froment *Industrial & Engineering Chemistry Research*. **1988**, 27, 1966-1971.
[25] T. Turanyi, A. S. Tomlin, et al. *The Journal of Physical Chemistry*. **1993**, 97, 163-172.
[26] P. A. Reyniers, C. M. Schietekat, et al. *Industrial & Engineering Chemistry Research*. **2015**, 54, 12270-12282.
[27] W. H. Green in RMG - Reaction Mechanism Generator v4.0.1, Vol. (Ed.^Eds.: Editor), City, **2013**.
[28] in OpenFOAM - The Open Source CFD Toolbox - User's Guide, Vol. 2.2.x edition (Ed.^Eds.: Editor), OpenCFD Ltd, City, **2013**.
[29] P. M. Plehiers, Rijksuniversiteit Gent, **1989**.

443828 Analysis of the Particle Reynolds Number in a Gas-Solid Down-Flow Reactive System Using Computational Fluid Dynamics

Abstract:

This study examines how the particle Reynolds number evolves in a gas-solid down-flow reactive system, using fluid catalytic cracking of gasoil in a downer reactor as case study. In heterogeneous catalytic processes such as fluid catalytic cracking (FCC), in which particles act as catalyst, the contact between gas and particles determines the overall performance. This contact depends on the turbulent conditions around the particles. Depending on the flow conditions, the gas layer around particles may behave differently, particularly when chemical reactions (kinetics) are present. This study explores how the particle Reynolds number (Re_p) characterizes the hydrodynamics in a gas-solid down-flow reactive system.

The particle Reynolds number is a dimensionless variable that relates the physical properties of the carrier fluid and the particles. It gives information about the flow conditions around the particle under specific conditions of velocity. The present work used computational fluid dynamics (CFD) coupled with Large Eddy Simulation (LES) to conduct the multiphase flow simulations. The hydrodynamic behavior of the system was modeled under an Euler-Lagrange framework. 200000 particles were injected to a three-dimensional down-flow system to perform the hydrodynamic study. Mono-dispersed, 70-micron size particles with a particle density of 1500 kg/m^3 were used for the simulations that were conducted with and without chemical kinetics. The catalytic cracking mechanism proposed by Gianetto et al., 1994 was employed for the simulations of the reactive system.

In the simulations in absence of chemical reactions, three different zones were observed along the longitudinal coordinate in the down-flow system: in the first one, near the entrance, the Re_p was high (>1) due to the large slip velocity between gas and particles. A second zone was observed as the particles were accelerated by the action of the drag force and gravity. In this zone, the slip velocities reach similar values and Re_p decreases, and becomes less than 1, a value that indicates the existence of creeping flow regime around the particles. As particles velocity increases, Re_p increases again, passing to a third zone. The first and third zone operate in an intermediate or transition regime around the particle between laminar and turbulent flow as Re_p is higher than 1.

The three different zones, above described, were also observed in the simulations including chemical reactions. The variations on Re_p of the first and second zone were more pronounced in comparison to the non-reactive system. The expansive effect produced by the action of cracking molecules increases the gas velocity, generating higher slip velocities between gas and solids in the first zone of the reactor. In consequence, higher Re_p numbers were obtained. The results indicate that the presence of creeping flow regime is reduced when the catalytic cracking reactions are included. As the accelerating effect is stronger, the extent of the second zone where Re_p is lower than 1, becomes shorter. The present study allows to have a guide on how flow regimes around particles evolve in a down-flow reactive system.

*** Formatting Issue - Best Available Paper/Graphic ***

444218 - Using All the Tools in the Toolbox – Case Studies in Process Improvement

Kenric A. Marshall[1], Sadeka Onam[2], Amarnath Singh[1]
The Dow Chemical Company
1. Freeport, Texas 2. Midland , Michigan
1. KAMarshall@dow.com, 2. Sonam@dow.com

Keywords: Statistical Analysis, Kinetics, Catalysts

Abstract

At commissioning, a chemical manufacturing plant represents the end results of years, and sometimes decades, of research and continuous improvement in design. Each plant is built with a goal of maximizing the economic return for that investment. However, it is doubtful if any plant has ever been built that did not contain an opportunity to reduce cost or otherwise improve performance at some point in its lifetime. Even if a plant were to be built that represented the state of the art in performance by every measure, the reality that markets are not static and scientific innovation inexorably advances means that opportunities to maximize the value from these expensive assets will continue to be created.

Factors that create opportunities for value creation or preservation will be presented. A challenge for process researchers is that project timing and resource availability invariably complicate the problem-solving process. A series of examples will be discussed that demonstrate the use of available tools (modeling, lab-scale experimentation, pilot-scale development) to achieve process improvement goals.

1. Introduction

Any research endeavor, whether in academia or industry, must at a minimum be based on principles of sound scientific research. Chemical engineering, as an applied science, has as its mission the delivery of products needed by a society, using the materials available to it, all at a cost (economic or otherwise) that the society can afford to bear. In the industrial realm, the constraints implicit in this statement are realized as the need to do all of the above while providing a return to the investor while meeting schedules demanded by the market place. A successful project must therefore be based not only on sound science, but also be timely and profitable. The application of some of tools helpful in meeting the expectations of stakeholders is explored in this presentation.

2. Opportunities for Continuous Process Improvement

In the chemical industry, processes must be continuously improved to maintain economic competitiveness. These opportunities present themselves throughout the life cycle of a plant. It begins when ideas are developed for a new product or a novel process. Many companies have innovation programs that are used to develop and evaluate concepts at an early stage. The final idea that may be moved to development may bear little resemblance to the original concept.

Additionally, in the design stage the interaction between pilot operations and designers will continually improve the concept. Even in new plants, there are inevitably improvement opportunities that are recognized. These come about because of factors such as the impossibility of testing every possible alternative in a pilot program, scale-up issues, and many others.

This discussion will focus mainly on opportunities for improving existing plants to maintain their economic competitiveness.

2.1 Drivers for continuous improvement of existing processes

Some of the drivers for continuous improvement are found in the following table.

Drivers	
Competitive process introduced	– Competitor or internal – Better – Different
Shifting market demand	– Different balance of products – Higher purity products
Changing regulatory requirements	– Reduce waste – Reduce targeted impurities – Substitutions
Elimination of Process Hazards / Improve Process Safety	– Reactive relief scenarios – Hazard recognition
Advancing Technology	– Improved catalysts – New unit operations – Process additives – Process analytical – Modeling capability

For multi-product plants, changes in market demand are frequently a driver for re-optimization of an existing plant. Some products may be increasing demand, while demand for others may be falling. This can be problematic, especially when options appear to be constrained by nature and/or physical limitations of an existing facility. When this situation is encountered, it is time to open up the chemical engineers' toolbox to begin to resolve the issues at hand.

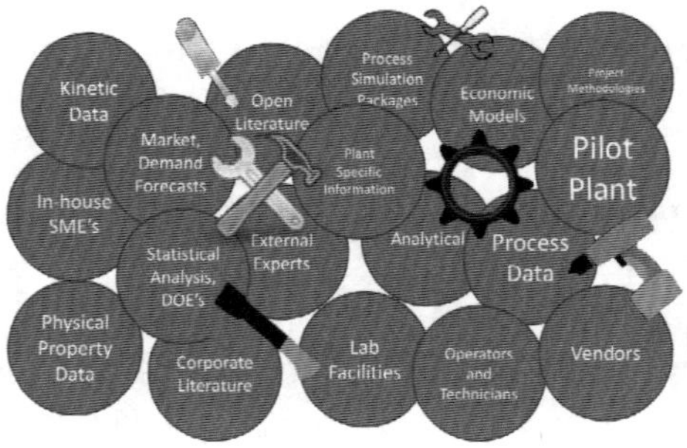

***** Formatting Issue - Best Available Paper/Graphic *****

3. Case Study: Shifting Product Demand

For the case presented for this discussion, market analysis determined that a significant shift in demand was occurring for the products from an existing multi-product plant. The data in the Table 1 suggests that we need to decrease selectivity of A by 13% lower from the current, and this decrease in selectivity of A will need to be balanced by increases in B, C and D selectivity. No change in selectivity of E is expected. So we basically need to meet 5 different selectivity goals which each have unique responses to changes in process variables.

Table 1: Current and Target Product Ratio Distribution

Product	Current Selectivity (%)	New Target (% Change from Current)
A	S_A	13% Lower
B	S_B	63% Higher
C	S_C	18% Higher
D	S_D	33% Higher
E	S_E	No Change

Information about the plant's historical production capability was collected, and a best case scenario was determined for meeting production demand based on the demonstrated capability of the plant. The gap between this performance and the expected market demand was determined, and the value of closing this gap was determined to be significant.

Options to close the production gap were assessed. In this case, the process is catalytic. Subject Matter Experts (SME's) within the company believed that a modification in the formulation of the existing catalyst could provide a solution to the dilemma. However, the shift in market demand will occur before the next opportunity for a catalyst recharge, creating a significant opportunity for value preservation in the meantime.

On this basis, a multi-pronged solution approach was proposed. Catalyst development would begin, but at the same time a project was initiated to expand the knowledge of the capability of the incumbent catalyst to move to production ratios that had not been demonstrated

Leverage Existing Knowledge

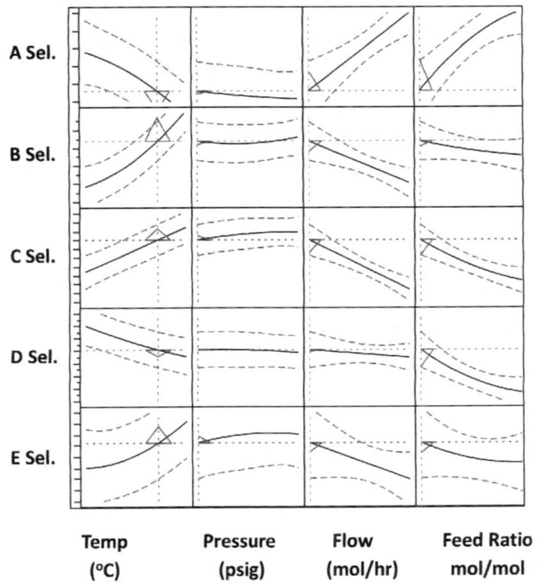

stand what we know about the we investigated an empirical l using data from a series of nents. These experiments were ; a continuous pilot reactor 4 .e need for this project was ıre 1 represents the prediction nodel which basically suggests of the product A decreases th decreased feed ratio and ow.

rical Statistical Model for the ern

*** Formatting Issue - Best Available Paper/Graphic ***

However, the predicted optimum operating ranges were outside the operating range of the existing plant. As a result, it was important to verify how the process would behave at conditions scalable to an existing plant facility. Hence, this historical data directed our research towards the study of some process region that was never investigated before. This is an example of using a storehouse of data developed for other purposes and combining it with process knowledge to identify opportunities for improvement.

3.2 Identify Key Variables

To understand effect of process variables, we started with a study of the response of two variables, feed ratio and pressure, that could be readily changed in the existing process. Baseline operating conditions were temperature T, pressure P, feed ratio R, feed flow F, and gas flow, G. The plot (Figure 2) here shows the order of magnitude of the effect on the selectivity of A as a function of conversion as feed ratio and operating pressure are reduced. Hence, with the combined effect of decreased feed ratio and decreased pressure, selectivity of A is now very

Identify Key Variables

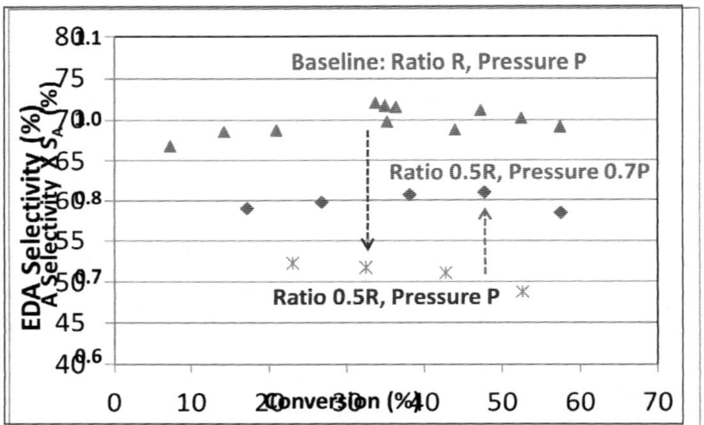

Figure 2: Effect of Pro[illegible]roduct Ratio

Results looked promising enough to meet target via process condition optimization: Need DOE

Table 2: Progress towards Goal at Changed Feed Ratio and Pressure

	% Selectivity at Ratio 0.5R & Pressure 0.7P	% Progress Towards Goal
A	$0.8S_A$	95
B	$1.8S_B$	99
C	$1.4S_C$	76
D	$0.8S_D$	75
E	$1.6S_E$	82

This result is very promising as it indicates that it might be possible to meet the goal with process optimization. However, targets for two key components were not met. Since there are multiple variables that can be optimized, the team proposed a 5 variable Designed Set of Experiments (DOE).

3.3 Develop Statistical Model

The design study required 63 individual experiments in a continuous pilot reactor. It took one day to complete each experiment to establish steady state. For this new statistical analysis, the range of the operating variables was chosen to reflect scalable conditions for the existing facility. We choose a Box Behnken response surface design and generated a significant model for conversion and selectivity of A, B, C, D and E [1,2]. The table (Table 3) summarizes the statistics for the empirical model. The statistical models were estimated for a 95% confidence interval. The P-values for the models were much lower than 0.05, suggesting a very good fit. The R2 adjusted values are very close to the R2 value which indicates model significance.

Table 3: Fit Model Statistics

Model Output	R^2	R^2 Adjusted	Analysis of Variance
			(p-value)
Conversion	0.97	0.95	<0.0001
A Selectivity	0.97	0.96	<0.0001
B Selectivity	0.97	0.96	<0.0001
C Selectivity	0.96	0.95	<0.0001
D Selectivity	0.96	0.95	<0.0001
E Selectivity	0.95	0.93	<0.0001

3.4 Optimization with Statistical Model

The result of the statistical analysis of the design set of data was a quadratic model [3]. The model predicts the outputs (conversion and selectivity) based on any value of the five input variables within the studied range.

The plot or the collection of small plots in Figure 3 is called a prediction profiler which presents a graphical representation of how the conversion and selectivity change with the operating conditions within the studied range. The prediction profiler is a very useful optimization tool in predicting the operating conditions required for the process to generate a target conversion and selectivity to products.

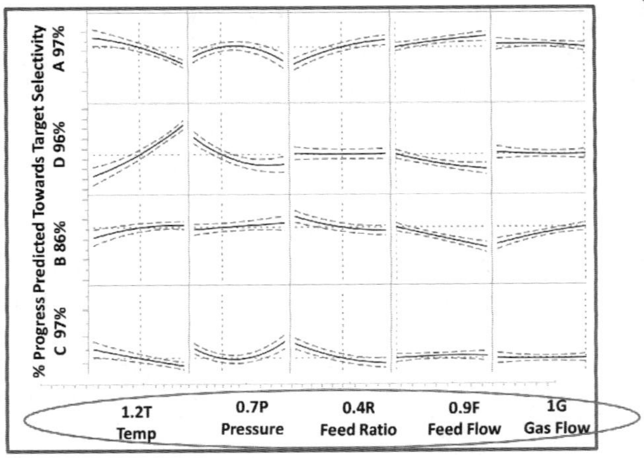

re 3: Statistical Model Prediction ler

The profiler (see Figure 3) ests that if we operate at 1.2T erature, about 0.7P pressure, 0.4R ratio, 0.9F feed flow and 1G gas then we expect to achieve a product that is very close to the new target. example, the product mix is cted to contain product A, B, C and th selectivity that are 97, 86, 97 and of the new target.

*** Formatting Issue - Best Available Paper/Graphic ***

The prediction profiler tool allowed a desired outcome to be set, which then manipulated the five variables to reach the target. In order to verify the predicted optimum, we operated our continuous pilot reactor at the predicted operating conditions and the actual results are tabulated in Table 4 that allows us to compare actual result with the model predicted optimum.

Table 4: Verification of Model Predicted Optimum

	Target Result	Model Prediction Towards Target	Actual Result from Lab	Model Predicted Deviation Range
A Selectivity	$0.9S_A$	97%	97%	+5.9 - 5.9%
C Selectivity	$1.2S_C$	97%	92%	+19 - 12%
B Selectivity	$1.6S_B$	86%	95%	+8.8 - 8.8%
D Selectivity	$1.3S_D$	96%	97%	+14 - 16%
E Selectivity	$1.0S_E$	100%	77%	+73 - 73%

The verification process strengthened the confidence in the validity of the predictive model. This empirical model serves as a process solution (with the existing catalyst) which can be used as a guide for changing the product ratio in the plant.

The pilot reactor was operated at similar plant conditions to the production plant. Differences in product selectivities were detected. These were attributed to known differences between the pilot and production-scale reactors (e.g. isothermal vs. non-isothermal, differences in aspect ratio). To mitigate this issue, the research team proposed to develop a kinetic model which can account for the differences in the two sets of reactors.

3.5 Develop Kinetic Model

A kinetic model was developed utilizing the data used to generate the statistical model [4]. The fundamental model accounts for non-isothermal behavior and other non-idealities of the plant reactor. This helps to minimize scale up risk from pilot to plant reactor associated with using an empirical model developed for an isothermal reactor system. Figure 4 includes parity plots for the model that compares model prediction with the actual plant result and shows an acceptable level of agreement.

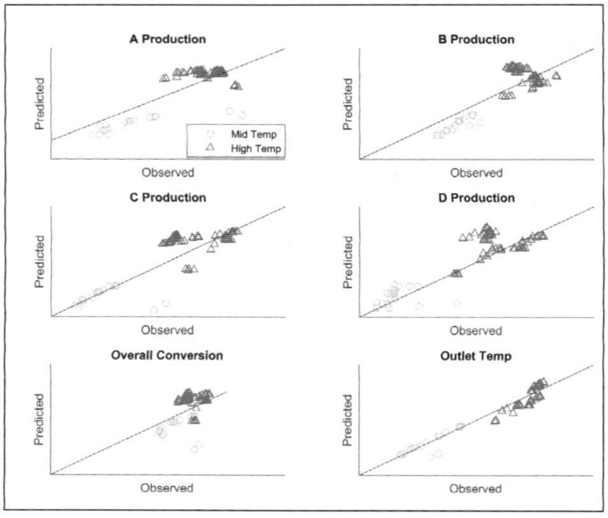

Figure 4: Plant Model Parity Plots

Both the statistical and kinetic model proposed solutions with the existing catalyst which successfully met target selectivity for the products A, B, C and D but not for the product E (only 77% of E selectivity target). The solution may or may not be acceptable to the plant and this greatly depends on the relative importance of meeting productivity target for individual product which is also a function of time. The research team decided to continue the catalyst development program to find a catalyst that will meet selectivity target for all components over the range of predicted production requirements.

3.6 Catalyst Development

Many approaches have been successfully adopted to develop new catalysts, from the empirical to modern combinatorial methods [5,6]. The chemistry of this system is well defined, so a breakthrough in catalyst performance is not required. In this system there are competing reactions that may be controlled by different aspects of the catalyst. A successful catalyst case requires modifying the relative influence of these aspects, rather than the enablement of new pathways.

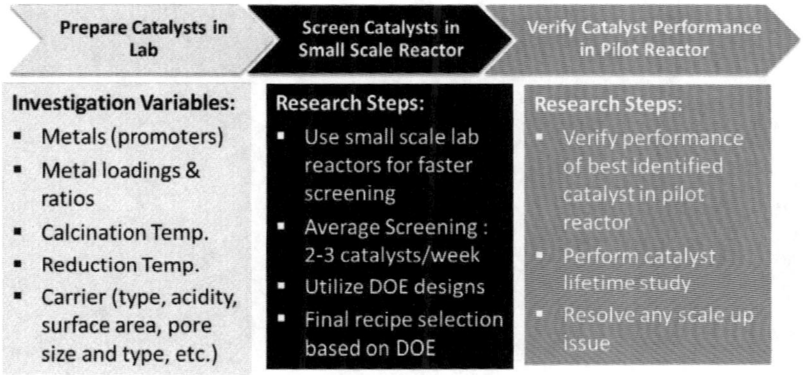

Figure 5: Catalyst Design Strategy

The strategy for developing an improved catalyst formulation is shown in Figure 5. Design variables were selected, and catalysts are screened in a lab-scale reactor. Promising catalysts are scaled to the pilot facility for more extensive testing.

Before full scale screening commenced, the measurement capability of the lab system was verified by a Gage R&D study [6]. This study determines the relative contributions to the measurement error from the operator, the instrument, and the samples themselves. This study validated that the measurement error was low enough that the required changes in catalyst performance could be distinguished from the noise of the system.

A second study looked at the stability of the lab process itself. In this study, a series of samples were collected over a period of time to determine if the process remained in statistical control over that period. This was important to validate that any changes in sample composition over the course of a day could be attributed to changes in process variables, rather than instability in the system.

A number of designed experiments have been conducted to determine the contributions of various factors to catalyst performance. An example is a 4 factor DOE used to determine the effects of catalyst concentration, promoter, and promoter concentration within a series of catalyst carriers representing a range of specific carrier properties. Four carriers and 4 promoters (categorical variables) were paired with two levels of catalyst and promoter concentration (continuous variables). The design was condensed to a total of 16 samples. Three replicates were included in the study.

***** Formatting Issue - Best Available Paper/Graphic *****

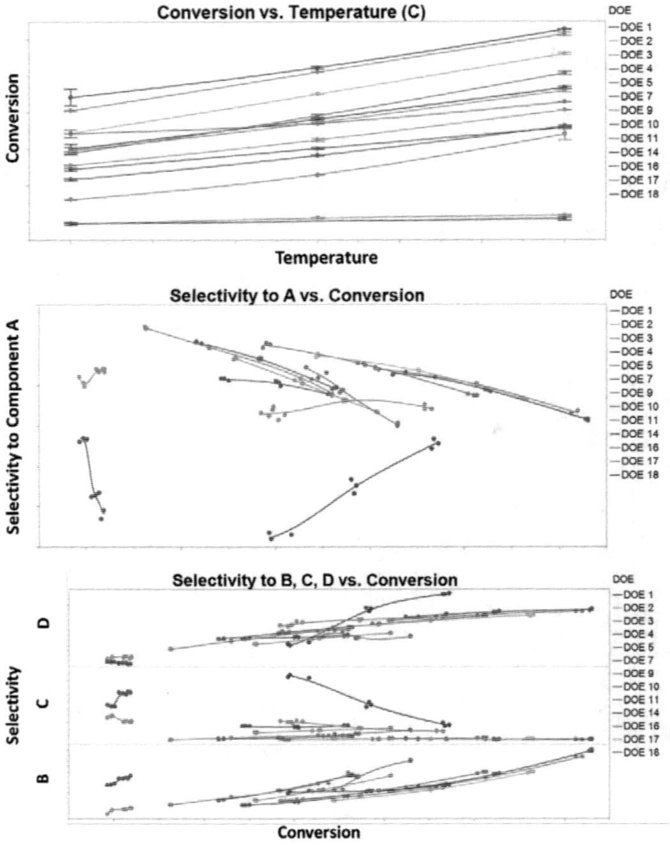

Figure 6: Catalyst performance, representative DOE study

Figure 6 shows results from this particular study. Catalyst formulations were identified that delivered a range of activity (as measured by the conversion as a function of temperature). More importantly, levers were identified that enable selectivities to be manipulated over a large range. One of the catalysts from this study was in fact selected for further investigation in the pilot-scale reactor system.

4 Conclusion

Factors that create opportunities for value creation or preservation have been presented. A case study has been presented that demonstrates the use of available tools (modeling, lab-scale experimentation, pilot-scale development) to achieve process improvement goals for a process of industrial significance.

5. References

[1] Box, G.E.P., and Behnken, D.W., "Some New Three Level Designs for the Study of Quantitative variables," *Technometrics*, Volume 2, pp. 455-475, 1960.

[2] Mandel, John. (1964)., "Statistical Analysis of Experimental Data.", Dover Publications. Online version available at: http://app.knovel.com/hotlink/toc/id:kpSAED0005/statistical-analysis/statistical-analysis

*** Formatting Issue - Best Available Paper/Graphic ***

[3] Mathews, Paul G.. (2005). Design of Experiments with MINITAB. American Society for Quality (ASQ). Online version available at: http://app.knovel.com/hotlink/toc/id:kpDEMINIT1/design-experiments-with/design-experiments-with

[4] Davis, Mark E. Davis, Robert J.. (2003). Fundamentals of Chemical Reaction Engineering. Dover Publications. Online version available at: http://app.knovel.com/hotlink/toc/id:kpFCRE0002/fundamentals-chemical/fundamentals-chemical

[5] Hager, Thomas. *The Alchemy of Air.* Harmony Books, Random House Inc. New York, NY. 2008

[6] Miyazaki, T., Osturk, S., Onal, I., Senken, S, "Selective oxidation of propylene to propylene oxide using combinatorial methodologies," Catalysis Today, Volume 81, pp. 473-484, 2003

[7] Sloop, Rick, "Understand Gage R&R," Quality Magazine, www.qualitymag.com/articles/86819-understand-gage-r-r (accessed March 16, 2016)

Discovery Driven Planning – A Tool to Quantify and Report on Uncertainty

A.R. Cartolano
Vice President of Process Engineering
SPEC Process Engineering & Construction, Inc.
Burlington, MA

R. Pierantozzi
Managing Partner
Cameron & Associates
Blue Bell, PA

ABSTRACT

As a new process or product is developed there are many turning points along the way, any of which could de-rail the program if the wrong decision is made. These decisions can be technical decisions about process options and/or business decisions about products and markets. Early in the project, there is usually uncertainty about elements of these aspects, so the project team will make assumptions, make decisions, and move forward based on those assumptions.

There are techniques for evaluating the uncertainties associated with the assumptions made and for driving decisions based on the sensitivity of the desired outcome to key assumptions. One of these tools is known as Discovery Driven Planning, developed at Wharton Business School. Although the tool was developed primarily to address uncertainty about a business plan, it can be used to analyze the technical side of a project where uncertainty exists and assumptions need to be made. This can be uncertainty about a product cost strip, capital investment, or any technical aspect of a development program.

We present the basic tenets of a Discovery Driven Plan, and outline the steps needed for any project team to perform it. The plan results in the development of a model that produces objective functions dependent on the assumptions made. We also present a simulation tool that allows the team to test any objective function against the assumptions made, resulting in predictions of the objective function(s), and sensitivity of the function to the assumptions. Finally, a new-product cost strip is presented as an example, and the management decisions that resulted from this analysis are discussed.

Discovery Driven Planning – A Tool to Quantify and Report on Uncertainty

INTRODUCTION

During a development program, whether to produce a new product, or change a process, there is always uncertainty around the outcome. Various studies are then undertaken to resolve this uncertainty, such as laboratory studies, pilot plant runs, engineering analysis, cost estimates etc. Each of these activities involves cost and time and scarce resources, so the organization needs to choose the efforts that will reduce the uncertainty (risk) at the lowest cost possible. There are tools that allow the project team to focus their efforts on the items that have the most impact on the process, thereby reducing project risk. Discovery Driven Planning (DDP), developed at The Wharton Business School (MacGrath 1995), is one such tool that allows the team to identify the biggest unknowns and develop plans to focus on and reduce those risks.

There are numerous assumptions made along the way as a development program evolves. The essential element of the DDP process is to identify these assumptions, to differentiate those from facts and assess what the most important assumptions are. Some items that might be considered facts would be the published cost of oil, or of other commodity chemicals, the price of stainless steel, or the labor index for a region. These are substantiated numbers that can be used as inputs to an analysis without questioning. However, what those prices might be 5 or 10 years down the road are assumptions. For example, who would have thought 5 years ago that the price of oil would be below $30/bbl.? If a business' profits depended on oil being $110/bbl, then there are tough times ahead, unless that possibility was considered and a plan was developed to deal with it. There are numerous examples of projects that went awry because key assumptions were not validated or understood. See McGrath 2009 on how to develop and use DDP to support growth initiatives, including examples of implementation across various companies and industries.

The DDP process starts by having the project team sit down and discuss the assumptions made around key parameters in the project. This could be: size of the business, manpower needs to support the business, capital investment, raw material costs, etc. Then, the team agrees to the most likely value for the parameters, with a range of uncertainty for each. Objective functions are then defined, such as manufacturing costs, business revenue, profit, ROI, etc. A Monte Carlo simulation is then performed on the objective function that varies each of the input values according to the uncertainty ranges agreed to by the team. Such tools as Crystal Ball from Oracle, are add-ins to Microsoft Excel and relatively easy to use for such analyses. Typically, the simulation is run for thousands of trials, to obtain a smooth probability distribution of the objective function.

The result of the analysis presents two key outcomes to consider. The first is the distribution of the objective function, as shown in Figure 1 below. A full statistical analysis of this distribution is also generated, which is useful in the example analysis presented below. This allows the team to determine the likelihood that the objective function will fall within the desired outcome tolerances. The second is the sensitivity analysis that shows the sensitivity of the variability of the objective function to the parameters input by the team, shown in Figure 2. This is very informative, and directs the team to focus their efforts on reducing the uncertainty of the key parameters, allowing for efficient use of resources.

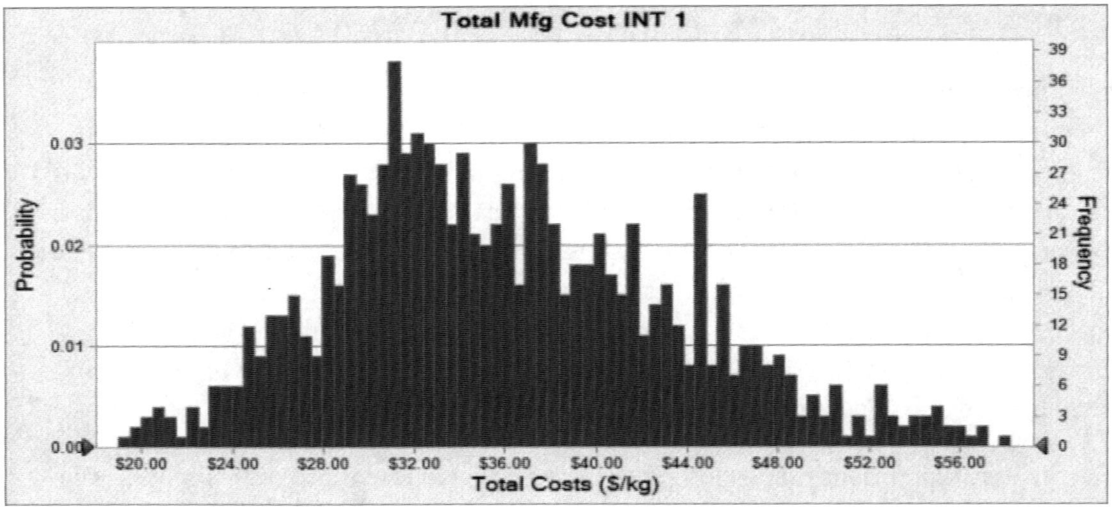

Figure 1: Typical Objective Function Distribution

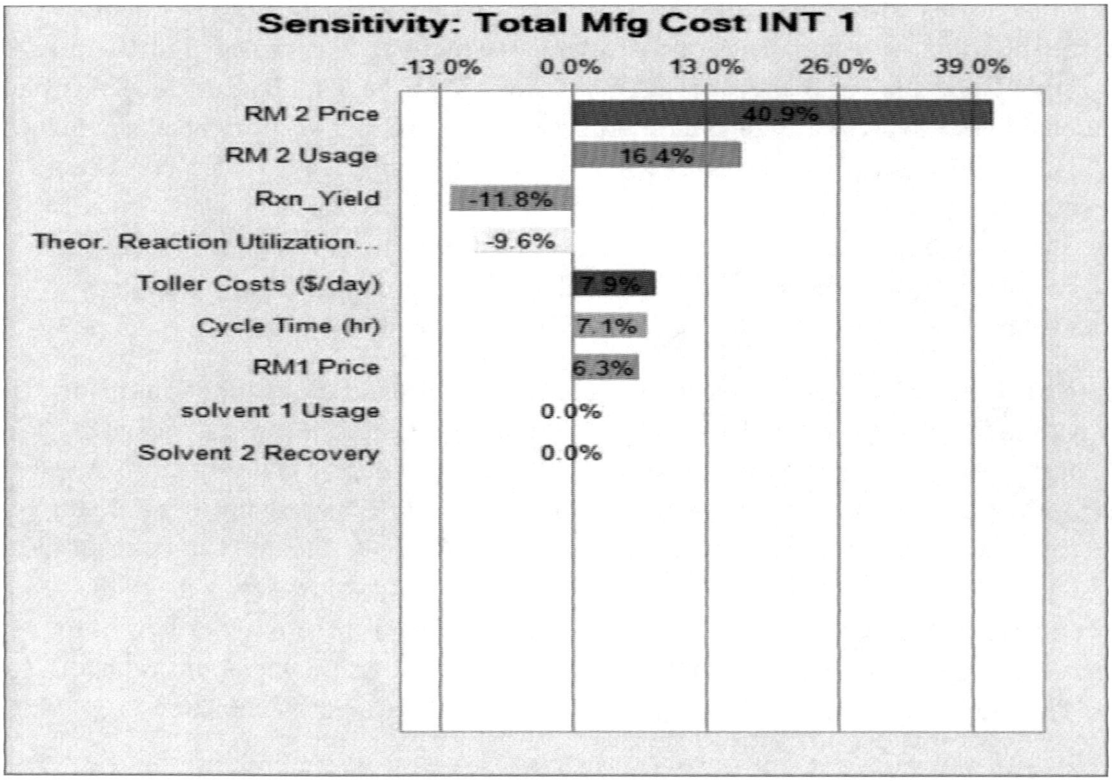

Figure 2: Typical Sensitivity Chart Resulting from a DDP Analysis

APPLICATION AND ANALYSIS

The use of the technique for helping to guide a materials synthesis program will help to illustrate the power of the technique. As part of a conductive polymers program, a specialty monomer (FINAL) was required as a key intermediate. It was proposed that the monomer be manufactured in house for security of supply. An effort was undertaken to evaluate the feasibility of manufacturing FINAL on a contract basis to supply development and initial commercialization of the polymer program. A proposed route to FINAL was evaluated in the laboratory based on chemistry published in the open literature. At the same time, an economic analysis was conducted to understand if the cost position provided by contract manufacture allowed for a viable supply strategy. A Discovery Driven Planning (DDP) type analysis of the manufacturing cost strip was conducted to support the economic evaluation. This type of analysis provided valuable insights into the drivers of the cost position and helped to set a supply strategy for the conductive polymers program.

The route to FINAL that was investigated involved a three-step process, starting with commercially available raw materials.

1. The first step involved the reaction of RM1 with RM2 in Solvent 1 to produce INT1.
2. The second step involved converting INT1 to INT2 by generating RM5 in situ from RM3 and RM4, and then reacting with INT1 in Solvent2. The acid released during the reaction would then need to be scrubbed in a RM6 scrubber.
3. The final step is the conversion of INT2 to FINAL by reaction with RM7 in Solvent 3 using RM8 as a catalyst. Purification of FINAL to electronic grade is then required.

To complete this analysis, a series of 3 nested DDP analyses were conducted. First, the cost to produce INT1 was predicted using assumptions made by the team around possible yields and cycle times. Contract manufacturing costs were assumed based on previous similar project costs. The cost strip used to generate this analysis is shown in Figure 3. Note that assumptions are in green, and objective functions are in blue. Note that all of the ranges are listed in the spreadsheet for clarity and future reference. In this case, the team looked at the variable costs, fixed costs, and total manufacturing costs. For this discussion, we will concentrate on the total manufacturing costs.

Figure 1 above shows the distribution of manufacturing costs for INT1, and Figure 2 shows the sensitivity analysis generated at the same time. Note that the uncertainty around the key raw material cost, and concurrently the reaction yield, involves key parameters that need to be investigated further. To reduce the uncertainty likely requires effort by two members of the team, a business or purchasing function, and the laboratory chemist.

	MW			
Product: INT 1	114		Assumption Number	
Reaction Yield		65.0%	1	Range from 50-75%
Theoretical Usage n(kg/kg)				
RM 1	86	0.75		
RM 2 Usage	106	1.86	10	Range from 0% to 300% excess
Acid	98	0.02		
ADH	115	0.001		
HQ	110	0.05		
solvent 1 Usage	32	1.75	2	Range from 1.5 to 2.5
Yields (%)				
RM1		65.0%		
RM 2		65.0%		
ACID		65.0%		
ADH		65.0%		
HQ		65.0%		
Solvent 1 Recovery		80.0%	3	Range from 50-90% recovery
Usages (kg/kg)				
RM 1		1.16		
RM 2		2.86		
Acid		0.02		
ADH		0.00		
HQ		0.07		
Solvent 1		0.54		
Raw Material Prices ($/kg)				
RM1 Price		$5.00	4	Range from $2-$10/kg
RM 2 Price		$5.00	5	Range from $2-$10/kg
Acid		$0.11		
ADH		$0.55		
HQ		$2.00	6	Range from $5-$20/kg
Solvent 1		$0.33		
Raw Material Variable Costs ($/kg)				
RM 1		$5.80		
RM 2		$14.31		
Acid		$0.00		
ADH		$0.00		
HQ		$0.15		
Solvent 1		$0.18		
Total RM Variable Costs ($/kg)		$20.44		
Theor. Reaction Utilization (kg/liter/batch)		0.15	7	Range from .12 to .25
Cycle Time (hr)		18	8	Range from 12 to 24
Reactor Size (liter)		24000		
Toller Costs ($/day)		$36,000	9	$10000 to $20000/reactor set, need 3 8000 liter reactor sets
Manufacturing Costs ($/kg) INT 1		$11.54		
Total Costs ($/kg) INT 1		$31.98		

Figure 3: Cost strip generated for INT1 with assumptions listed

A full report is generated by the simulation software that documents both inputs and outputs. For example, in this analysis, a triangular distribution was assumed for the ranges of assumptions, which is typical for most of these analyses. It is not useful to assume any other type of assumption at this level, because of the level of uncertainty around any distribution function. A typical assumption distribution is shown in Figure 4 below.

Worksheet: [Intermediate 1_11_15.xlsx]Cost Strip Int 1

Cell: C48

Triangular distribution with parameters:

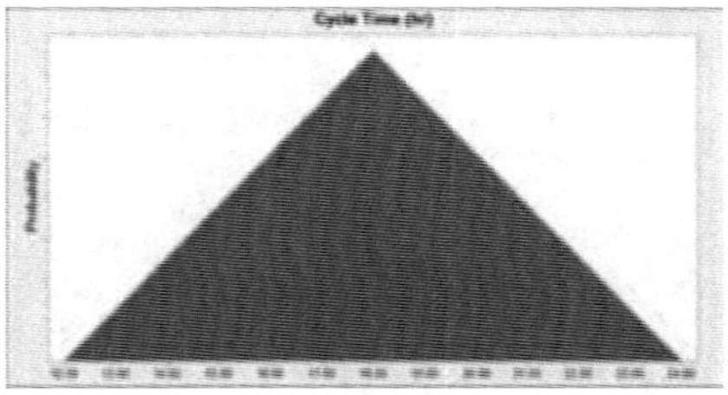

Minimum	12.00
Likeliest	18.00
Maximum	24.00

Figure 4: Simulation Report: Example Triangular Distribution of Assumption

The report also generates statistical information about the objective function as shown in Table I below. This gives a good idea of the variation of the objective function, and the likelihood it will lie within target values based on the assumptions used. For this example, this data was used in the next step of the analysis.

TABLE I: Statistical Information from Monte Carlo Simulation

Forecast: Total Mfg Cost INT 1

Summary:

Entire range is from $19.00 to $67.42
Base case is $31.98
After 1,000 trials, the std. error of the mean is $0.25

Statistics:

	Forecast values
Trials	1,000
Base Case	$31.98
Mean	$36.24
Median	$35.22
Mode	---
Standard Deviation	$7.85
Variance	$61.57
Skewness	0.6421
Kurtosis	3.46
Coeff. of Variation	0.2165
Minimum	$19.00
Maximum	$67.42
Range Width	$48.42
Mean Std. Error	$0.25

The cost for INT1 was then input to a second DDP spreadsheet to predict the cost to manufacture INT2. The INT1 cost was inserted as a range ($20.75-$51.83/kg) based on the mean ($36.24/kg) predicted by the first analysis plus or minus 2 standard deviations. The other assumptions were generated in a similar fashion as used in the cost strip for INT1. The cost to manufacture INT2 is shown in Figure 4, and the sensitivity analysis associated with INT2 is shown in Figure 5. Note that raw material costs start to dominate the cost structure.

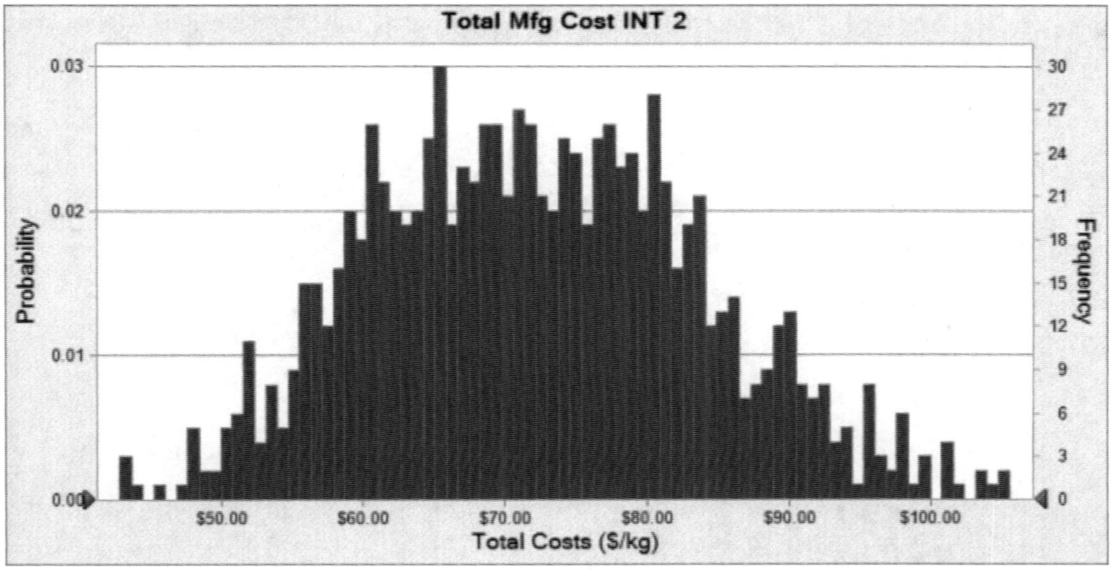

Figure 5: Total Manufacturing Cost Prediction for INT2

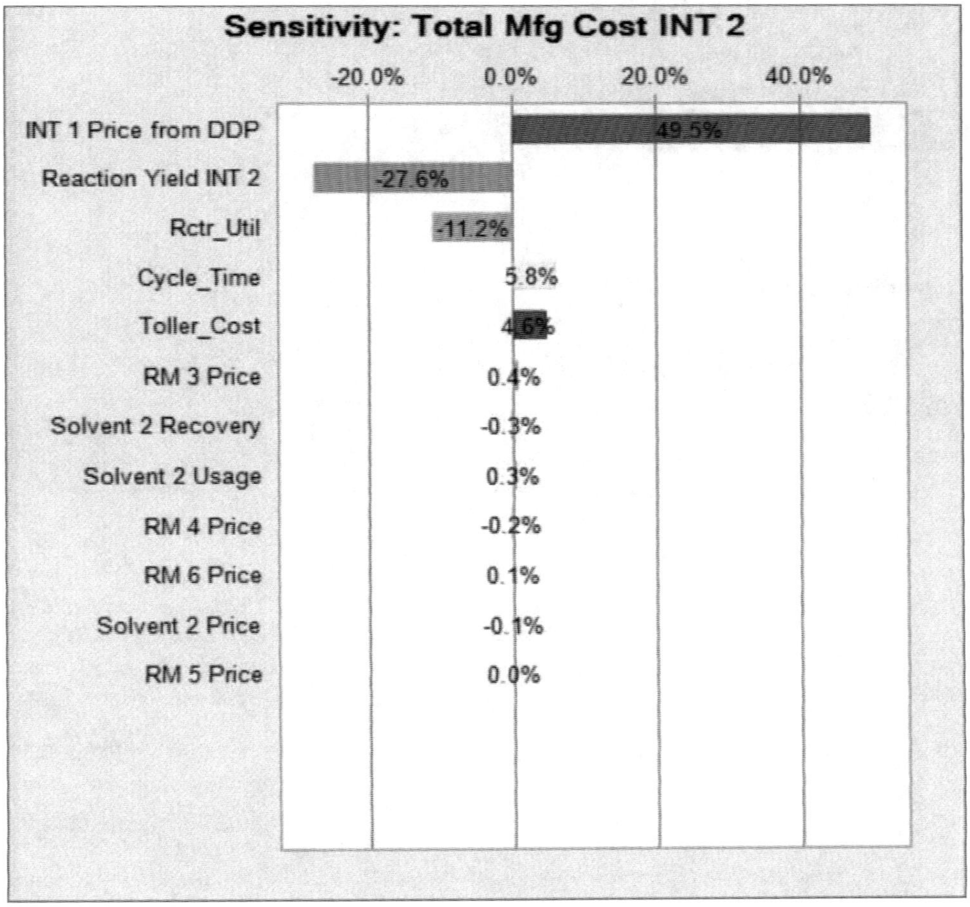

Figure 6: Sensitivity Chart for INT2 Manufacturing Cost

Following the same methodology, the cost to manufacture FINAL was predicted using the INT2 cost generated by the second DDP spreadsheet.

The results are shown in the attached Figures 6 and 7. The key part of the analysis for the team and project managers was the sensitivity charts generated during the simulation. As the synthesis proceeded, the raw material inputs began to dominate the manufacturing cost. Also, the key to reducing the cost of the entire process was to improve the yields, at each step, but more importantly, at the first step, that is the manufacture of INT1.

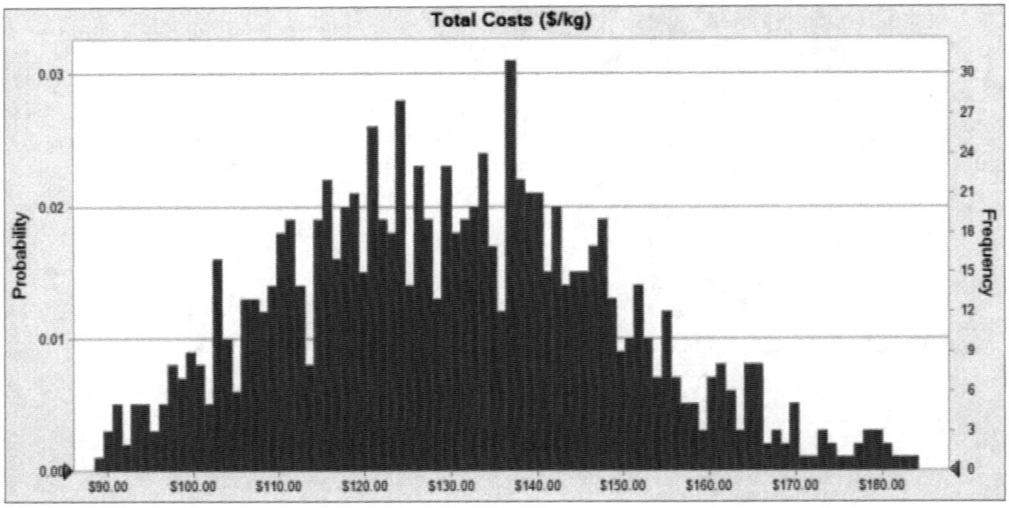

Figure 7: Total Manufacturing Cost Prediction for FINAL

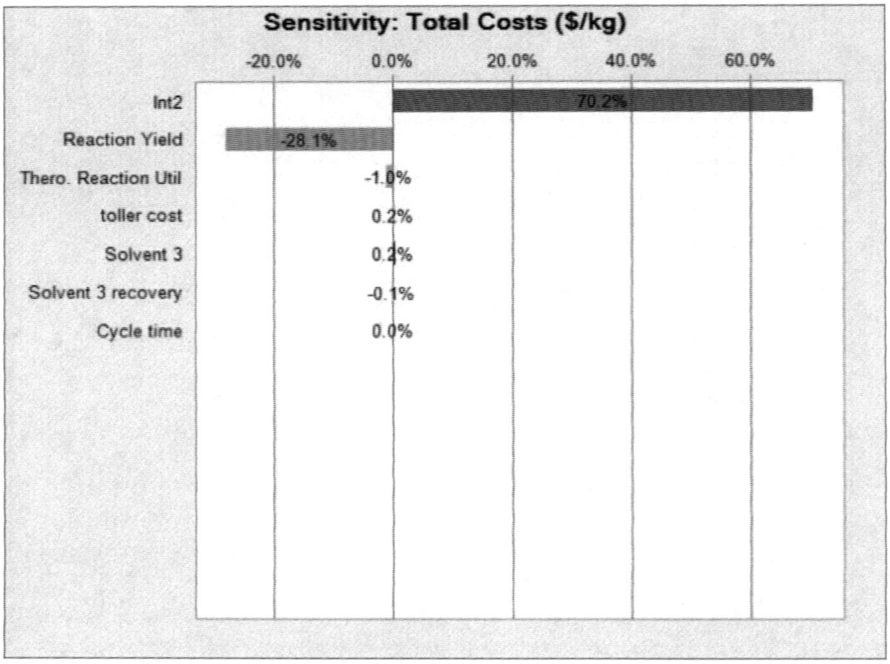

Figure 8: Sensitivity Analysis for Cost to Manufacture FINAL

This leads to the conclusion that substantial process development efforts were needed on a yield improvement program for each step of the process. As a result the project team elected to purchase FINAL to keep the development program going and even into the initial commercialization phases, until a synthesis development program would be justified based on the project's viability.

CONCLUSION

The Discovery Driven Planning Process has shown to be a useful tool during a new process or product development project to help the development team quantify the uncertainties of assumptions, and then develop plans to address the most influential assumptions. This type of analysis can be incorporated into many of the gated work processes in use today, by allowing the team to list the activities that have the highest impact on the project. This allows the development of meaningful milestones, and guides the development team to perform activities in a cost-effective manner. This analysis is repeated at each stage, so that as uncertainty around key assumption(s) can be reduced to acceptable levels, the team can focus on the next level of variables that affect project success. This in no way guarantees success of any project, but does allow for more efficient use of resources in a systematic, documented fashion.

REFERENCES

1. McGrath, R.G. & MacMillan, I.C. 1995. Discovery Driven Planning. *Harvard Business Review,* 73(4):44-54.
2. McGrath, R.G. & MacMillan, I.C. 2009. *Discovery Driven Growth: a breakthrough process to reduce risk and seize opportunity.* Boston: Harvard Business Publishing.

444265 Friend Today, Foe Tomorrow: How to Protect Your Intellectual Property When Entering into an Oil & Gas Industry Joint Venture

Abstract:

Exploration and development of oil and gas reserves increasingly requires costly, tailored technologies and enhanced safety and environmental measures. This is due in part to greater desire to develop unconventional oil and gas resources, such as shale gas and tight oil, and to develop reserves in more remote and environmentally-sensitive areas. Exploration and production companies have been developing deepwater reserves, shale plays, and other unconventional sources, but many lack the infrastructure to gather and process the resulting oil and gas. As a result, the industry has seen a rise in joint ventures between upstream and midstream companies that have the capacity to transport and store gas or crude oil, as well as joint ventures between two or more upstream companies.

Although joint ventures can help companies spread capital costs, overcome technological limitations, and mitigate risks, they also create potential pitfalls for companies with intellectual property assets. For example, a company may need to grant its joint venture partners a license to use its patented technologies or even share its trade secrets and other know-how with those partners, who may be industry competitors. Additionally, joint ventures can lead to development of new innovations, and disputes may arise as to who owns the resulting intellectual property and has the right to use it. Protecting these valuable intellectual property assets is an aspect of managing joint venture and joint development projects that is often overlooked. This paper will outline potential perils of joint ventures in the energy industry and discuss strategies to protect intellectual property portfolios when entering these agreements.

444082 CFD Modeling of Fouling in Crude Oil Refinery Heat Exchangers: Effect on Heat Transfer and Fluid Flow

Abstract:

The complex phenomenon of crude oil fouling costs millions of dollars in maintenance for refineries per year, globally. The equipment which is affected the most is crude oil pre-heat exchangers. Fouling increases the fuel requirement, compromises the heat recovery, causes undesirable production line shutdowns and other safety concerns. It is important to have an approach to determine the degree of fouling in heat exchangers, and monitor their health. This paper focuses on the use of CFD modeling to get a good estimate of fouling inside crude oil heat exchangers. CFD analysis will also show how the fouling particles are deposited on the surfaces, decreasing the heat transfer and creating higher pressure drop in the system.

443645 A Different CFD Approach to Model Sloshing of Offshore Multiphase Separators

Abstract:

Liquid sloshing in offshore multiphase separators has been one of the predicaments for reduced efficiency of process equipments. This led to intense investigations for last several decades both by experiments and computer simulations. For sloshing calculations, the most commonly used multiphase model is the Volume of Fluid (VOF) model. This method assumes the fluids (oil, water and gas) being immiscible and therefore no interpenetration and intermixing occurs among phases. This method is very useful to see the impact of sloshing on the vessel, but it is not so useful for the estimation of the actual separation. Hence, a different approach is needed. In this study sloshing calculation are run using the Eulerian multiphase model in combination with the multi-fluid VOF scheme (for accurate interface tracking). ANSYS FLUENT CFD software is used for all calculations with a user defined function to include the six degree sloshing motions. The primary goal of this study is to validate this new method for sloshing and separation of the oil/water/gas. This approach will help estimate in-situ carry over and carry under more accurately during the sloshing motion.

445012 An Efficient Hybrid CFD-DEM Modeling Framework for Industrial Drum Coating Operations

Abstract:

Tablet coating is a common unit operation in pharmaceutical manufacturing. The coating process depends on a number of process variables – spray parameters, coater geometry, mass loading, drying air flow etc. Process engineers in the pharmaceutical industry are interested in optimizing coater operation from both process and product perspectives. Performing physical experiments with coaters to determine the optimal set of process parameters is, more often than not, impractical. Digital experiments using computer simulations of tablet coating offer a viable alternative to experiments. We present here an efficient hybrid CFD-DEM model framework that enables a relatively rapid and efficient optimization of the coating process. The approach described here is an effective alternative to full scale CFD/DEM co-simulation and is designed to provide useful design guidance to process engineers at a significantly reduced computational expense.

443848 Practical Considerations of Melt Crystallization

Abstract:

INTRODUCTION:

Crystallization involves purification of a chemical from a liquid mixture by solidification of the desired component. Crystallization falls into two main categories: solution crystallization and crystallization from the melt (also referred to as “fractional crystallization”). Solution crystallization means a material is crystallized from a mixture containing a solvent. Crystals are formed by either concentrating the desired component by evaporation or cooling or both, and the solid phase is typically formed at a temperature well below its pure component freezing point. In melt crystallization, no solvent is added. The crystals are generated by cooling of the “melt” and are formed at a temperature near the pure-component freezing point. The product is collected in its melt form, and solids handling can be largely avoided for some process designs particularly if the product is desired in a liquid form.

APPLICATIONS:

Melt crystallization technology is somewhat uncommon, but can be very effective in particular applications. It is a separation technique that offers certain advantages over traditional separation methods including distillation or solution crystallization. It is capable of producing an extremely high product purity of >99.99%. Producing a material which is a liquid melt can be a great advantage for processes in which a solid product is not required. Typical challenges associated with handling a solid can be altogether avoided (except within the crystallizer itself). In addition, a cost savings may be realized since solid-liquid separation and drying equipment are unnecessary. Relative to distillation, energy usage can be low since the heat of fusion is generally smaller that the heat of evaporation for organic compounds. It may be desirable for separation of materials which are thermally unstable and for separating components which have low relative volatilities (such as isomers) or that form azeotropes. Finally, crystallizing from the melt rather than from solution means that the use of solvents and subsequent concentration or solvent removal steps are unnecessary. Without the addition of solvents, volumes of process streams are smaller and associated equipment is smaller as well.

This presentation will provide an overview of this process technology and describe the variations in available equipment for employing this technology. Criteria indicating whether melt crystallization is appropriate for processing a given stream will be presented. Current uses of this technology in industry will also be discussed.

FEASIBILITY AND SCALE-UP STUDIES:

MATRIC has constructed laboratory scale falling film melt crystallizers and has experience in conducting feasibility studies to evaluate the use of melt crystallization for a given purification process. Small scale testing can provide valuable information at an earlier stage in a project when large quantities of material are not available. The work process employed by MATRIC consists first of conducting a simple “bottle test” (in which the process stream is subjected to a static crystallization) as a preliminary indicator of performance. A partition coefficient (K) is calculated based on the impurity concentrations in the solids and mother liquor. A K value of <1 indicates that the impurities preferentially stay in the liquid phase, and systems with K ≤0.25 are considered ideal candidates for melt crystallization technology. At this stage, phase diagrams are also obtained and assessed to determine what recoveries are possible and whether the operating temperatures are practical.

Next, crystallization trials are carried out in batch mode in the falling film crystallizer, where crystal formation can be easily observed through the glass crystallizer walls. Crystallization, sweating and melting steps are carried out, and the resulting fractions are collected and analyzed. Individual and overall interfacial distribution coefficients are calculated. These data can be used to construct a material balance and estimate the number of stages which would be required to reach a specified purity and recovery. The technique for subcooling the solution can be determined through this exercise. Observations can be made regarding how well crystals adhere to the tube wall and whether seeding is required. If seeding is required, special start-up techniques may be necessary. Recommendations for sweating can also be made.

443832 Drivers for Operating Pilot Plant Processes and Associated Design Considerations

Abstract:

There are a number of reasons a company may be interested in operating pilot plants. Among these reasons are scale-up, process validation, continuous improvement of existing commercial facilities, and production of sample materials and/or market development quantities. Each of these reasons introduce a different set of design and operational considerations.

MATRIC operates pilot plant facilities under contract for a broad spectrum of clients, each with a different business need and reasons for wanting to complete piloting work. The business drivers define the technical objectives of the pilot plant, the budget constraints, and the design considerations that go into the pilot plant.

This presentation will discuss various business and technical drivers behind piloting processes and the design considerations that should be considered depending on what is motivating the need for piloting. These considerations include the sizing basis, handling recycle and purge streams, temperature loss management at small scales, and budget considerations. The discussion will also cover the need for clear owners' project objectives so that the needs and drivers of the sponsor are well understood and are properly incorporated into the pilot plant design.

AUTHOR INDEX

AUTHOR INDEX

AUTHOR INDEX

AIChE
120 Wall Street, FL 23
New York, NY 10005-4020

ISBN 978-1-5108-2497-3

CURRAN ASSOCIATES INC.
proceedings.com